Determinism, Holism, and Complexity

Determinism, Holism, and Complexity

Edited by

Vieri Benci
University of Pisa
Pisa, Italy

Paola Cerrai
University of Pisa
Pisa, Italy

Paolo Freguglia
University of L'Aquila
L'Aquila, Italy

Giorgio Israel
University of Rome "La Sapienza"
Rome, Italy

and

Claudio Pellegrini
University of California at Los Angeles
Los Angeles, California

Kluwer Academic / Plenum Publishers
New York, Boston, Dordrecht, London, Moscow

Library of Congress Cataloging-in-Publication Data

Determinism, holism, and complexity/edited by Vieri Benci ... [et al.].
p. cm.
"This proceedings includes papers presented at the meeting and some of the results of the discussions that took place during the workshop held in Arcidosso (GR) Italy in September 3–8 2001."
Includes bibliographical references and index.
ISBN 0-306-47472-7
1. Mathematical physics—Philosophy—Congresses. 2. Galilei, Galileo, 1564–1642—Congresses. I. Benci, V. (Vieri)

QC20.6 .D48 2003
530'.01—dc21

2002040666

Proceedings of Determinism, Holism and Complexity, held at Castello Aldobranesco di Arcidosso, Italy, on September 3–8, 2001

ISBN 0-306-47472-7

233 Spring Street, New York, New York 10013

http://www.wkap.nl

10 9 8 7 6 5 4 3 2 1

A C.I.P. record for this book is available from the Library of Congress

Printed in the United States of America

Contents

Preface ix

Contributing Authors xiii

Part I Physics

Complexity and emergence of meaning 3
Fortunato Tito Arecchi

A geometric optics experiment to simulate the betatronic motion 27
Armando Bazzani, Paolo Freguglia, Leone Fronzoni, Giorgio Turchetti

Some remarks on the arrow of time and the notion of information 41
Vieri Benci

How real is the quantum world? 53
Marcello Cini

Decoherence and Classical Behaviour in Quantum Mechanics 63
Gianfausto Dell'Antonio, Rodolfo Figari, Alessandro Teta

Scaling Laws: Microscopic and Macroscopic Behavior 79
Raffaele Esposito

Measure of diffusion entropy of weak turbulence 87
Laura Galeotti, Giulia Menconi, Leone Fronzoni

Complexity in physics of an adhesive tape 95
Bruno Giorgini, Enrico Lunedei, Matteo Ciccotti

Reflections about the time arrow 105
Antonio Lepschy

The Big Computer. Complexity and Computability in Physical Universe 117
Ignazio Licata

On the Uniqueness or Multiplicity of Physical Theories 125
Claudio Pellegrini

An interplay between determinism and one-parameter semigroups 133
Silvia Romanelli

From dynamical systems to complex systems 141
Giorgio Turchetti

Part II Biology

Shape and size in biology and medicine 153
Vincenzo Capasso

Assessment of the quality of waters and the environment 171
Nicola Ceccopieri, Rosalba Banchetti

Synchronization of neocortical interneurons 181
Santi Chillemi, Michele Barbi, Angelo Di Garbo

The fractal borderland 191
Giuseppe Damiani

Emergent properties and complexity for biological theories 199
Paolo Freguglia

Ignoring complex interactions in natural ecosystems 211
Manuela Giovannetti

A compression algorithm as a complexity measure on DNA sequences 221
Giulia Menconi

Reductionism and history: the biology between Scylla and Charybdis 231
Renzo Morchio

A characterization for a set of trinucleotides to be a circular code 241
Giuseppe Pirillo

Deterministic and random components of over time evolution 245
Giuseppe Pulina, Corrado Di Mauro, Niccolò Macciotta, Aldo Cappio-Borlino

Toward creating life in a test-tube 257
Martino Rizzotti

Tubulin and the complexity of life 273
Giovanni Santangelo, Giampaolo Magagnini

Phylogenies and the new evolutionary synthesis 283
Francesco Santini

Cell system complexity and biological evolution 293

Michele Sarà

Self-organization and prebiotic environment 305
Silvano Traverso

Part III History and Philosophy of Sciences

James and Freud on psychical determinism 313
Paolo Casini

Probabilistic aspects in George D. Birkhoff's work 327
Luca Dell'Aglio

The metamorphosis of holism 339
Elena Gagliasso

Early approaches to the management of complexity 349
Ana Millán Gasca

The dignity of the natural sciences 359
Pietro Omodeo

Holism: some historical aspects 379
Silvana Procacci

Towards a history of complexity 387
Tito M. Tonietti

Preface

Determinism, holism and complexity: three epistemological attitudes that have easily identifiable historical origins and developments. Galileo believed that it was necessary to "prune the impediments" to extract the mathematical essence of physical phenomena, to identify the mathematical structures representing the underlying laws.

This Galilean method was the key element in the development of Physics, with its extraordinary successes. Nevertheless the method was later criticized because it led to a view of nature as essentially "simple and orderly", and thus by choosing not to investigate several characteristics considered as an "impediment", several essential aspects of the phenomenon under investigation might be left out. The Galilean point of view also contains an acknowledgement of the central role played by the causal nexus among phenomena. The mechanistic-deterministic description of reality – for instance, à la Laplace – although acknowledging that it is not possible to predict phenomena exactly owing to unavoidable measurement error, is based on the recognition of the their causal nature, even in an ontological sense. Consequently, deterministic prediction became the methodological fulcrum of mathematical physics. But although mechanistic determinism has had and, in many cases, still has, considerable success in Physics, in other branches of science this situation is much less favourable. This leads to the tendency – at least expressed at the paradigmatic level – to oppose the consideration of a small number of "simple" descriptive elements of the phenomenon against the consideration of its totality (holism), or rather, the emphasizing of the need to take into account as many of the phenomenon's characteristics as possible, even those that are apparently irrelevant. Above all in Biology, but also in Economics and other sectors of contemporary Physics the aim of, or rather the hope for, an overall, and even exhaustive, description, achieved using cognitive, but not necessary mathematical techniques (such as metaphors, protocol analysis, etc.) sought to establish itself from the second half of the nineteenth century on, although in a variety of different ways.

This happened also in several non negligible cases of failure of mathematical modeling in a number of sectors. In particular, the use of analogies with Physics, and with Mechanics in particular, did not always prove productive or satisfactory. Nevertheless, in modern science, it is increasingly difficult to do without the mathematical tool, above all in view of the sophisticated techniques that have been developed. Modern science must now solve the problem of using the mathematical tool to study "complicated" or "complex" phenomena. Indeed, while the term "complication" used to be used to refer exclusively to the existence in a given system of a large number of components and of the interactions among them, the reference to "complexity" now increasingly stresses the holistic and integrated view of the system. A new approach thus takes on shape and substance – namely, that of the "science of complexity". However, in actual fact, to what extent, at the present state of the art, are the approaches leading to the construction of models or of mathematical theories in accordance with the science of complexity, truly different, and not just in their approach, to those of Galilean reductionism? The reader will find a partial answer precisely in the articles contained in the present book. One has the feeling, or rather the confirmation, that it is an uphill path that is actually proceeding gradually side by side with the gradual perfecting of the techniques and methods.

This proceedings includes papers presented at the meeting and some of the results of the discussions that took place during the workshop held in Arcidosso (GR) Italy in September 3–8 2001.

We wish to express our gratitude to Daniela Visetti for all her work, which has been essential for the successful publication of these proceedings. We also want to thank the editors of Kluwer Academic/Plenum Publishers for their patience and constant help, and in particular Marissa Kraft.

Our thanks to the following institutions:

— Centro Interdipartimentale di Ricerca in Metodologia delle Scienze, Università di Roma "La Sapienza"

— Centro Interdipartimentale "L. Galvani" per Studi Integrati di Bioinformatica, Biofisica e Biocomplessita (CIG), Universita di Bologna

— Centro Interdisciplinare per lo Studio dei Sistemi Complessi (CISSC), Universita di Pisa

— Coordinamento Nazionale di Biologia Teorica

— Department of Physics and Astronomy, University of California Los Angeles (UCLA)

— Domus Galilaeana, Pisa

— Rivista di Biologia/Biology Forum

— Società Italiana di Storia della Scienza
and moreover:
— Amministrazione Comunale di Arcidosso
— Amministrazione Provinciale di Grosseto
— Banca Monte dei Paschi di Siena
— Comunità Montana del Monte Amiata

VIERI BENCI

PAOLA CERRAI

PAOLO FREGUGLIA

GIORGIO ISRAEL

CLAUDIO PELLEGRINI

Contributing Authors

Fortunato Tito Arecchi, Università di Firenze and Istituto Nazionale di Ottica Applicata, Italy.

Rosalba Banchetti, Dipartimento di etologia, ecologia ed evoluzione, Pisa, Italy.

Michele Barbi, Istituto di Biofisica CNR, Pisa, Italy.

Armando Bazzani, Dipartimento di Fisica, INFN and CIG, Università di Bologna, Bologna, Italy.

Vieri Benci, Dipartimento di Matematica Applicata "U. Dini", Università di Pisa, Pisa, Italy.

Vincenzo Capasso, MIRIAM and Dipartimento di Matematica, Università di Milano, Milano, Italy.

Aldo Cappio-Borlino, Dipartimento di Scienze Zootecniche, Università di Sassari, Sassari, Italy.

Paolo Casini, Dipartimento di Pianificazione Territoriale e Urbanistica, Università "La Sapienza", Roma, Italy.

Nicola Ceccopieri, Dipartimento di etologia, ecologia ed evoluzione, Pisa, Italy.

Santi Chillemi, Istituto di Biofisica CNR, Pisa, Italy.

Matteo Ciccotti, Università di Bologna and INFN-sezione di Bologna, Italy.

Marcello Cini, Dipartimento di Fisica, Università La Sapienza and Istituto Nazionale di Fisica della Materia, Roma, Italy.

Giuseppe Damiani, IDVGA–CNR, Segrate (MI), Italy.

Luca Dell'Aglio, Dipartimento di Matematica, Università della Calabria, Arcavacata di Rende (CS), Italy.

Gianfausto Dell'Antonio, Dipartimento di Matematica, Università di Roma "La Sapienza" and Laboratorio Interdisciplinare SISSA-ISAS, Trieste, Italy.

Angelo Di Garbo, Istituto di Biofisica CNR, Pisa and Istituto Nazionale di Ottica Applicata, Firenze, Italy.

Corrado Di Mauro, Dipartimento di Scienze Zootecniche, Università di Sassari, Sassari, Italy.

Raffaele Esposito, Dipartimento di Matematica Pura ed Applicata, Università degli Studi dell'Aquila and Centro di Ricerche Linceo Interdisciplinare "Beniamino Segre", Italy.

Rodolfo Figari, Dipartimento di Scienze Fisiche, Università di Napoli "Federico II", Napoli, Italy.

Paolo Freguglia, Dipartimento di Matematica Pura ed Applicata, Università degli Studi dell'Aquila and Domus Galilaeana, Pisa, Italy.

Leone Fronzoni, Dipartimento di Fisica, Università di Pisa, INFM and CISSC, Università di Pisa, Pisa, Italy.

Elena Gagliasso, Dipartimento di Studi Filosofici ed Epistemologici, Università degli Studi di Roma "La Sapienza", Roma, Italy.

Laura Galeotti, Dipartimento di Fisica dell'Università di Pisa, Pisa, Italy.

Bruno Giorgini, Università di Bologna and INFN-sezione di Bologna, Italy.

Manuela Giovannetti, Dipartimento di Chimica e Biotecnologie Agrarie, Università di Pisa and Center for Soil Microbiology, C.N.R., Pisa, Italy.

Antonio Lepschy, Dipartimento di Elettronica ed Informatica, Università di Padova, Padova, Italy.

Ignazio Licata, Istituto di Cibernetica Non-Lineare per lo Studio dei Sistemi Complessi, Marsala (TP), Italy.

Enrico Lunedei, Università di Bologna and INFN-sezione di Bologna, Italy.

Niccolò Macciotta, Dipartimento di Scienze Zootecniche, Università di Sassari, Sassari, Italy.

Giampaolo Magagnini, Dipartimento di Etologia, Ecologia ed Evoluzione, Pisa, Italy.

Giulia Menconi, C.I.S.S.C. Centro Interdisciplinare per lo Studio dei Sistemi Complessi, Università di Pisa, Pisa, Italy.

Ana Millán Gasca, Dipartimento di Matematica, Università di Roma "La Sapienza", Roma, Italy.

Renzo Morchio, Dipartimento di Fisica, Università di Genova, Italy.

Pietro Omodeo, Dipartimento di Biologia Evolutiva, Università di Siena, Siena, Italy.

Claudio Pellegrini, Department of Physics and Astronomy, UCLA.

Giuseppe Pirillo, IAMI CNR, Firenze, Italy and Université de Marne-la-Vallée, Marne-la-Vallée, France.

Silvana Procacci, Dipartimento di Scienze filosofiche, Università degli Studi di Perugia, Perugia, Italy.

Giuseppe Pulina, Dipartimento di Scienze Zootecniche, Università di Sassari, Sassari, Italy.

Martino Rizzotti, Dipartimento di Biologia, Università di Padova, Padova, Italy.

Silvia Romanelli, Dipartimento Interuniversitario di Matematica, Università di Bari, Bari, Italy.

Giovanni Santangelo, Dipartimento di Etologia, Ecologia ed Evoluzione, Pisa, Italy.

Francesco Santini, Department of Zoology, University of Toronto, Toronto, Canada.

Michele Sarà, Dipartimento per lo studio del Territorio e delle sue Risorse, Genova, Italy.

Alessandro Teta, Dipartimento di Matematica Pura e Applicata, Università di L'Aquila, Italy.

Tito M. Tonietti, Dipartimento di Matematica, Università di Pisa, Pisa, Italy.

Silvano Traverso, Rivista di Biologia / Biology Forum, Genova, Italy.

Giorgio Turchetti, Dipartimento di Fisica, INFN and CIG, Università di Bologna, Bologna, Italy.

Determinism, Holism, and Complexity

I

PHYSICS

COMPLEXITY AND EMERGENCE OF MEANING: THE FUNDAMENTAL LEVEL OF NEUROPHYSICS

Fortunato Tito Arecchi
Università di Firenze and
Istituto Nazionale di Ottica Applicata
arecchi@ino.it

Abstract The central issue of cognitive science is how a large collection of coupled neurons does not limit to automatic responses to environmental inputs, as done by brainless lower animals, but combines external data with internal memories into new coherent patterns of meaning. Based on recent laboratory investigations of homoclinic chaotic systems, and how they mutually synchronize by weak coupling,a novel conjecture on the dynamics of the single neuron is formulated. Homoclinic chaos appears as the easiest way to code information in time by a code consisting of trains of equal spikes occurring at erratic times; synchronization of trains of different individual neurons is the basis od a coherent perception. The percept space P can be given a metric structure by introducing a distance measure. The distance in P space is conjugate to the duration time in the sense that a quantum uncertainy relation in percept space is associated with time limited perceptions.

1. Introduction

In a debate on "Complexity and emergence", we should first of all provide clear definitions for these terms, in order to ascertain how much of what we say depends on our cultural bias, is an artifact of our linguistic tools, and how much corresponds to hard facts, to our embedding in an open environment, whose features, even though actively elaborated by our semantic memory, can not be taken as sheer "autopoiesis", but are grounded on an ontology.

This inquiry is done from the point of view of a physicist who has been active for decades in investigating the formation of collective or coherent processes out of a large amount of otherwise separate individuals, pointing out the critical appearance (*emergence*) of new world configurations and the elements of *novelty* of this emergence, which make this phenomenon *complex*. By complex we do not mean the trivial fact that the computational cost of their description is high (in such a case I would rather call them *complicate*) but the fact that available knowledge stored in well established models is not sufficient to predict reliably the emergence, and one must integrate the deductive chains with extra information which introduces a historical flavour into the scientific procedure.

This presentation is organized as follows.

The core of the paper goes to the very ground of perceptive processes. If we accept – as proper of complex systems – to organize our knowledge as different and mutually irreducible hierarchical levels, each one with its own rules and language, then the most fundamental one in cognitive processes is the physical description of how external stimuli (light, sound, pressure, chemicals) are transformed into sensorial perceptions.

Without entering those levels which for a physicist are metalevels rather foreign to his/her competence, already at the neurodynamical level, we come across a quantum limitation, here presented explicitly for the first time, which forbids the brain operations to be fully simulated by a universal computing machine. By purpose, I said "brain" since I do not wish to enter the debates on "mind", "consciousness" etc.

I have called "neurophysics" the combination of neurodynamical facts, whereby neurons are treated as physical objects to be compared to lasers or other nonlinear dynamical systems, and the quantum limitation emerging from the peculiar spike synchronization strategy selected in course of the natural evolution as the optimal strategy to elaborate information into relevant cognitive processes.

As for the Reference list, I have often replaced the specific mention of an article or a book by a Web site, where one can conveniently browse for a more satisfactory answer. I think that time is ripe to consider this reference tool as a standard one.

2. What is neurophysics?

It is by now firmly established that a holistic perception emerges, out of separate stimuli entering different receptive fields, by synchronizing the corresponding spike trains of neural action potentials [30, 26].

We recall that action potentials play a crucial role for communication between neurons [13]. They are steep variations in the electric potential across a cell's membrane, and they propagate in essentially constant shape from the soma (neuron's body) along axons toward synaptic connections with other neurons. At the synapses they release an amount of neurotransmitter molecules depending upon the temporal sequences of spikes, thus transforming the electrical into a chemical carrier.

As a fact, neural communication is based on a temporal code whereby different cortical areas which have to contribute to the same percept P synchronize their spikes. Spike emission from a nonlinear threshold dynamical system results as a trade off between bottom-up stimuli to the higher cortical regions (arriving through the LGN (lateral geniculate nucleus) from the sensory detectors, video or audio) and threshold modulation due to top-down readjustment mediated by glial cells [20].

It is then plausible to hypothesize, as in ART (adaptive resonance theory [10]) or other computational models of perception [9] that a stable cortical pattern is the result of a Darwinian competition among different percepts with different strength. The winning pattern must be confirmed by some matching procedures between bottom-up and top-down signals.

It is the aim of this section to present two fundamentals aspects of percept formation, namely,

i) The neurodynamics of spike formation

ii) A quantum limitation in information encoding/decoding through spike trains

As for the first aspect, a saddle point instability separates in parameter space an *excitable* region, where axons are silent, from a *periodic* region, where the spike train is periodic (equal interspike intervals). If a control parameter is tuned at the saddle point, the corresponding dynamical behavior (homoclinic chaos) consists of a frequent return to the instability [1]. This manifests as a train of geometrical identical spikes, which however occur at erratic times (chaotic interspike intervals). Around the saddle point the system displays a large susceptibility to an external stimulus, hence it is easily adjustable and prone to respond to an input, provided this is at sufficiently low frequencies; this means that such a system is robust against broadband noise as discussed later.

As for the second aspect, the temporal coding requires a sufficiently long sequence of synchronized spikes, in order to realize a specific percept. If the sequence is interrupted by the arrival of new uncorrelated stimuli, then a fundamental uncertainty ΔP emerges in the percept space P. This is related to the finite duration ΔT allotted for the code processing by a fundamental uncertainty relation

$$\Delta P \cdot \Delta T \geq C$$

where C is a positive dimensional quantity whose non zero value represents a quantum constraint on the coding. This constraint implies that the percepts are not set-theoretical objects, that is, objects belonging to separate domains, but there are overlap regions where it is impossible to discriminate one percept from another. We will discuss later the occurrence of this new class of time dependent perceptual illusions.

We call "neurophysics" the combination of i) and ii), by analogy with "econophysics" which as extracted some general physical from economic phenomena [16].

Neurophysics is distinct from neurodynamics, which is the investigation of dynamical models of neuron behaviors, as well from neurophysiology, which explores the coupling of different brain areas. Neurophysics is restricted to the two above items, and it is rather model-indipendent, so that it provides a general ground upon which different models can be built and compared.

3. Neurodynamics

The neurodynamic aspect has been dealt with in a preliminary series of reports, that here I recapitulate as the following chain of linked facts.

1 A single spike in a 3D dynamics corresponds to a quasihomoclinic trajectory around a saddle point (fixed point with 1 (2) stable direction and 2 (1) unstable ones); the trajectory leaves the saddle and returns to it (Fig. 1).

2 A train of spike corresponds to the sequential return to, and escape from, the saddle point. A control parameter can be set at a value B_C for which this return is erratic (chaotic interspike interval) even though there is a finite average frequency. As the control parameter is set above or below B_C, the system moves from *excitable* (single spike triggered by an input signal) to *periodic* (yielding a regular sequence of spikes without need for an input), with a frequency monotonically increasing with the separation ΔB from B_C (Fig. 2) [17].

3 Around the saddle point the dynamical system has a high susceptibility. This means that a tiny disturbance applied there provides a large response. Thus the homoclinic spike trains can be synchronized by a periodic sequence of small disturbances (Fig. 3). However each disturbance has to be applied for a minimal time, below which it is no longer effective; this means that the system is insensitive to broadband noise, which is a random collection of fast positive and negative signals.

4 The above considerations lay the floor for the use of mutual synchronization as the most convenient way to let different neurons

respond coherently to the same stimulus, organizing as a space pattern. In the case of a single dynamical system, it can be fed back by its own delayed signal. As the delay is long enough the system is decorrelated with itself and this is equivalent to feeding an independent system. This process allows to store meaningful sequences of spikes as necessary for a long term memory [6].

4. The role of synchronization in neural communications

The role of elementary feature detectors has been extensively studied in the past decades. Let us refer to the visual system [12]; by now we know that some neurons are specialized in detecting exclusively vertical or horizontal bars, a specific luminance contrast, etc. However the problem arises: how elementary detectors contribute to a holistic (Gestalt) perception? A hint is provided by figure 4 [26]. Both the woman and the cat are made of the same visual elements, horizontal and vertical contour bars, different degrees of luminance, etc. What are then the neural correlates of the identification of separate individual objects? We have one million fibers connecting the retina to the visual cortex, through the LGN. Each fiber results from the merging of approximately 100 retinal detectors (rods and cones) and as a result it has its own receptive field which is about 3.5 angular degrees wide. Each receptive field isolate a specific detail of an object (e.g. a vertical bar). We thus split an image into a mosaic of adjacent receptive fields, as indicated in the figure by white circles for the woman and black circles for the cat.

Now the "feature binding" hypothesis consists of assuming that all the neurons whose receptive fields are pointing to a specific object (either the woman or the cat) synchronize the spikes as shown in the right of the figure. Here each vertical bar, of duration 1 millisec, correspond to a single spike, and there are two distinct spike trains for the two objects.

Direct experimental evidence of this synchronization is obtained by insertion of microelectrodes in the cortical tissue of animals just sensing the single neuron [26]. Indirect evidence of synchronization has been reached for human beings as well, by processing the EEG (electroencephalogram) data [23].

The advantage of such a temporal coding scheme, as compared to traditional rate based codes, which are sensitive to the average pulse rate over a time interval and which have been exploited in communication engineering, has been discussed in a recent paper [27].

Based on the neurodynamical facts reported above, we can understand how this occurs [10, 14]. In figure 5 the central cloud represents the higher cortical stages where synchronization takes place. It has two inputs. One (bottom-up) comes from the sensory detectors via the early stages which classify elementary features. This single input is insufficient, because it would provide the same signal for e.g. horizontal bars belonging indifferently to the woman or to the cat. However, as we said already, each neuron is a nonlinear system passing close to a saddle point, and the application to a suitable perturbation can stretch or shrink the interval of time spent around, and thus lengthen or shorten the interspike interval. The perturbation consists of top-down signals corresponding to conjecture made by the semantic memory.

In other words, the perception process is not like the passive imprinting of a camera film, but it is an active process whereby the external stimuli are interpreted in terms of past memories. A focal attention mechanism assures that a matching is eventually reached. This matching consists of resonant or coherent behavior between bottom-up and top-down signals; that is why it has received the name ART as introduced by Grossberg (1976) and later specified in term of synchronization of the spike positions by Von der Malsburg has tested by Singer and his school. If matching does not occur, different memories are tried, until the matching is realized. In presence of a fully new image without memorized correlates, then the brain has to accept the fact that it is exposed to a new experience.

Notice the advantage of this time dependent use of neurons, which become available to be active in different perceptions at different times, as compared to the computer paradigm of fixed memory elements which store a specific object and are not available for others (the so called "grandmother neuron" hypothesis).

5. The self-organizing character of synchronized patterns

We have presented above qualitative reasons why the degree of synchronization represents the perceptual salience of an object. Synchronization of neurons located even far away from each other yields a space pattern on the sensory cortex, which can be as wide as a few millimeter-square, involving one million neurons. The winning pattern is determined by dynamic competition (the so-called "winner takes all" dynamics).

This model has an early formulation in ART and has been later substantiated by the synchronization mechanisms. Perceptual knowledge appears as a complex self-organizing process. We show how this approach overcomes earlier approaches of A.I. (Artificial Intelligence) and PDP (Parallel Distributed Processing) models.

Classical accounts of knowing and learning, influenced by the information processing paradigm, hold that procedural and declarative knowledge reside as information in long-term memory and are assembled during problem solving to produce appropriate solution. The underlying assumption is that any cognitive agent possesses some language-like sign system to represent the world; cognition is said to occur when these sign systems are manipulated according to rules with IF ... THEN ... structure [2].

This classical approach to cognition, which posits some sign system as the necessary and sufficient condition for any cognitive agent to exhibit intelligence, is known as the physical symbol system hypothesis [18]. However, this approach, in which learning is conceived of as the rule-governed updating of a system of sentences encountered numerous failures to account for empirical data as [8]:

- the preanalytic human judgements of credibility which is the basis of any account of large scale conceptual change as the acceptance of a new scientific paradigm;
- the perceptual discriminations;
- the connections between conceptual and material practices which are the basis of manual activities;
- the speed with which human beings construct appropriate frames relevant to explanatory problems (the so-called "frame problem").

In the PDP models the connectivity is realized by three layers consisting of input, hidden and output units (Fig. 6).

In contrast with the traditional computer models such networks are *not* programmed to contain procedural and declarative knowledge but are *trained* to do specific things [8]. The network is exposed to a large set of examples. Each time the output is compared to the "correct" answer, the difference is used to readjust the connection weights through the network. This is a form of training known as "back propagation". However, a strong criticism [11] is that connections are fixed apriori and there is no self-organizing behavior as instead it occurs in the dynamical formation of a synchronized pattern.

A possible model for a patterned coupling of neurons across the sensorial cortex has been suggested by Calvin [7] in analogy with the pattern formation at the top of a fluid layer heated from below. In both cases if the excitation range is fixed (in our case by the axon length), then the most likely configuration is an equilateral triangle, since each vertex confirms the other two. But each vertex can also be coupled to other equilateral triangles, yielding an overall mosaic pattern which looks as a floor of connected hexagons.

6. A metric in percept space *P*

We discuss two proposal of metrics of spike trains. The first one that we call class A [29] considers each spike as very short, almost like a Dirac delta-function and each coincidence as an instantaneous event with no time uncertainty. The metric spans a large, yet discrete, space and it can be programmed on a standard computer.

A more recent proposal [24] accounts for the physical fact that each spike is spread in time by filtering process, hence the overlap takes a time breadth t_c and any coincidence is a smooth process. Indeed, as discussed above, in neurodynamics the high sensitivity region around the saddle point has a finite time width (unpublished results of my research group). We call B the class of smooth metrics.

The discrete metrics A satisfy a set-theoretical expectation and are programmable in a computer. The continuous metrics B fit better neurodynamical facts; they however imply a quantum limitation which then seems to be intrinsic of cognitive processes.

6.1 Class A metrics

Victor and Purpura [29] have introduced several families of metrics between spike trains as a tool to study the nature and precision of temporal coding. Each metric defines the distance between two spike trains as the minimal "cost" required to transform one spike train into the other via a sequence of allowed elementary steps, such as inserting or deleting a spike, shifting a spike in time, or changing an interspike interval length.

The geometries corresponding to these metrics are in general not Euclidean, and distinct families of cost-based metrics typically correspond to distinct topologies on the space of spike trains. Each metric, in essence, represents a candidate temporal code in which similar stimuli produce responses which are close and dissimilar stimuli produce responses which are more distant.

Spike trains are considered to be points in an abstract topological space. A spike train metric is a rule which assigns a non-negative number $D(S_a, S_b)$ to pairs of spike trains S_a and S_b which expresses how dissimilar they are.

A metric D is essentially an abstract distance. By definition, metrics have the following properties:

- $D(S_a, S_a) = 0$
- Symmetry: $D(S_a, S_b) = D(S_b, S_a)$
- Triangle inequality: $D(S_a, S_c) \leq D(S_a, S_b) + D(S_b, S_c)$
- Non-negativity: $D(S_a, S_b) > 0$ unless $S_a = S_b$,

The metrics may be used in a variety of ways – for example, one can construct a neural response space via multidimensional scaling of the pairwise distances, and one can assess coding characteristics via comparison of stimulus-dependent clustering across a range of metrics.

Cost-based metrics
Cost-based metrics fulfill the general definition of a metric, and are constructed with the following ingredients:

- a list of allowed elementary steps (allowed transformations of spike trains)
- an assignment of non-negative costs to each elementary step

For any such set of choices one can define a metric $D(S_a, S_a)$ as the least total cost of any allowed transformation from S_a to S_b via any sequence of spike trains S_a, S_1, $S_2, \ldots, S_n, S_b$.

The spike count distance D^{count}
This is a cost-based metric in which the elementary steps and associated costs are:

- insert a spike: cost $= 1$
- delete a spike: cost $= 1$
- shift a spike in time: cost $= 0$

The distance between two spike trains is the difference in the number of spikes, independent of when they occur. That is, this metric formalizes the notion that the only significant aspect of a spike train is the number of spikes it contains.

The spike time distance $D^{\text{spike}}[q]$
This is a family of cost-based metrics, parametrized by a "cost per unit time" q (units of 1/sec). The elementary steps and associated costs are:

- insert a spike: cost = 1
- delete a spike: cost = 1
- shift a spike by an amount of time t: cost $= qt$

If q is very small, this becomes the spike count distance. If q is very large, all spike trains are far apart from each other, unless they are nearly identical. For intermediate values of q, the distance between two spike trains is small if they have a similar number of spikes, occurring at similar times.

The motivation for this construction is that neurons which act like coincidence detectors should care about this metric. The value of q corresponds to the temporal precision $1/q$ of the coincidence detector.

This distance can be calculated efficiently by a dynamic programming algorithm.

The spike interval distance $D^{\text{interval}}[q]$
In this family of cost-based metrics, the elementary steps and associated costs are:

- insert a spike (along with an interspike interval): cost = 1
- delete a spike (along with an interspike interval): cost = 1
- change an interspike interval by an amount of time t: cost $= qt$

If q is very small, this becomes the spike count distance. If q is very large, all spike trains are far apart from each other, unless they are nearly identical. For intermediate values of q, the distance between two spike trains is small if they have a similar number of spikes, occurring with similar intervals.

Although this method has been applied successfully [15], the calculation of the full cost function is quite involved. The reason is that it is not always clear where a displaced spike came from, and if the number of spikes in the trains is unequal, it can be difficult to determine which spike was inserted/deleted.

6.2 A class B metric

Rossum has introduced an Euclidean distance measure that computes the dissimilarity between two spike trains [24]. First of all, filter both spikes trains giving to each spike a duration t_c. In our language, this would be the time extension of the saddle point region of high susceptibility, where perturbations are effective. To calculate the distance, evaluate the integrated squared difference of the two trains. The simplicity of the distance allows for an analytical treatment of simple cases. Numerical implementation is straightforward and fast.

The distance interpolates between, on the one hand, counting non-coincident spikes and, on the other hand, counting the squared difference in total spike count. In order to compare spike trains with different rates, total spike count can be used (large t_c). However, for spike trains with similar rates, the difference in total spike number is not useful and coincidence detection is sensitive to noise.

The distance uses a convolution with the exponential function. This has an interpretation in physiological terms.

As an alternative measure, one could convolve the spikes with a square window. In that case the situation becomes somewhat similar to binning followed by calculating the Euclidean distance between the number of spikes in the bins. But in standard binning the bins are fixed on the time axis, therefore two different spike trains yield identical binning patterns as long as spikes fall in the same bin. However, with the proposed distance (convolving with either square or exponential) this does not happen; the distance is zero only if the two spike trains are fully identical (assuming tc is finite).

The distance is loosely related to the distance measure introduced by Victor and Purpura. The distance introduced here is explicitly embedded in Euclidean space, which makes it less general but easier to analyze than the class A distance.

Interestingly, the distance is related to stimulus reconstruction techniques, where convolving the spike train with the spike triggered average yields a first order reconstruction of the stimulus [22]. Here the exponential corresponds roughly to the spike triggered average and the filtered spike trains correspond to the stimulus. The distance thus approximately measures the difference in the reconstructed stimuli. This might well explain the linearity of the measure for intermediate tc.

7. Role of duration T in perceptual definition: a quantum aspect

How does a synchronized pattern of neuronal action potentials become a relevant perception? This is an active area of investigation which may be split into many hierarchical levels. At the present level of knowledge, we think that not only the different receptive fields of the visual system, but also other sensory channels as auditory, olfactory, etc. integrate via feature binding into a holistic perception. Its meaning is "decided" in the PCF (pre frontal cortex) which is a kind of arrival station from the sensory areas and departure for signals going to the motor areas. On the basis of the perceived information, actions are started, including linguistic utterances.

Sticking to the neurodynamical level, which is the most fundamental one, and leaving to other sciences, from neurophysiology to psychophysics, the investigation of what goes on at higher levels of organization, we stress here a fundamental temporal limitation.

Taking into account that each spike lasts about 1 msec, that the minimal interspike separation is 3 msec, and that the average decision time at the PCF level is about $T = 240$ msec, we can split T into $240/3 = 80$ bins of 3 msec duration, which are designated by 1 or 0 depending whether they have a spike or not. Thus the total number of messages which can be transmitted is

$$2^{80} \approx 10^{27}$$

that is, well beyond the information capacity of present computers. Even though this number is large, we are still within a finitistic realm. Provided we have time enough to ascertain which one of the 10^{27} different messages we are dealing with, we can classify it with the accuracy of a digital processor, without residual error.

But suppose we expose the cognitive agent to fast changing scenes, for instance by presenting in sequence unrelated video frames with a time separation less than 240 msec. While small gradual changes induce the sense of motion as in movies, big differences imply completely different subsequent spike trains. Here any spike train gets interrupted after a duration ΔT less than the canonical T. This means that the PCF cannot decide among *all* perceptions coded by the neural systems and having the same structure up to ΔT, but different afterwards. How many are they: the remaining time is $\tau = T - \Delta T$. To make a numerical example, take a time separation of the video frames $\Delta T = T/2$, then $\tau = T/2$. Thus in spike space an interval ΔP comprising

$$2^{\tau/3} \approx 2^{40} \approx 10^{13}$$

different perceptual patterns is uncertain.

As we increase ΔT, ΔP reduces, thus we have an uncertainty principle

$$\Delta P \cdot \Delta T \geq C$$

If we adopt the Victor Purpura metrics, with the distances measured by costs in nondimensional units, then C has the dimensions of a time.

If we adopt the Rossum metric then the distance ΔP is the square of a voltage and therefore an energy, so that C has the dimensions of an action, has Planck's constant in the quantum energy-time uncertainty relation.

The problem faced thus far in the scientific literature, of an abstract comparison of two spike trains without accounting for the available time for such a comparison, is rather unrealistic. A finite available time ΔT places a crucial role in any decision, either if we are trying to identify an object within a fast sequence of different perceptions or if we are scanning trough memorized patterns in order to decide about an action.

As a result the perceptual space P per se is meaningless. What is relevant for cognition is the joint (P, T) space, since "in vivo" we have always to face a limited time ΔT which may truncate the whole spike sequence upon which a given perception has been coded. Only "in vitro" we allot to each perception all the time necessary to classify it.

A limited ΔT is not only due to the temporal crowding of sequential images, as reported clinically in behavioral disturbances in teenagers exposed to fast video games, but also to sequential conjectures that the semantic memory essays via different top-down signals. Thus in the metrical space (P, T), while the isolated localization of a percept P (however long is T) or of a time T (however spread is the perceptual interval ΔP) have a sense, a joint localization both in percept and time has an ultimate limit when the corresponding domain is less than the quantum area C.

Let us consider the following thought experiment. Take two percepts $P1$ and $P2$ which for long processing times appear as the two stable states of a bistable optical illusion, e.g the Necker cube. If we let only a limited observation time ΔT then the two uncertainty areas overlap. The contours drawn in figure 7 have only a qualitative meaning. The situation is logically equivalent to the non commutative coordinate-momentum space of a single quantum particle. In this case it is well known [32] that the quasiprobability Wigner function has strong non classical oscillations in the overlap region. We cannot split the coordinate-momentum space into two disjoint domains (sets) to which we can apply a Boolean logic or a classical Kolmogorov probability. This is the reason why the Bell inequalities are violated in an experiment dealing with such a situation [19]. The Wigner function formalism derives from a Schroedinger wavefunction treatment for pure state, and corresponding density matrix for mixed states.

In the perceptual (P, T) space no Schroedinger treatment is yet available, but we can apply a reverse logical path as follows.

The uncertainty relation $\Delta P \cdot \Delta T \geq C$ forbids a partition of the (P, T) space into sets. Therefore the (P, T) space is non commutative. Thus it must be susceptible of a Wigner function treatment and we can consider the contours of figure 7 as fully equivalent to isolevel cuts of a Wigner function. Hence we can introduce Schroedinger cat states and violations of Bell inequalities exactly as in quantum physics but with a reverse logical process, as illustrated in figure 8.

The equivalent of a superposition state should be a bistable situation observed for a time shorter than whole decision time. An experimental test is in preparation in my research group. Such a test should provide an estimation of the C value, which plausibly changes from individual to individual, and for a single one may be age and motivation dependent.

Thus, in neurophysics, time occurs under two completely different meanings, that is, as the ordering parameter to classify the position of successive events and as the useful duration of a relevant spike sequence, that is the duration of a synchronized train. In the second meaning, time T is a variable conjugate to perception P.

The quantum character has emerged as a necessity from the analysis of an interrupted spike train in a perceptual process. It follows that the (P, T) space cannot be partitioned into disjoint sets to which a Boolean yes/not relation is applicable and hence where ensembles obeying a classical probability can be considered. A set-theoretical partition is the condition to apply the Church-Turing thesis, which establishes the equivalence between recursive functions on a set and operations of a universal computer machine.

The evidence of quantum entanglement of overlapping perceptions should rule out in principle a finitistic character of the perceptual processes. This should be the negative answer to the Turing 1950 question whether the mental processes can be simulated by a universal computer [28].

Among other things, the characterization of the "concept" or "category" as the limit of a recursvive operation on a sequence of individual related perceptions gets rather shaky, since recursive relations imply a set structure.

Quantum limitations were also put forward by Penrose [21] but on a completely different basis. In his proposal, the quantum character was attributed to the physical behavior of the "microtubules" which are microscopic components of the neurons playing a central role in the synaptic activity. However, speaking of quantum coherence in biological processes is very hard to accept, if one accounts for the extreme vulnerability of any quantum system due to "decoherence" processes, which make quantum superposition effects observable only in extremely controlled laboratory situations.

References

[1] ALLARIA E., ARECCHI F.T., DI GARBO A., MEUCCI R., *Synchronization of homoclinic chaos*, Phys. Rev. Lett. **86** (2001), 791.

[2] ANDERSON J.R., *Cognitive Psychology and its Implications*, Freeman, San Francisco 1985.

[3] ARECCHI F.T., FARINI A., *Lexicon of Complexity*, Studio Editoriale Fiorentino, Firenze 1996.

[4] ARECCHI F.T., *Complexity and adaptation: a strategy common to scientific modeling and perception*, Cognitive Processing **1** (2000), 23.

[5] ARECCHI F.T., *Complexity versus complex system: a new approach to scientific discovery*, Nonlin. Dynamics, Psychology, and Life Sciences **5** (2001), 21.

[6] ARECCHI F.T., MEUCCI R., ALLARIA E., DI GARBO A., TSIMRING L.S., *Delayed self-synchronization in homoclinic chaos*, submitted for publication (2001).

[7] CALVIN W.H., *The Cerebral Code: Thinking a Thought in the Mosaics of the Mind*, MIT Press, Cambridge Mass 1996.

[8] CHURCHLAND P.S., SEJNOWSKI T.J., *The computational Brain*, MIT Press, Cambridge, Mass 1992.

[9] EDELMAN G.M., TONONI G., *Neural Darwinism: The brain as a selectional system* in "Nature's Imagination: The frontiers of scientific vision", J. Cornwell ed., Oxford University Press, New York 1995, 78–100.

[10] GROSSBERG S., *The attentive brain*, The American Scientist **83** (1995), 439.

[11] GROSSBERG S., *Review of the book by F. Crick: The astonishing hypothesis: The scientific search for a soul* 83 (n. 1) (1995).

[12] HUBEL D.H., *Eye, brain and vision*, Scientific American Library 22, W.H. Freeman, New York 1995.

[13] IZHIKEVICH E.M., *Neural Excitability, Spiking, and Bursting*, Int. J. of Bifurcation and Chaos **10** (2000), 1171.

[14] JULESZ B., *Early vision and focal attention*, Reviews of Modern Physics **63** (1991), 735–772.

[15] MACLEOD K., BACKER A., LAURENT G., *Who reads temporal information contained across synchronized and oscillatory spike trains?*, Nature **395** (1998), 693–698.

[16] MANTEGNA R.N., STANLEY H.E., *An Introduction to Econophysics*, Cambridge University Press, Cambridge UK 2000.

[17] MEUCCI R., DI GARBO A., ALLARIA E., ARECCHI F.T., *Autonomous Bursting in a Homoclinic System*, submitted for pubblication (2001).

[18] NEWELL A., SIMON H.A., *Computer science as empirical inquiry*, in "Mind Design", J. Haugeland ed., MIT Press, Cambridge 1976, 35–66.

[19] OMNÈS R., *The interpretation of Quantum Mechanics*, Princeton University Press, Princeton NJ 1994.

[20] PARPURA V., HAYDON P.G., *Physiological astrocytic calcium levels stimulate glutamate release to modulate adjacent neurons*, Proc. Nat. Aca. Sci. **97** (2000), 8629.

[21] PENROSE R., *Shadows of the Mind*, Oxford University Press, New York 1994.

[22] RIEKE F., WARLAND D., DE RUYTER VAN STEVENINCK R., BIALEK W., *Spikes: Exploring the neural code*, MIT Press, Cambridge Mass 1996.

[23] RODRIGUEZ E., GEORGE N., LACHAUX J.P., MARTINERIE J., RENAULT B., VARELA F., *Perception's shadow: Long-distance synchronization in the human brain*, Nature **397** (1999), 340–343.

[24] ROSSUM VAN M., *A novel spike distance*, Neural Computation **13** (2001), 751.

[25] SIMON H., *The architecture of complexity*, Proc. Amer. Philos. Soc. **106** (1982), 467–485.

[26] SINGER W.E., GRAY C.M., *Visual feature integration and the temporal correlation hypothesis*, Annu. Rev. Neurosci. **18** (1995), 555.

[27] SOFTKY W., *Simple codes versus efficient codes*, Current Opinions in Neurobiology **5** (1995), 239.

[28] TURING A., *Computing Machinery and Intelligence*, Mind **59** (1950), 433.

[29] VICTOR J.D., PURPURA K.P., *Metricspace analysis of spike trains: theory, algorithms and application*, Network: Comput. Neural Syst. **8** (1997), 127–164.

[30] VON DER MALSBURG C., *The correlation theory of brain function*, reprinted in "Models of neural networks II", E. Domani, J.L. Van Hemmen and K. Schulten eds., Springer, Berlin 1981.

[31] WEBER M., WELLING M., PERONA P., *Unsupervised Learning of Models for Recognition*,in "Proc. 6th Europ. Conf. Comp. Vis., ECCV2000", Dublin, Ireland 2000.

[32] ZUREK W.H., *Decoherence and the transition from quantum to classical*, Phys. Today, October issue (1991), 36.

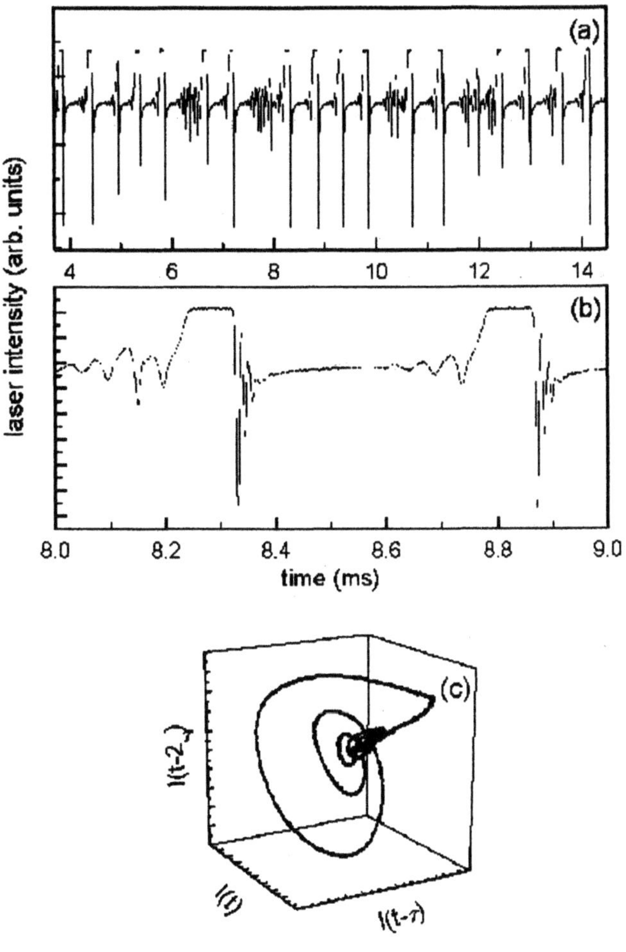

Figure 1. (a) Experimental time series of the laser intensity for a CO2 laser with feedback in the regime of homoclinic chaos. (b) Time expansion of a single orbit. (c) Phase space trajectory built by an embedding technique with appropriate delays (from [1]).

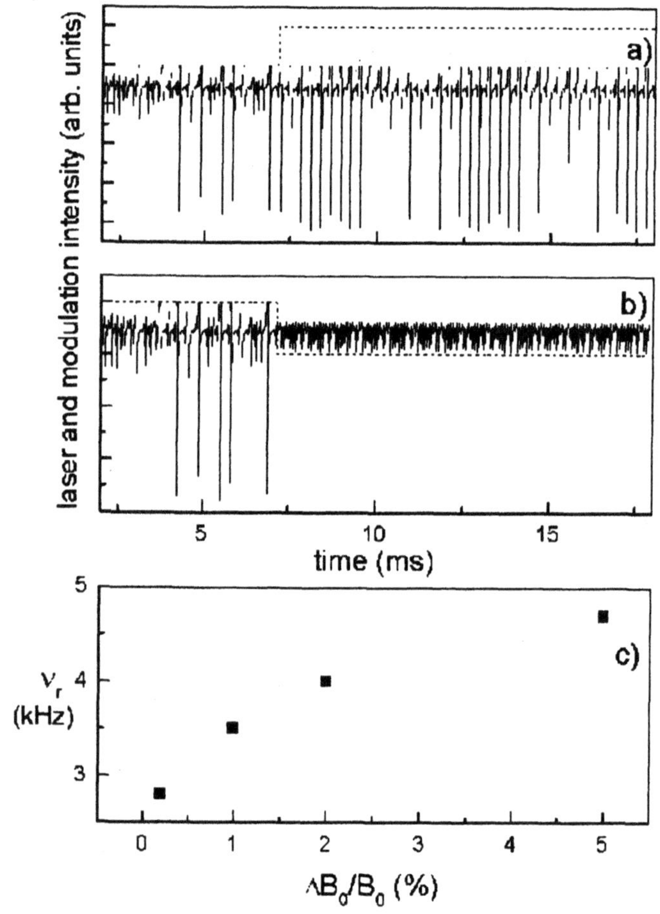

Figure 2. Stepwise increase a) and decrease b) of control parameter B0 by +/ − 1% brings the system from homoclinic to periodic or excitable behavior. c) In case a) the frequency ν_r of the spikes increases monotonically with ΔB_0 (from [17]).

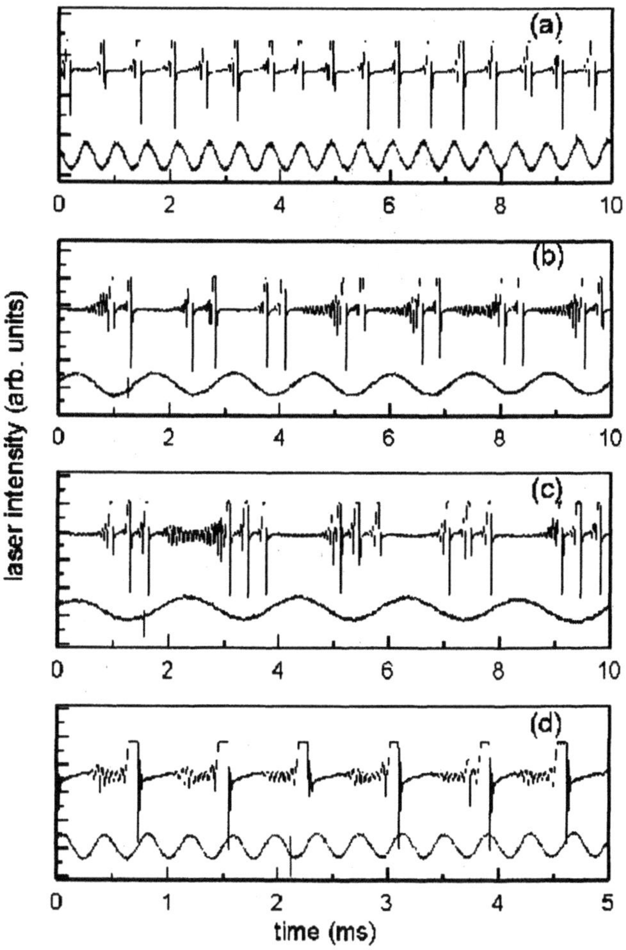

Figure 3. Experimental time series for different synchronization induced by periodic changes of the control parameter. (a) 1:1 locking, (b) 1:2, (c) 1:3, (d) 2:1 (from [1]).

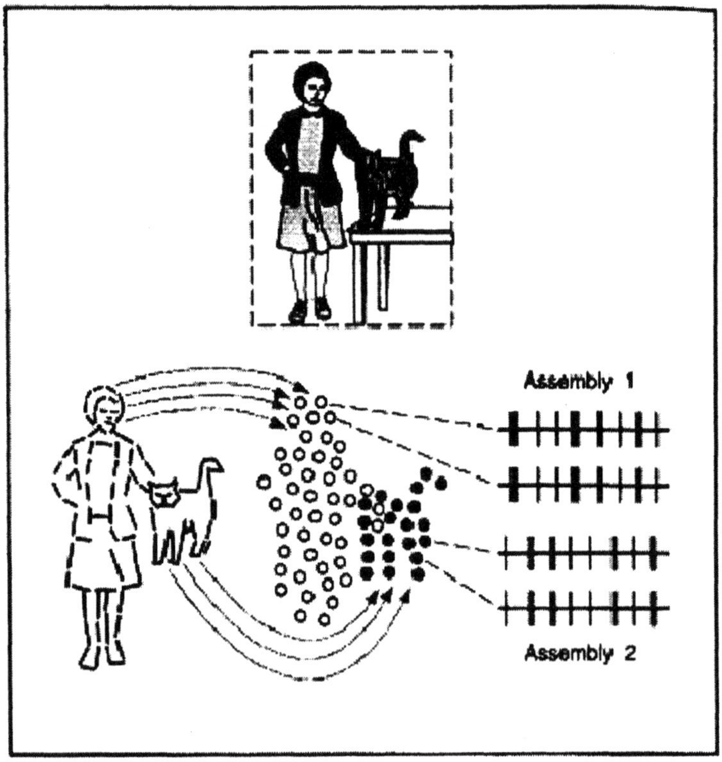

Figure 4. Feature binding: the lady and the cat are respectively represented by the mosaic of empty and filled circles, each one representing the receptive field of a neuron group in the visual cortex. Within each circle the processing refers to a specific detail (e.g. contour orientation). The relations between details are coded by the temporal correlation among neurons, as shown by the same sequences of electrical pulses for two filled circles or two empty circles. Neurons referring to the same individual (e.g. the cat) have synchronous discharges, whereas their spikes are uncorrelated with those referring to another individual (the lady) (from [26]).

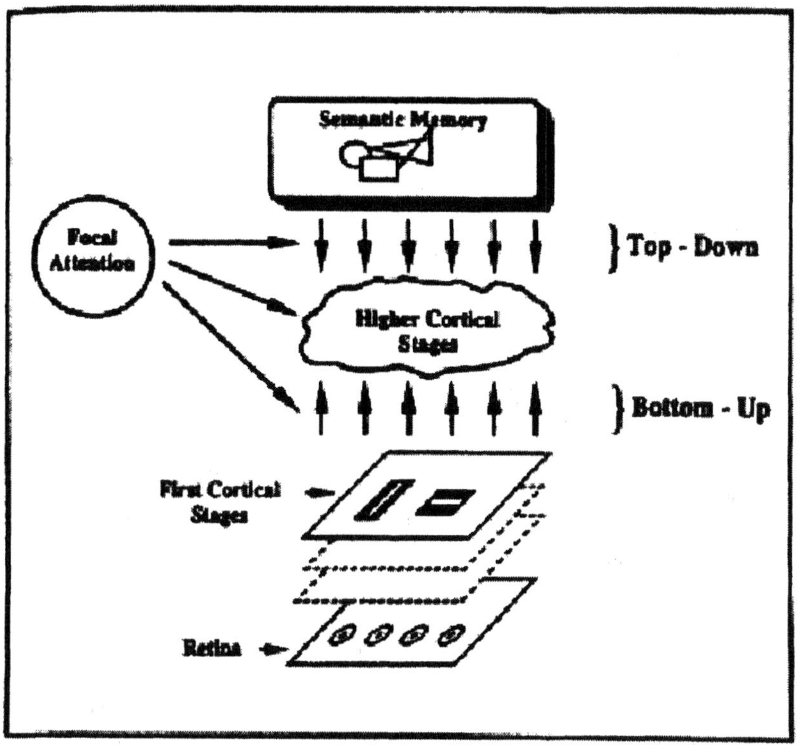

Figure 5. ART = Adaptive Resonance Theory. Role of bottom-up stimuli from the early visual stages an top-down signals due to expectations formulated by the semantic memory. The focal attention assures the matching (resonance) between the two streams (from [14]).

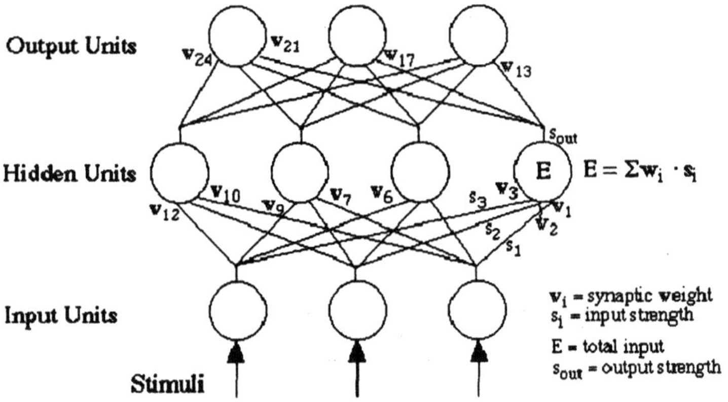

Figure 6. A simple network including one example of a neuronlike processing unit. The state of each neural unit is calculated as the sum of the weighted inputs from all neural units at a lower level that connect to it.

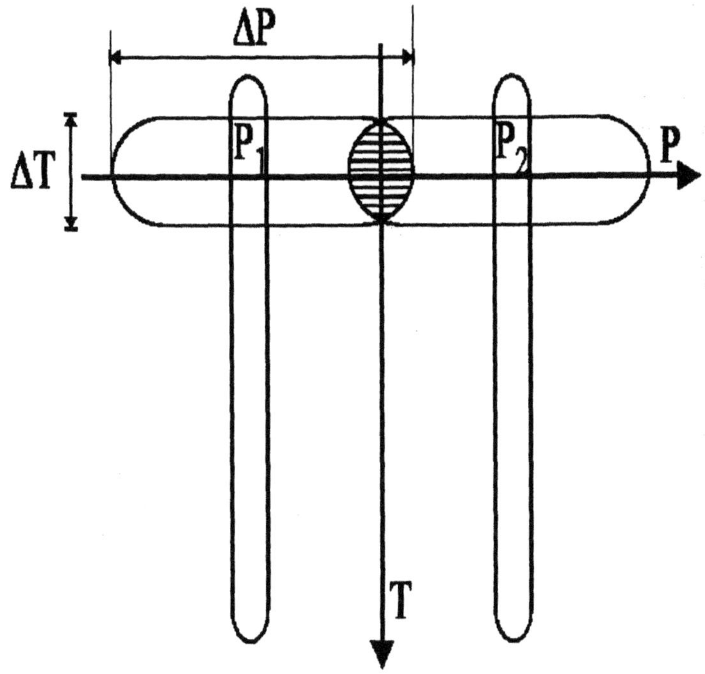

Figure 7. Uncertainty areas of two perceptions P_1 and P_2 for two different durations of the spike trains. In the case of short ΔT, the overlap region is represented by a Wigner function with strong positive and negative oscillations which go as $\cos \frac{\Delta P}{C} T$ along the T axis; therefore with a frequency given by the ratio of the percept separation $\Delta P = P_2 - P_1$ to the perceptual "Planck's constant" C.

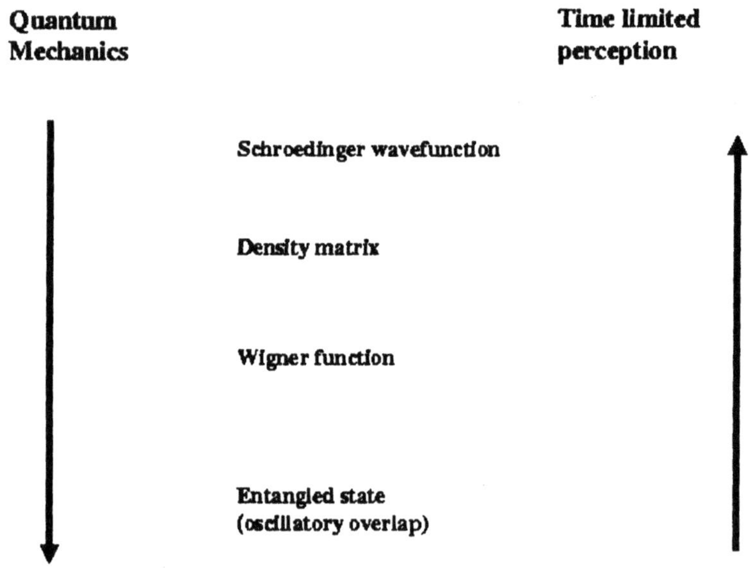

Figure 8. Direction of the logical processes which lead from wavefunction to entangled states or viceversa.

A GEOMETRIC OPTICS EXPERIMENT TO SIMULATE THE BETATRONIC MOTION

Armando Bazzani
Dipartimento di Fisica,
INFN e CIG dell'Università di Bologna,
Via Irnerio 46, 40126 BOLOGNA - ITALY
bazzani@bo.infn.it

Paolo Freguglia
Dipartimento di Matematica Pura ed Applicata,
Università degli Studi dell'Aquila
and
Domus Galilaeiana, PISA - ITALY

Leone Fronzoni
Dipartimento di Fisica, Università di Pisa,
INFM e CISSC dell'Università di Pisa, PISA - ITALY

Giorgio Turchetti
Dipartimento di Fisica,
INFN e CIG dell'Università di Bologna,
Via Irnerio 46, 40126 BOLOGNA - ITALY

Abstract We consider the analogy between geometric optics and mechanics to simulate the betatronic motion in a particle accelerator by using an optic experiment. Some experimental results are discussed and the main difficulties to improve the actual performances are briefly presented. We also introduce a representation of a ray trajectory by means of the quaternion numbers that can be used to study the geometrical properties.

1. Introduction

The Hamiltonian dynamics [1] has been developed from the Fermat's principle of geometric optics using the analogy between the light propagation and the particle motion. The Hamilton's formulation of the equations of a mechanical system has been applied from long time to the problems of celestial mechanics and rigid body dynamics and it has yielded a great number of fundamental mathematical and physical results [2].

The application of the Hamiltonian mechanics to the everyday physics is difficult due to the unavoidable presence of dissipative effects and stochastic external perturbations. In the recent years the particle accelerator physics has provided new fields of application: a proton bunch confined in a vacuum pipe by an external magnetic field is a very good example of Hamiltonian system and the problem of the transverse confinement of a particle can be studied by the methods and the results of Hamiltonian dynamics [3]. Moreover the particle accelerators could be used as analogic experiments to test the physical applicability of the theoretical results. The dynamics on a transverse section of a particle accelerator is called as betatronic motion and it is based on the FODO magnetic structure [4] where two subsequent quadrupole magnets produce a global focusing effect on both the transverse directions. The non-linear effects present in the real accelerators perturbs the linear stability of the motion so that the definition of dynamics aperture is a very difficult problem [3].

Very recently we have proposed an optical experiment of accelerator physics based on the analogy between geometric optics and mechanical systems [5]. Our goal was to explore the possibility and the limits of an optical FODO cell to simulate the betatronic motion in a particle accelerator [6]. In this paper we compare the first experimental results with the theoretical simulations and we discuss the future developments. We also take advantage from the Hamilton's application of the quaternionic numbers [7] to dynamical problems [8] and we propose a representation of a light ray orbit by means of a quaternions sequence [5].

In spite of the intensity reduction of a light beam across an optical device, the dynamics in the geometric optics approximation preserves the Hamiltonian character. Therefore we remark the possibility of the geometric optics to be used for the project of simple (and not expensive) experiments of Hamiltonian dynamics that point out some interesting features to be interpreted by the theoretical simulations.

2. Dynamics of a magnetically confined charge particle

The dynamics of a charge particle in presence of a magnetic field satisfies the Lorentz equation

$$\frac{d\mathbf{p}}{dt} = \frac{e}{c}\mathbf{v} \times \mathbf{B} \tag{1}$$

where $\mathbf{p} = m\gamma\mathbf{v}$ is the mechanical momentum and $\mathbf{B}$ is the magnetic field. It is possible to write equation (1) in the canonical form by introducing the Hamilton function [10]

$$H = \sqrt{m^2c^4 + c^2\left(\mathbf{P} - \frac{e}{c}\mathbf{A}\right)^2} \tag{2}$$

where $\mathbf{A}$ is the vector potential associated to the magnetic field (rot$\mathbf{A} = \mathbf{B}$) and $\mathbf{P} = \mathbf{p} + \mathbf{A}e/c$ is the generalized momentum. Then the distribution function $\rho(\mathbf{x}, \mathbf{P}, t)$ of an ensemble of particles in the phase space $(\mathbf{x}, \mathbf{P})$ satisfies the Liouville equation

$$\frac{\partial \rho}{\partial t} + \frac{\partial \rho}{\partial \mathbf{x}}\frac{\partial H}{\partial \mathbf{P}} - \frac{\partial \rho}{\partial \mathbf{P}}\frac{\partial H}{\partial \mathbf{x}} = 0 \tag{3}$$

In the modern accelerators the magnetic confinement of a particle in the vicinity of e reference orbit (central orbit) is realized by using magnetic elements that introduce a vector potential $\mathbf{A} = A(x, y, s)\hat{s}$ where $\hat{s}$ is the direction of the reference orbit, s is the arc length of the reference orbit and (x, y) define the transverse plane to the orbit. Taking into account the Maxwell equations [10], the vector potential $\mathbf{A}$ can be written in the form

$$A(x, y, s) = \mathrm{Re}\sum_{k\geq 2} \frac{B_k(s)}{k}(x + iy)^k \tag{4}$$

where i is the imaginary unit. The coefficients B_k define the k-pole components of the magnetic field and are associated with the quadrupole, sextupole... magnetic elements. The Hamiltonian (2) can be simplified if one uses the coordinate s as parameter, and the new Hamiltonian reads [3]

$$H_s = -\sqrt{1 - p_x^2 - p_y^2} - \frac{e}{p_0 c}A(x, y; s) \tag{5}$$

where p_0 is the total mechanical momentum and p_x (resp. p_y) is the ratio between the transverse momentum along the $\hat{x}$ (resp. $\hat{y}$) direction and the total momentum. Taking into account the physical hypothesis $x, y, p_x, p_y \ll 1$ the Hamiltonian (5) can be approximated in the form

$$H_s = \frac{1}{2}(p_x^2 + p_y^2) - \frac{eB_2(s)}{2p_0c}(x^2 - y^2) + \dots \quad (6)$$

that it is the Hamiltonian of the betatronic motion in a particle accelerator. We remark that the dependence on s of the Hamiltonian (6) is discontinuous due to the discontinuous structure of the magnetic lattice. For this reason it is convenient to study the dynamics using a map formalism instead of differential equation. The basic idea is to associate to each magnetic element a transfer map M_k that gives the correspondence between the dynamical status of a particle at fixed transverse sections of the reference orbit that separate the various magnetic elements (see Fig. 3); the composition of the transverse maps allow to construct the orbit in the phase space.
The basic magnetic structure that is used to achieve the linear stability of the transverse dynamics in the vicinity of the reference orbit is the FODO magnetic structure [4]. A FODO cell is composed by two quadrupole magnets of different polarity (focusing and defocusing) separated by two empty spaces (drifts) (see Fig. 3). A focusing quadrupole is defined by the condition $eB_2/p_0c < 0$ and introduces an attractive elastic force along the x direction and a repulsive elastic force along the y direction; the converse happens for the defocusing quadrupole. We observe that according to the definition (4) it is not possible to achieve an attractive force on both the transverse directions, but with a suitable choice of the parameters (strength of the quadrupole field and length of the empty spaces) the global effect of a FODO cell is a linear focalization in the transverse plane. In such a case the dynamics of a particle in the phase space is a rotation along invariant ellipses with fixed rotation angles that are called *betatronic tunes.*

3. Geometric Optics

The propagation of the light is described by the wave equation for the electro-magnetic field [10]

$$\frac{\partial^2 E}{\partial \mathbf{x}^2} - \frac{n^2}{c^2}\frac{\partial^2 E}{\partial t^2} = 0 \quad (7)$$

where n is the refraction index of the medium; the ratio c/n is the light velocity in the medium. In the limit of small wave length we look for a solution in the form $E(\mathbf{x}, t) = e(\mathbf{x}) \exp(ik_0 S(\mathbf{x}) - i\omega t)$ for $k_0 \to 0$. The

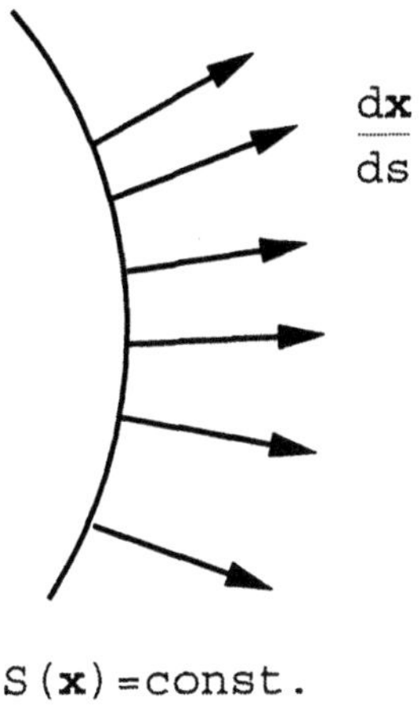

Figure 1. Wave front propagation according to the Geometric Optics.

existence of a non-trivial solution implies the eikonal equation

$$\left(\frac{\partial S}{\partial \mathbf{x}}\right)^2 = n^2(\mathbf{x}) \tag{8}$$

where the neglected terms are of order $O(1/k_0)$ time the derivatives of the field intensity and the refraction index: therefore equation (8) cannot be used when the various functions change in a space scale of the order of the wavelength. The eikonal equation is the basic equation to study the light propagation in the limit of geometric optics. The wave propagates parallel to $\partial S/\partial \mathbf{x}$ so that we can introduce the light rays equation

$$n\frac{d\mathbf{x}}{ds} = \frac{\partial S}{\partial \mathbf{x}} \tag{9}$$

where s is the arc length of the trajectory and the surfaces $S(\mathbf{x}) = const.$ define the wave front (see Fig. 1) Following the optics-mechanics analogy of W. Hamilton [1] one recognizes that the eikonal equation has the structure of a Hamilton-Jacobi equation of mechanical systems [2]

$$H\left(\mathbf{x}, \frac{\partial S}{\partial \mathbf{x}}\right) = E \tag{10}$$

where the energy E is 0 and the Hamiltonian reads

$$H = \mathbf{p}^2 - n^2(\mathbf{x}) \tag{11}$$

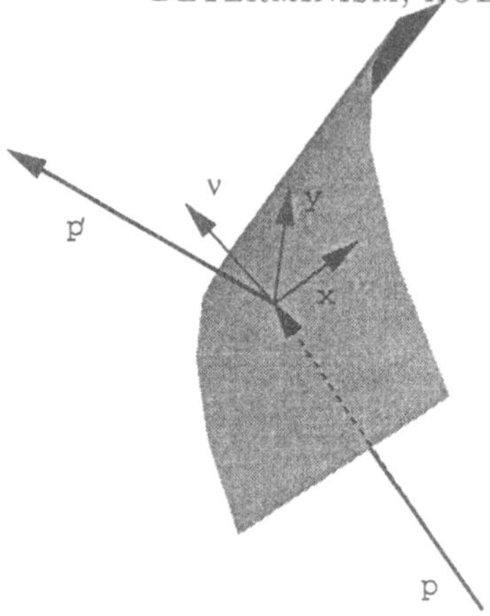

Figure 2. Change of the ray direction at a discontinuity surface (Snell's law).

Only the trajectories with $H = 0$ are physical so that we have the constraint $\|\mathbf{p}\| = n$. We shall consider the possibility of simulating the dynamics described by the Hamiltonian (6) with an Hamiltonian of the form (11) taking into account that in optics the discontinuity surfaces of the refraction index are very easy to realize so that a map formalism is more suitable to describe the ray dynamics. At a discontinuity surface we have to use the *Snell's law* of refraction that can written in the form

$$\mathbf{p} \cdot \mathbf{t} = \mathbf{p}' \cdot \mathbf{t} \tag{12}$$

$\forall\ \mathbf{t}$ tangent to the surface where $\mathbf{p}$ (resp. $\mathbf{p}'$) is the momentum of the incoming (resp. out-coming) ray (see Fig. 2). The Snell's law means that the component of the ray momentum tangent to the surface is continuous whereas the normal component changes according to $n' \sin\theta' = n \sin\theta$ where θ is the angle between the ray direction and the normal to the surface. From a geometric point of view the direction of the light ray is rotated in the plane defined by the normal and the incoming direction with a rotation angle that depends on the incident point [5].

We explicitly consider the case of a symmetric lens whose surfaces are defined by the formula $s = f(x, y)$ when the ray direction is almost parallel to $\hat{s}$ and the s-axis itself is a physical trajectory (reference orbit); (x, y, s) is a fixed reference system in the space. In the limit $p_x, p_y \ll 1$ the Snell's law can be approximated according to

$$p'_x = p_x + \left(\sqrt{n'^2 - n^2(\nu_x^2 + \nu_y^2)} - n\sqrt{1 - (\nu_x^2 + \nu_y^2)}\right)\nu_x \tag{13}$$

$$p'_y = p_y + \left(\sqrt{n'^2 - n^2(\nu_x^2 + \nu_y^2)} - n\sqrt{1 - (\nu_x^2 + \nu_y^2)}\right)\nu_y \tag{14}$$

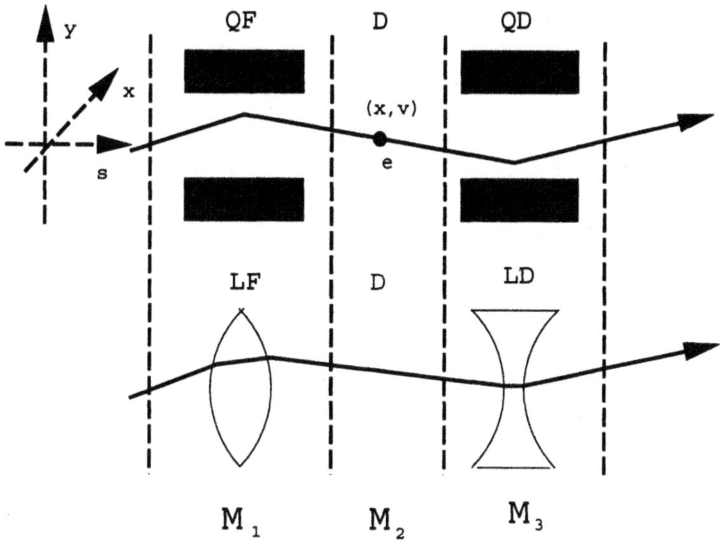

Figure 3. Top figure: sketch of the trajectory of a particle along a FODO cell: QF, QD denote the focusing and defocusing quadrupoles and D denotes the drift space. Bottom figure: scheme of an optical FODO cell; LF and LD denote spherical lenses with opposite convexity. The dashed vertical lines define the transverse sections of the reference orbit used to compute the transfer maps M_k.

where

$$\nu_{x,y} = -\frac{\frac{\partial f}{\partial x,y}}{\sqrt{1 + \left(\frac{\partial f}{\partial x}\right)^2 + \left(\frac{\partial f}{\partial y}\right)^2}} \tag{15}$$

are the components of the normal versor to the surface. It is not difficult to prove that for $x, y, \ll 1$ (linearized dynamics) the effect of a focusing quadrupole can be simulated by a symmetric spherical lens of different convexity along the x and y directions ($f(x, y) = R - \sqrt{R^2 - x^2 + y^2}$) and with a correspondence $eB_2/p_0c = 2(n' - n)/R$ among the parameters. The defocusing effect is achieved by exchanging the role of the x and y coordinates. Therefore we can in principle construct an optical FODO-like cell by using two spherical lenses of opposite convexity according to figure 3. The dynamics can be studied introducing the transfer map M_k for each element.

4. Optics and Quaternions

The quaternion numbers has been defined by W. Hamilton [7] to introduce the division operation among the vectors in $\mathbb{R}^3$. The quaternion numbers are a generalization of the complex numbers and they contain three imaginary units i, j, k which obey to the algebraic rules

$$i^2 = j^2 = k^2 = -1 \qquad ij = k \qquad ji = -k \tag{16}$$

As a consequence any quaternion number can be written in the form

$$q = q_0 + q_1 i + q_2 j + q_3 k \qquad q \in \mathbb{R}^4 \tag{17}$$

The product between quaternions is defined according to the rules (16) and it is distributive, associative and anticommunative. We introduce the norm of a quaternion q according to

$$\|q\|^2 = q\overline{q} = q_0^2 + q_1^2 + q_2^2 + q_3^2 \tag{18}$$

where $\overline{q} = q_0 - q_1 i - q_2 j - q_3 k$ is the conjugate quaternion. The idea of Hamilton was to give to the quaternions in $\mathbb{R}^3$ the same role as the complex numbers in $\mathbb{R}^2$. Let $\mathbf{x} \in \mathbb{R}^3$ any vector, we define the quaternion $q_\mathbf{x} = \mathbf{i}\cdot\mathbf{x}$ where $\mathbf{i} = (i, j, k)$ is the vector of the imaginary units. It is also possible to use a antisymmetric matrix representation of the quaternion numbers which is a homeomorphism between the quaternion algebra and the matrix algebra [9]. Moreover following Hamilton, we associate a unitary quaternion to a unitary vector $\hat{v}$ and an angle α according to

$$q_{\alpha,\hat{v}} = \cos\frac{\alpha}{2} + \mathbf{i}\cdot\hat{v}\sin\frac{\alpha}{2} \tag{19}$$

The algebra of quaternion number has then a simple geometrical interpretation: the product

$$q_{\alpha,\hat{v}}\, q_\mathbf{x}\, \overline{q}_{\alpha,\hat{v}} = q_{\mathbf{x}'} \tag{20}$$

represents a rotation of the vector $\mathbf{x}$ by an angle α around the axis $\hat{v}$ [7]. As a consequence the quaternions algebra allows a non-singular representation of the rotations group $SO(3)$. As a corollary of the previous result if $\hat{v}\cdot\mathbf{x} = 0$ then equation (20) can be simplified

$$q_{\alpha,\hat{v}}\, q_\mathbf{x}\, \overline{q}_{\alpha,\hat{v}} = q_{2\alpha,\hat{v}}\, q_\mathbf{x} = q_{\mathbf{x}'} \tag{21}$$

Equation (21) is easily inverted $q_{2\alpha,\hat{v}} = q_{\mathbf{x}'}\overline{q}_\mathbf{x}/\|\mathbf{x}\|^2$. We can use the quaternion algebra to get a geometrical description of the ray orbit in the geometric optics approximation. At any point the ray direction is defined by the momentum vector $\mathbf{p}$ with the constraint $\|\mathbf{p}\| = n$; since

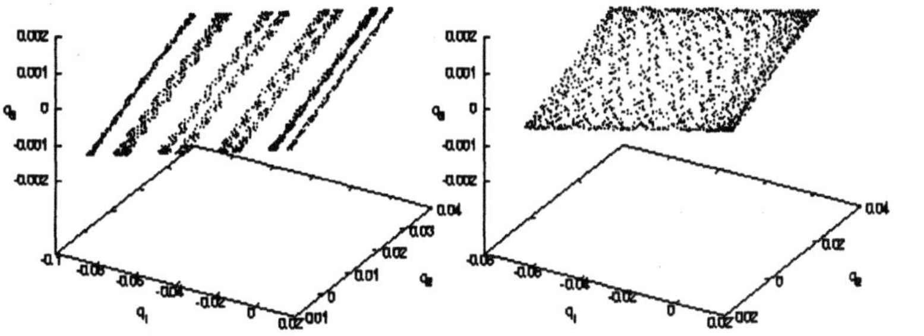

Figure 4. Geometric representation of an optical path by using a sequence of quaternion numbers; the right picture show the imaginary part of the quaternions associated to a regular orbit in the FODO like cell; the left part refers to a resonant orbit.

in a region of constant refraction index n the light propagates along straight lines, the optical path is completely determine by an initial points and a sequence of vectors $\mathbf{p}_k$ that gives the change of direction at the discontinuity surfaces of the refraction index. Moreover for any k there exists a unique quaternion $q_{2\alpha_k,\hat{v}}$ such that

$$q_{2\alpha_k,\hat{v}}\, q_{\mathbf{p}_0} = q_{\mathbf{p}_k} \Rightarrow q_{2\alpha_k,\hat{v}} = q_{\mathbf{p}_k}\, \overline{q}_{\mathbf{p}_0} \tag{22}$$

An algebraic calculation gives the solution of equation (22) in the form

$$q_{2\alpha_k,\hat{v}} = \mathbf{p}_k \cdot \mathbf{p}_0 - \mathbf{i}(\mathbf{p}_k \times \mathbf{p}_0) \tag{23}$$

where $\times$ is the usual vector product. The quaternion sequence $q_{2\alpha_k,\hat{v}}$ is a representation of the optical orbits. We have applied such idea to the optical path in the FODO structure of figure 3 [5]. Due to our hypotheses the trajectory presents only small deviations with respect to the reference orbit and the real component of the quaternions $q_{2\alpha_k,\hat{v}}$ is much bigger than the imaginary components. Therefore it is enough to plot the imaginary part to have information on the geometry of the trajectory. The results are shown in the figure 4: the right plot refers to a regular orbit and the left plot refers to a resonant orbit where the imaginary parts of $q_{2\alpha_k,\hat{v}}$ clusterize on separated sets as it is expected from the Birkhoff theorem [2]. According to equation (23) the quaternion representation of an orbit is related to a Lagrangian representation in the momentum space instead of the usual coordinate space. The quaternions $q_{2\alpha_k,\hat{v}}$ belongs to the 3D unitary sphere $\|q\| = 1$, but the

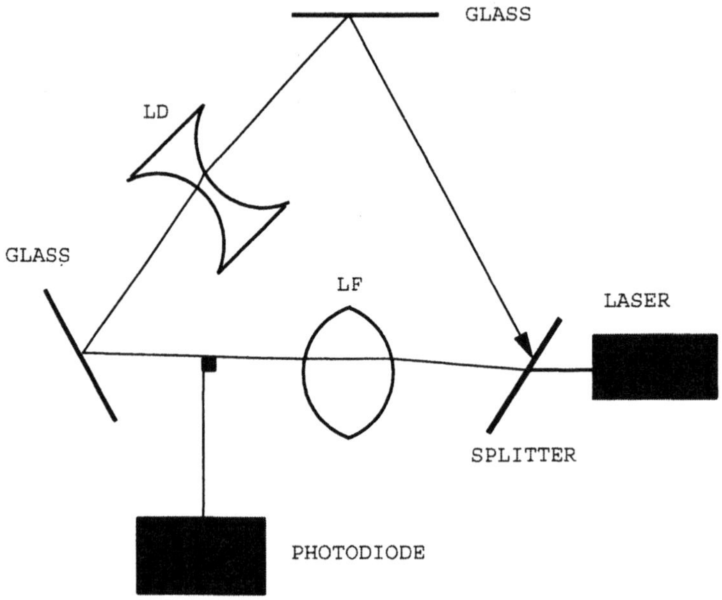

Figure 5. Scheme of the experimental set up of an optical FODO cell.

real component is much bigger then the imaginary components and the last component is negligible with respect to the others. As a consequence from a geometrical point of view the sequence $q_{2\alpha_k,\hat{v}}$ apparently explores of a small region of a 2D sphere ; this is the reason why the points in the figure 4 seem to lie on a plane.

5. Experimental results

We have used the previous idea to construct an optical-like FODO cell according to the scheme described in figure 5. A laser beam is injected in the device by using a beam splitter with a high reflection coefficient ($\simeq 90\%$) and it is confined by means of two glasses and two spherical lenses of opposite convexity. The distance between the lenses is .2 m. and the refraction index is $n = 1.4$ that corresponds to a focusing length of .15 m. These parameters give to a *betatronic tune* of $\simeq 90°$ for the linearized dynamics and we expect that the beam returns near to the initial position after 4 turns of the device. The light intensity is

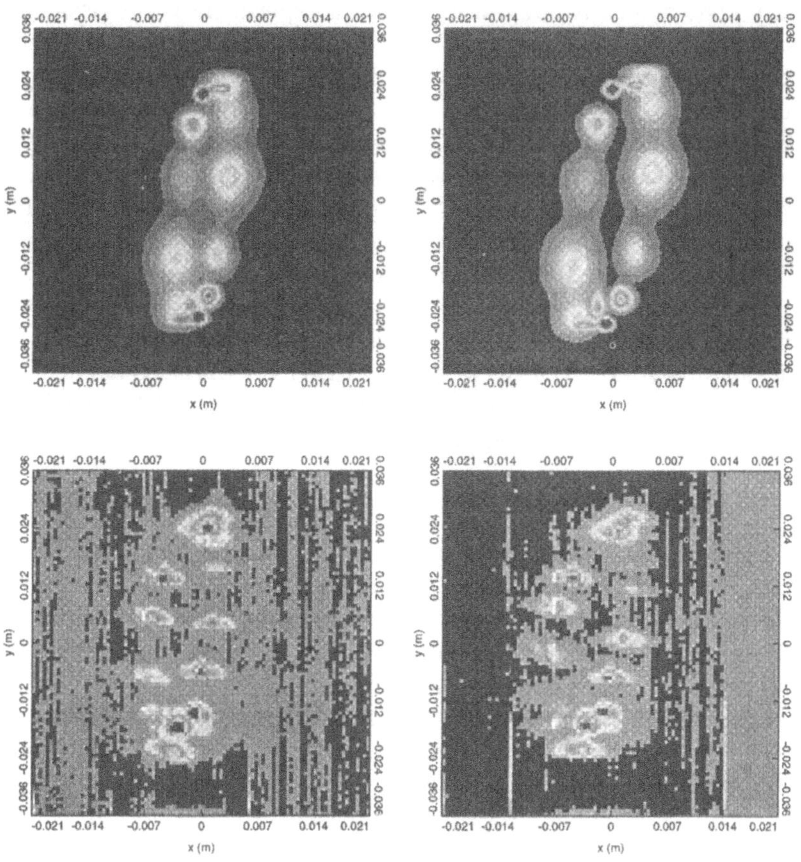

Figure 6. Comparison between the experimental measures (bottom pictures) of the light intensity (the scale of colors from blue to red corresponds to a logarithmic scale in the light intensity) the integration of the Liouville (upper pictures) equation for the optical FODO cell; the initial conditions is a gaussian distribution in the phase space centered at $x = 3$ mm. $p_x = 0$ $y = 7.02$ mm. $p_y = .042$ (left part) and $x = 3$ mm. $p_x = 0$ $y = 7.41$ mm. $p_y = .045$ (right part); the number of spots is $\simeq 15$ and gives to the number of turns of the device.

registered by a photo-diode at a fixed section of the reference orbit.
The purpose of this experiment is to control the real possibility of using an optical device to analogically simulate the betatronic motion in a particle accelerator. The experimental results have pointed out that the intensity losses reduce drastically the number of turns which are detectable by the photo-diode and that the diffraction destroys the beam

quality. The main advantages of an optical experiment is the possibility of simulating the dynamics of the whole beam and not only the single particle dynamics and of preserving the Hamiltonian character of the dynamics without friction effects. We have compared the experimental results with the solution of the Liouville equation (3) for a gaussian initial distribution of light intensity centered at different initial (x, y, p_x, p_y) coordinates at the beam splitter. In figure 6 we plot the experimental light intensity (bottom part) for initial conditions $x = 3$ mm. $p_x = 0$ $y = 7.02$ mm. $p_y = .042$ (left part) and $x = 3$ mm. $p_x = 0$ $y = 7.41$ mm. $p_y = .045$ (right part). We can see a good agreement between theory and experiment even if some differences are present probably due to the alignment errors of the lenses and the errors in the measure of the lenses parameters. The number of turns is of the order of 15 that is not sufficient to detect non-linear effects due to aberration of the lenses.

6. Conclusions

A optical FODO structure can be used to simulate the betatronic dynamics in a particle accelerator. The main problem is the very limited number of turns due to the losses of light intensity. The possible advantages are the analogic simulation of the whole beam dynamics and the preservation of the Hamiltonian character.
A representation of the orbit by using a sequence of quaternion numbers is able to give information on the geometric properties and further studies are necessary.

References

[1] HAMILTON W.R., *Theory of systems of rays*, Trans. Of the Royal Irish Academy **15** (1828), 69–174.

[2] ARNOLD V.I., *Méthod mathématiques de la mécanique classique*, MIR, Moscow 1976.

[3] BAZZANI A., TODESCO E., TURCHETTI G., *A normal form approach to the theory of the nonlinear betatronic motion*, CERN 94-02 (1994).

[4] COURANT E., SNYDER H., *Theory of the alternating-gradient synchrotron*, Ann. Phys. (N.Y.) **3** (1958), 1–48.

[5] FREGUGLIA P., TURCHETTI G. EDS., *Mechanics and Geometry, some topics*, Quattroventi, Urbino 2002.

[6] BAZZANI A., FREGUGLIA P., FRONZONI L., TURCHETTI G., *An optical experiment towards the analogic simulation of the betatronic motion*, AIP Conference Proceedings **581** (2001), 211–220.

[7] HAMILTON W.R., *On quaternions*, Trans. Of the Royal Irish Academy **3** (1847), 1–16.

[8] HAMILTON W.R., *On the application of the method of quaternions to some dynamical questions*, Trans. Of the Royal Irish Academy **3** (1847) Appendix, XXXXVI–L.

[9] KUIPERS J.B., *Quaternions and Rotation Sequences*, Princeton University Press, Princeton 1999.

[10] LANDAU L.D., LIFSITS E.M., *Teoria dei campi*, Editori Riuniti, Roma 1981.

SOME REMARKS ON THE ARROW OF TIME AND THE NOTION OF INFORMATION

Vieri Benci
Dipartimento di Matematica Applicata "U. Dini",
Università degli studi di Pisa,
Via Bonanno 25/b, 56121 PISA - ITALY
benci@dma.unipi.it

Abstract We discuss the relation between the notions of information content and entropy and their relations with the problem of irreversibility. The notion of information content is not uniquely defined and we analyze various definitions and their physical meaning. We think that this analysis shed new light on the problem of irreversibility.

1. The arrows of time

The problems related to time are as old as human thinking. One of the most fascinating and unsettling problem is the arrow of time. In modern Science the most meaningful indicators of flowing of time and of its direction are essentially two:

(II law) The second law of thermodynamics: the passing of time destroys information. Time is Cronos who devours his offspring. Everything is consumed by the flow of time, and, at the very end, the universe will be an undifferentiated mass where even light and darkness will be hopelessly mixed together.

(Evolution) Historical, biological, cosmological evolution: the passing of time creates information. In cosmological evolution, light is separated from darkness; galaxies, stars, and planets take their form; in biological evolution life arises from mud and bacteria, protista, fungi, plants and animals evolve in always more complex forms; and then intelligence appears and evolution continues in history and gives origin to more and more complex civilizations.

These two aspects of the arrow of time are apparently contradictory with each other. But there is more to say: both of them are in contradiction with the fundamental law of physics.

In fact the fundamental laws of physics are reversible: they do not distinguish past and future. From a mathematical point of view, the laws of physics are expressed by differential equations (e.g. the equations of Hamilton, Maxwell, Schroedinger, Einstein etc.). The "state" of any physical system at time t is described by a function $u(t)$ which solves the equations involved. A peculiarity of all these equations lies in the fact that if $u(t)$ is a solution, also $u(-t)$ is a solution. This is the meaning of the word "reversibility", at least in this context. This fact, translated in the physical world, has the following meaning: if $u(t)$ describes the evolution of a physical system, also $u(-t)$ represents a possible evolution of the same system (with different initial conditions). This fact is in evident contradiction with experience. Nobody is born old and gets younger until becoming a child and finally disappearing in an egg.

Today, these apparent contradictions are understood reasonably well. The relation between the second law and the irreversibility of the fundamental law of physics has been explained by Boltzmann. His theory has received and receives many objections, but it is essentially correct (at least in the opinion of most scientists). The evolution arrow and its apparent contradiction with the second law, in recent times, has been object of a lot of attention and the study of dissipative systems explains reasonably well the underlying mechanisms. Nevertheless, still there are many subtle questions to be settled from many points of view: philosophical, physical and mathematical. One of this question is related to the meaning of "information" and its relation with the many notions of "entropy". This is the point which will be discussed in this paper.

2. Information Content

The first mathematical notion of "information content" of a string is due to Shannon. However the Shannon notion of information relies strongly on the notion of probability and this is very disappointing for the aims of this paper for the following reasons:

- we think that from an epistemological point of view the notion of probability is quite obscure and will not help to clarify the nature of notions such as "entropy" and irreversibility

- we think that the notion of information is primitive and that the notion of probability should be derived from it

- our goal is to give a definition which can be applied also to cellular automata and to computer simulations and this objects are strictly deterministic; thus the notion of probability should be avoided at least as primitive concept.

However there are measures of information which depend intrinsically on the string and not on its probability within a given context. We will give a general definition of information content which apply to many different contexts.

Let $\mathcal{A}$ be an alphabet, namely a finite collection of "letters" and let $\mathcal{A}^*$ denote the set of finite strings in this alphabet, namely

$$\mathcal{A}^* = \bigcup_{n=1}^{\infty} \mathcal{A}^n$$

Given a string $\sigma \in \mathcal{A}^*$, $|\sigma|$ will denote its length. If $\sigma \in \mathcal{A}^*$, we set

$$S_0(\sigma) = \log W(\sigma)^1 \tag{1}$$

where $W(\sigma)$ is the number of strings which can be obtained by σ permuting its letters. Notice that

$$S_0(\sigma) \leq |\sigma| \cdot \log |\mathcal{A}|$$

where $|\mathcal{A}|$ denotes the number of letters of $\mathcal{A}$. Moreover $S_0(\sigma) = 0$ iff σ is constant.

We will call parsing of σ a partition of σ in shorter strings w which we will call words. For example if

$$\sigma = \text{“}betubetube\text{''}, \tag{2}$$

two parsings of σ are given by

$$\begin{aligned} \alpha_1 &= (be, tube, tube) \\ \alpha_2 &= (bet, u, bet, u, e) \end{aligned}$$

Given a parsing α, we will denote by $W(\alpha)$ the number of strings which can be obtained permuting the words of α. In our example we have

$$\begin{aligned} W(\alpha_1) &= 3; \\ W(\alpha_2) &= \frac{5!}{2! \cdot 2! \cdot 1!} = 30. \end{aligned}$$

[1]In all this paper, "log" will denote the base 2 logarithm.

Given a parsing α, we will call dictionary of α the set $V(\alpha)$ of words w (with $|w| > 1$) which appear in α. In our example we have

$$\begin{aligned} V(\alpha_1) &= \{be, tube\} \\ V(\alpha_2) &= \{bet\} \end{aligned}$$

We set

$$S_{com}(\sigma) = \min_{\alpha} \left[\log W(\alpha) + \sum_{w \in V(\alpha)} S_0(w) \right] \tag{3}$$

Notice that $S_{com}(\sigma) \leq S_0(\sigma)$. In fact, if α contains only one word, we have $S_0(\alpha) = 0$ and $\sum_{w \in V(\alpha)} S_0(w) = S_0(\sigma)$. The function $S_{com}(\sigma)$ will be called the combinatorial entropy of the string σ. In our example, we have

$$\begin{aligned} S_{com}(\sigma) &= \log W(\alpha_1) + S_0(be) + S_0(tube) \\ &= \log 3 + \log 2 + \log 4! \cong 7.169 \end{aligned}$$

Definition 1. *Let*

$$U : \mathcal{A}^* \to \{0,1\}^*$$

be an injective map and set

$$I_U(\sigma) = |U(\sigma)|$$

The function

$$I_U : \mathcal{A}^* \to \mathbb{N}$$

is called information function relative to U if, for any infinitely long string ω, we have

$$\lim_{n \to \infty} \frac{I_U(\omega^n)}{n} \leq \lim_{n \to \infty} \frac{S_{com}(\omega^n)}{n} \tag{4}$$

whenever these limits exist. The number $I_U(\sigma)$ will be called U-information content of σ.

$U(\sigma)$ can be thought as a coding of the string σ in binary alphabet. Equation (4) represents a kind of *optimality* of the coding U.

3. Algorithmic Information Content

One of the most important of the information functions is the Algorithmic Information Content (AIC). In order to define it, it is necessary to define the notion of partial recursive function. We limit ourselves to give an intuitive idea which is very close to the formal definition. We can consider a partial recursive function as a computer C which takes a program P (namely a binary string) as an input, performs some computations and gives a string $\sigma = C(P)$, written in the given alphabet $\mathcal{A}$, as an output. The AIC of a string σ is defined as the shortest binary program P which gives σ as output, namely

$$I_{AIC}(\sigma, C) = \min\{|P| : C(P) = \sigma\}$$

In this case the function $U(\sigma)$ of Definition 1 is just the shortest program which produces σ. We require that our computer is a universal computing machine. Roughly speaking, a computing machine is called *universal* if it can simulate any other machine. In particular every real computer is a universal computing machine, provided that we assume that it has virtually infinite memory. For a precise definition see e.g. [5] or [4]. We have the following theorem due to Kolmogorov

Theorem 1. *If C and C' are universal computing machine then*

$$|I_{AIC}(\sigma, C) - I_{AIC}(\sigma, C')| \leq K\,(C, C')$$

where $K(C, C')$ is a constant which depends only on C and C' but not on σ.

This theorem implies that the complexity AIC of σ with respect to C depends only on σ up to a fixed constant and then its asymptotic behavior does not depend on the choice of C. For this reason from now on we will write $I_{AIC}(\sigma)$ instead of $I_{AIC}(\sigma, C)$.

The AIC allows to define the randomness in an intrinsic way.

Definition 2. *A string σ of length n is called AIC-random if*

$$I_{AIC}(\sigma) \geq n \log |\mathcal{A}|$$

It is possible to show that most of the string are AIC-random. Moreover, the shortest program P, which produces σ is a random string. In fact

$$I_{AIC}(\sigma) = |P| \tag{5}$$

and if there were a shorter program Q which would produce P, it would produce also σ and we would have

$$I_{AIC}(\sigma) \leq |Q| < |P|$$

and this contradicts (5).

Unfortunately the AIC cannot be computed by any algorithm (Chaitin Theorem[2]). This is a very deep statement and, in some sense, it is equivalent to the Turing halting problem or to the Goedel incompleteness theorem.

This fact has very deep consequences for our discussion on the arrow of time as we will see later. For the moment we can say that the AIC cannot be used as a reasonable physical quantity since it cannot be computed.

4. Computable Information Content

From an heuristic point of view, the shortest program P which outputs the string σ is a sort of optimal encoding of σ. The information that is necessary to reconstruct the string σ is contained in the program. Suppose that we have some lossless (reversible) coding procedure $Z : \mathcal{A}^* \to \{0,1\}^*$ such that from the coded string we can reconstruct the original string (for example the data compression algorithms used in any personal computer). Since the coded string contains all the information that is necessary to reconstruct the original string, we can consider the length of the coded string as an approximate measure of the quantity of information that is contained in the original string. We can define the information content of the string σ as the length of the compressed string $Z(\sigma)$, namely

$$I_Z(\sigma) = |Z(\sigma)|.$$

The advantage of using a Compression Algorithm lies in the fact that, in this way, the information content $I_Z(\sigma)$ turns out to be a computable function and hence it can be used in computer simulation and as measurable physical quantity.

There are many of such algorithms and they can be used for various purposes. (cf. e.g. [2]). It would be interesting to define the Computable Information Content (CIC) as the information quantity related to the "best" of such algorithms. We do not know if this is possible, since the notion of "best" is not well defined and, of course, it depends on what you want to do. Here, we will list the properties which such notion must satisfy for our purposes.

A function

$$I_{CIC} : \mathcal{A}^* \to \mathbb{N}$$

[2]Actually, the Chaitin theorem states a weaker statement: a procedure (computer program) which states that a string σ of length n can be produced by a program shorter than n, must be longer than n.

is called Computable Information Content if it satisfies the following properties:

(i) it is an information function in the sense of Definition 1.

(ii) it is computable.

(iii) $I_{CIC}(\sigma) = M_{CIC}(\sigma)+S_{CIC}(\sigma)$ where $S_{CIC}(\sigma)$ satisfy the following properties:

(S1) $S_{CIC}(\sigma) \leq \log W(\sigma)$

(S2) $S_{CIC}(\sigma\tau) \leq S_{CIC}(\sigma) + S_{CIC}(\tau)$

(S3) $S_{CIC}(\sigma) \geq I_{AIC}(\sigma) - o(|\sigma|)$.

The properties (i) and (ii) are satisfied by I_Z defined by any reasonable compression algorithm Z. The important peculiarity of the Computable Information Content lies in the possibility of decomposing the global quantity of information in two parts:

- $S_{CIC}(\sigma)$ which we will call computable entropy of σ and represents the *disordered* part of the information.
- $M_{CIC}(\sigma)$ which we will call macroinformation of σ which represent the *regular* part of the information.

The properties (S1), (S2) and (S3) of the entropy are chosen in order to fit our intuitive idea of measure of *disorder*. For example, by (S1), we deduce that a constant string has null entropy: no disorder. (S2) can be interpreted in the following way: the "disorder" of two strings is additive unless the two strings are correlated with each other in some way. (S3) gives a lower bound to the quantity of disordered information of a string. Since the best program P which produces σ must be random (in the sense of Definition 2), it contains a quantity of "disorder" at least equal to $|P| = I_{AIC}(\sigma)$ (plus a lower order term which cannot be eliminated by the request (S1)).

Moreover, these properties make the computable entropy close to the Boltzmann definition of entropy and this fact is very relevant for the interpretation of physical phenomena. Notice that $S_{CIC}(\sigma)$ is not an information function since there not exists a function U which satisfies the requirement of Definition 1. In fact, the set

$$A_n = \{\sigma \in \mathcal{A}^* \mid S_{CIC}(\sigma) \leq n\}$$

is infinite (it contains all the constants), while, for any injective function

$$U : \mathcal{A}^* \to \{0,1\}^*$$

the set

$$B_n = \{\sigma \in \mathcal{A}^* \mid U(\sigma) \leq n\}$$

is finite, actually $|B_n| \leq 2^n$.

Functions I_{CIC} which satisfy (i),(ii) and (iii) exist. For example, it is sufficient to take a compression algorithm U which satisfies (4) and set $I_{CIC}(\sigma) = I_U(\sigma)$, $S_{CIC}(\sigma) = S_{com}(\sigma)$ and $M_{CIC}(\sigma) = I_U(\sigma) - S_{com}(\sigma)$. In [1] there is a more involved construction of a function $I_{CIC}(\sigma)$ which satisfies some extra properties.

We think that this distinction between information and entropy is very important and the confusion of these notions is at the origin of many problems in the interpretations of statistical mechanics and its relations to information theory.

5. Dynamical systems

We assume to have a dynamical system consisting of many particles. As usual we divide the phase space X of each particle (μ-space) in L small cells. We can label any cell by a letter a_i of an ideal alphabet. Then the microstate of the system (with the accuracy given by the grain of our partition of the phase space) can be represented by a string of letters. Namely the string

$$\sigma = (a_1 a_2 a_3 \ldots\ldots)$$

represents the state in which particle 1 is in the cell a_1, particle 2 is in the cell a_2, and so on. The Boltzmann entropy of our microstate is given by $\log W$ where W is the set of all the configurations of our system. This entropy corresponds to the quantity $S_0(\sigma)$ defined by (1). If we are in a particular configuration (e.g. if the particles are concentrated in some region), the CIC-entropy $S_{CIC}(\sigma)$ is less than $S_0(\sigma) = \log W$ and $I_{CIC}(\sigma) = M_{CIC}(\sigma) + S_{CIC}(\sigma) < S_0(\sigma)$. If the system is chaotic, then during its evolution, $S_{CIC}(\sigma)$ will increase and $M_{CIC}(\sigma)$ will decrease. If the dynamics has no other constraints but the conservation of the number of particles, $S_{CIC}(\sigma)$ will reach the number $S_0(\sigma)$ and $M_{CIC}(\sigma)$ will become essentially 0. If there are few constraints (such as conservation laws), $S_{CIC}(\sigma)$ will reach the maximum compatible with the constraints and $M_{CIC}(\sigma)$ will reach the minimum compatible with these constrains. The reason is that the maximum entropy states fill most of the phase space. In this way we reach the thermodynamical equilibrium.

If in a physical dissipative system there is an input of energy from the outside, in general, stable equilibrium configurations cannot exist. In this case many phenomena may occur; stable periodic orbits, stable tori or even strange attractors. Sometimes, very interesting spatial structures may appear. Analogous phenomena occur in non reversible cellular automata. The most famous of them is the Conway game of life in which a lot of intriguing shapes appear in spite of the simplicity of the transition map.

From the point of view of Information theory, these are the systems which make the macroinformation to increase. If we start from an initial data with a low macroinformation content, the macroinformation will increase until reaching a limit value. If the system is infinite, the macroinformation will increase for ever. For example you may think of the game of life in an infinite grid with initial conditions having only a finite number of black cells (and thus you start with an initial condition which has finite information).

If we assume that the universal laws of the universe are reversible, isolated dissipative systems cannot exist. A dissipative system (Ω, f) must be considered as a part of a reversible system (Ω_1, f_1) such that $\Omega \subset \Omega_1$ and $f|_\Omega = f_1$. Thus, we can consider (Ω, f) as part of a reversible system of which we have ignored the variables external to the system and also the microscopic dynamic variables.

But we know that, in reversible systems, the entropy as well as the CIC-entropy increases (a more detailed discussion of this scenario is given in [1]). Thus the two main arrows of time, described in the introduction, will be present:

- the thermodynamic arrow defined by the increase of CIC-entropy;
- the evolution arrow defined by the increase of macroinformation.

However, in some sense, all the *information* is contained in the initial condition and in the transition map (this is the case of cellular automata for which the complete state of the system is described by finite information). But which information? The answer is simple: the AIC. If the state of the system at time t is described by a string $\sigma(t)$, then we have that

$$I_{AIC}(\sigma(t)) \leq I_{AIC}(\sigma(t_0)) + \log|t - t_0| + O(\log|t - t_0|)$$

and since the "log function" grows very slowly, the AIC can be considered essentially constant. If you look at your system from the point of view of the AIC, nothing new happens. But the basic fact lies in the non-computability of the AIC. The AIC, when applied to a real object, is a sort of metaphysical entity with no real meaning: it cannot be computed and hence it cannot be used to make predictions. On the contrary, any type of computable information, as e.g. the CIC, changes with time in an unpredictable way.

Concluding, if the universe is supposed to be strictly deterministic (such as a cellular automata), the AIC is essentially constant (up to the log of time); the algorithmic information cannot be created or destroyed. Past and future are contained in the present. However the AIC cannot be computed; it is only a mathematical abstraction of the *real information*, whatever is the meaning which you want to give to this notion. If you look at the CIC, both entropy and complexity, in general, increase. Moreover, even in a strictly deterministic universe, the future is unpredictable: this fact corresponds to the unpredictability of the halting of a Turing machine[3].

6. Conclusions

We think that the right description of the origin of irreversibility, complexity and the arrow of time lies in a good notion of "information content". A good notion must be independent of the notion of probability for the reasons described in section 2.

Moreover, we think that it is very important to distinguish two different meanings of the notion of information:

- the general *abstract* notion of information (such as the AIC) which in reversible system is (essentially) a constant of motion and exists only in the mind of God.

- a computable notion of information (such as the CIC) which is related to physical quantities; it changes in time and can be used to describe the observed phenomena.

The distinctions between these two notions is marked by the Turing halting theorem which, we think, is one of the deepest theorems discovered in last century (and whose consequences are not yet all completely understood).

Once we have agreed to consider the CIC (or any other "epistemologically" equivalent notion of information) as the relevant physical quantity, it is important to have a mathematical method to separate the CIC in two different components:

- the entropy, which corresponds to the old idea of "measure of disorder". From a physical point of view this information cannot be used to make exact deterministic previsions. It is the information dispersed in the chaos and cannot be recovered without a violation of the Turing halting theorem.

[3] In fact, there is a correspondence between Turing machines and cellular automata.

- the macroinformation, which is related to physical measurable quantities and can be used to make statistical previsions. Moreover, the macroinformation is strictly related to various indicators of complexity (see [1]).

Thus, in information theory, we have the distinction between macroinformation and CIC-entropy. This is similar to the distinction, in thermodynamics, between free energy $F = E - TS$ and degraded energy TS. The CIC-entropy cannot be used to make predictions, while TS cannot be used to perform any work. However, it is very important to underline that, in isolated systems at constant temperature, free energy and macroinformation behave in a quite different way: free energy always decreases, while macroinformation might increase. The development of life, in all its forms, determines a decrease of free energy and an increase of macroinformation. Probably there is a deep mathematical relation between the evolution of these two quantities. The study of the interplay between macroinformation, entropy and the other physical quantities is a good way to investigate the origin and the evolution of complex structures.

In [1], we have proposed a mathematical model which makes a distinction between macroinformation and entropy. This is not the only possible model and probably is not the best. However, we think that this is the right direction to investigate this kind of problems.

References

[1] BENCI V., *The Arrow of Time and Information Theory*, to appear.

[2] BENCI V., BONANNO C., GALATOLO S., MENCONI G., PONCHIO F., *Information, complexity and entropy: a new approach to theory and measurement methods*, `http://arXiv.org/abs/math.DS/0107067` (2001).

[3] BRILLOUIN L., *Scientific Uncertainty, and Information*, Academic Press, New York 1964.

[4] CHAITIN G.J., *Information, randomness and incompleteness, papers on algorithmic information theory*, World Scientific, Singapore 1987.

[5] LI M., VITANYI P., *An Introduction to Kolmogorov Complexity and Its Applications*, Springer-Verlag 1993.

HOW REAL IS THE QUANTUM WORLD?

Marcello Cini
Dipartimento di Fisica,
Università La Sapienza
and
Istituto Nazionale di Fisica della Materia, sez. di Roma,
Piazza A. Moro 2, 00185 ROMA - ITALY
marcello.cini@roma1.infn.it

Abstract In spite of recent spectacular progress of experimental techniques, which allow today the practical realization of the *gedankene experimente* envisaged by Einstein and Bohr in order to discuss the strange properties of the quantum mechanical world, its ontological status is far from being unanimously established, and the objective reality of its physical elements is generally denied. In this paper a realist interpretation of Quantum Mechanics is discussed, based on the concept that microscopic objects have "context dependent properties". This concept, in its turn, is grounded on the proof that the theory is compatible with the existence of macroscopic objects which behave classically for all practical purposes. From these two features of this interpretation also the peculiar long distance correlations of quantum particles can be understood, provided that one accepts that physical laws do not formulate detailed prescriptions about all that "must" happen in the world, but only provide constraints and express prohibitions about what "may" happen. This implies that random events just happen, provided they comply to these constraints and do not violate these prohibitions. Finally a new reformulation of Quantum Mechanics, in which the concept of probability waves is eliminated from the beginning, is presented. In this way, not only its whole structure in phase space is deduced from a classical formulation of the uncertainty principle and a single quantum postulate, but it is also shown that the introduction of q-numbers in quantum theory is not a mathematical postulate coming out of the blue, but is a consequence of well defined physical requirements.

1. Introduction

In spite of the fact that the recent extraordinary progresses of experimental techniques make us able to manipulate at will systems made of any small and well defined number of atoms, electrons and photons – allowing therefore the practical realization of the *gedankenexperimente* that Einstein and Bohr had imagined in order to support their opposite theses on the meaning of quantum mechanics – it does not seem that, after more than seventy years from the birth of the theory, a unanimous consensus has been reached in the physicist's community on the way of understanding the "strange" properties of the quantum world. Unfortunately, we cannot know whether Feynman would still insist in maintaining that "It is fair to say that nobody understands quantum mechanics". We can only discuss if, almost twenty years after, some progress towards this goal has been made. In my opinion this is the case, and I will try to show you why.

2. Context dependent properties of a quantum system

What does a realist interpretation of quantum mechanics consist of? In a recent work Mario Bunge (Bunge 2001) sets up to show that an interpretation of this kind, in spite of the conventional arguments against this convinction, is possible. I personally agree with his purpose, but I have to say that his arguments seem to me often too weak and insufficiently based on the present day knowledge.

As a first example, I join Bunge in criticizing the so called "orthodox Copenhagen interpretation" of Quantum Mechanics, but I do not think it is very useful to unify under this heading many different ways of looking at the phenomena of the quantum world which have been proposed (see e.g. [6, 14] in seventy years, and especially since the revival of the Bohr Einstein debate as a consequence of Bell's paper in 1964 [1]. I would limit myself to stress that *a realist interpretation of what happens in this world is possible*, based on the premise *that human, or extrahuman, minds have no influence whatsoever on the phenomena which occur outside us at the atomic and subatomic scale.*

This definition of realism does not exactly coincide with Bunge's one. It is necessary therefore at this stage to pin down where the difference lies. "Realists hold – I quote Bunge – that (a) the world external to the inquirer exists independently (ontological thesis); and (b) reality can and must be objectively described". I totally agree with (a), but I am puzzled about the meaning of (b). It all depends on what one intends with the word "reality". Does this mean that a *quanton* (this is the term he introduces to indicate any particle of the quantum world) is an *element* of reality possessing objective properties which do not depend on the surrounding matter? This statement is not true. I therefore understand that Bunge means by this that the physical properties of a quanton identified with the classical kynematic concepts of position and velocity objectively exist, but have generally blunt values and only occasionally, but not simultaneously, may acquire sharp values. But the question arises: when and how does this happen?

My answer is: when a quanton interacts with a suitable piece of matter which constrains it to assume, *at random but with a given probability*, a sharp value of either position or velocity. This answer implies that real microscopical objects with a definite mass, spin and charge never possess simultaneously the kynematic properties which would allow to describe their motion along a trajectory in space. It implies also that if a quanton interacts with a photographic plate we can "objectively describe" the system after the interaction as made of a photographic plate with a black spot and the quanton with a sharp position located where the black spot is. It implies furthermore that an analogous but complementary objective description of reality can be made when (a beam of) quantons interact(s) with a diffraction grating. (I wish to stress, however that an asymmetry exists between the two cases: we can describe a single quanton as having a definite position, but we need a beam of quantons to describe them as having a given value of the velocity. We will come back on this at the end). I prefer therefore to say that the quantons have "context dependent" properties. By saying that quantons have "context dependent properties" I do not at all deny realism: I only refuse to adopt an arbitrarily radical reductionist view of reality. It should be stressed that I deliberately did not speak of instruments, of measurements or of observers. I am speaking of physical objects.

My answer, however, introduces another problem. I have to add in fact that it is essential that the pieces of matter introduced in my definition of a realist interpretation of the quantum world should be macroscopical objects *with context independent properties.* Other quantons would not play the same role. *It becomes therefore necessary to prove that the existence of macroscopic pieces of matter with context independent properties is not a postulate but follows from the equations of quantum mechanics themselves.*

As is well known, the problem was tackled, in the early days of quantum mechanics, by Bohr, who *postulated* the existence of classical objects in order to explain how could the quantum objects acquire, in the interaction with them, sharp values of either position or velocity. However, as Max Jammer writes in his book *The Philosophy of Quantum Mechanics*: "The double nature of the macroscopic apparatus (on the one hand a classical object and on the other hand obeying quantum mechanical laws) remained a somewhat questionable or at least obscure feature in Bohr's conception of quantum mechanical measurements." I believe however, in spite of this, that Bohr's solution was appropriate at that time and did not conflict with realism. In fact, it was von Neumann's refusal of it which started all the nonsense about the role of the observer (and of his consciousness) in creating physical reality.

The fundamental question which should be answered in order to prove that a realist interpretation of the quantum world is possible, is therefore the following: How is it possible that classical objects with "objective" context independent values of their dynamical variables exist, given that the laws of Quantum Mechanics seem to deny this possibility?

3. The existence of classical objects

This question was investigated and answered by my group in Rome twenty years ago in two papers [4, 5], which at the time received some attention (*Nature* dedicated a whole page of comment to the second one). It is, however, fair to give credit to K. Gottfried (1966) for having correctly approached the problem many years before.

In these papers we proved that "macroscopic aggregates of particles in normal conditions never exhibit quantum behaviour because the phase relations of their microscopic constituent's wave functions are completely destroyed." We did in fact prove that the effects of quantum interference (the wavelike behaviour) tend to zero for a superposition of two macroscopically different states (technically wavefunctions or state vectors) of a macroscopic object. This implies that when a quanton P in a given state interacts with a suitable "instrument" S_q made of N quantons, the difference between the probabilistic predictions of quantum mechanics on the possible outcomes of this interaction and the predictions of classical statistical mechanics, for an ideal statistical ensemble in which a classical instrument S_c replaces S_q (with the same values of its macroscopic variables), tends to vanish when N becomes very large ($\gg 1$). This means that, after all, Bohr was right in assuming that classical bodies exist. Needless to say, our result proved also that Schrödinger's cat cannot be at the same time dead and alive, simply because it is a macroscopic "object".

However, once the existence of classical "objective" properties of macroscopic bodies is granted, "realists" still have to face another peculiar situation. Does the position of a single particle – one may ask – acquire its value only when it is "measured" even if its wave function is a linear combination of two wave functions localized in macroscopically distant space regions? The standard answer is "yes" but it is hard to swallow that an electron initially trapped in a double well which is later separated adiabatically in two single wells, one shipped to Rome and the other one to New York, materializes in one of them only when an "observer" decides to have look inside.

This last problem has been investigated, I believe with success, by Maurizio Serva and myself a few years ago [7, 8], and further clarified in collaboration with Philippe Blanchard [2]. Consider the above mentioned single particle state made of a superposition of two free wave packets localized in separate regions of space. In this case we can explicitly calculate the uncertainties Δx and Δp of position and momentum which appear in the well known general expression of the Heisenberg uncertainty principle

$$(\Delta x \Delta p) \geq (h/4\pi)$$

where h is Planck's constant. The result is that the above inequality becomes an equation whose right hand side is the sum of two terms. One is a classical term which expresses an epistemic uncertainty, while the other one is the minimum quantum term $(h/4\pi)$ expressing the irreducible nature of chance at this level. Their different physical nature is made clear by the fact that, while it is possible, without violating the Heisenberg principle, to reduce the classical term by simultaneously forcing both position and momentum of a particle to assume more precise values, the quantum term is irreducible because a localization of the position which reduces Δx necessarily leads to an increase of Δp and viceversa.

This interpretation of the uncertainty principle solves the paradox of the particle localization in one of the two distant single wells. It is the initial irreducible (ontological) indeterminacy of the particle's position in the double well which triggers a statistical (epistemical) uncertainty arising from our ignorance of its actual localization when the wells are sufficiently far apart. It is therefore appropriate to conclude that in this case the particle was already in one of them before measuring its position. It is worthwhile to stress that only this interpretation leads to a smooth transition from the quantum to the the classical domain and therefore to the universally accepted epistemical interpretation of probabilities in classical statistical mechanics.

4. Long distance correlation of quantum variables

I come now to the second counterintuitive quantum phenomenon, the famous EPR "paradox", whose solution, after the numerous experiments confirming the violation of Bell's inequalities, can only be expressed by saying that Einstein was wrong in believing that a description of the state of a pair of distant correlated particles, more detailed than the one provided by quantum mechanics, could be obtained by means of hidden variables, and in concluding that quantum mechanics is an incomplete theory.

The strange quantum behaviour of two particles whose variables remain correlated even after having been separated spacially once that their correlation has been established initially by means of a close interaction, is *not* that they remain correlated also when they are far away. This happens also to a pair of classical particles. Take for example an object with zero total angular momentum which is split in two parts which fly away in opposite directions. Obviously the sum of the angular momenta of the two parts remains zero even when they are far away. This means that *any* component of the angular momentum of one part is *always* equal and opposite to the *same* component of the other part. What is "paradoxical" is that this behaviour persists also in quantum mechanics in spite of the fact that in principle only *one* component of a particle's angular momentum *can have at a given time a sharp value.*

This is why people ask: how is it possible that, when the first particle acquires a sharp value of its angular momentum component during the interaction with its filter, the far away particle comes to "know" that it should acquire the same and opposite value of its own angular momentum component? I do not think that one can "explain" this counterintuitive behaviour, by minimizing the difference with its classical counterpart, because this difference has its roots, in my opinion, in the "ontological" (or irreducible) – not "epistemical" (or due to imperfect knowledge) – nature of the randomness of quantum events. If one takes seriously this postulate, in fact, one has to accept that physical laws do not formulate detailed prescriptions, enforced by concrete physical entities, about all that *must* happen in the world, but only provide constraints and express prohibitions about what *may* happen. Random events just happen, provided they comply to these constraints and do not violate these prohibitions.This means that the question "*how* can this happen?" becomes meaningless. One can only answer: "it happens *because* it is not forbidden".

With this in mind, one can try to understand *why* the far away particle, once the first one has acquired a sharp value of its angular momentum component, *has* to acquire an equal and opposite sharp value for the same component of its own angular momentum. The answer is: it *has* to do so, because, should it not behave this way, it would violate the law of conservation of angular momentum. This law is a consequence of a property of space, namely its homogeneity under rotations. The quantity "total angular momentum" *is* itself, by definition, a non-local quantity. Non locality therefore needs not to be enforced by a mysterious *action-at-a-distance.* In fact, the two filters are not two uncorrelated pieces of matter: they are two rigidly connected parts of one single piece of matter which "measures" this quantity. The non local constraint is therefore provided by the nature of the macroscopic "instrument". This entails that, once the quantum randomness has produced the first partial sharp result, there is no freedom left for the result of the final stage of the interaction: there is no source of angular momentum available to produce any other result except the equal and opposite sharp value needed to add up to zero for the total momentum.

5. Quantum mechanics in phase space

If chance has an irreducible origin the fundamental laws should allow for the occurrence of *different* events under *equal* conditions. The language of probability, suitably adapted to take into account all the relevant constraints, seems therefore to be the only language capable of expressing this fundamental role of chance.

The proper framework in which a solution of the conceptual problems discussed above should be looked for is therefore, after all, the birthplace of the quantum of action, namely phase space, where no probability amplitudes exist. It is of course clear that joint probabilities for both position and momentum having sharp given values cannot exist in phase space, because they would contradict the uncertainty principle. Wigner (1932) however, introduced the functions called after his name as pseudoprobabilities (which may assume also negative values) to represent Quantum Mechanics in phase space, and showed that by means of them one can compute any physically meaningful statistical property of quantum states. It seems reasonable therefore to consider these functions not only as useful tools for computations, but as a framework for looking at Quantum Mechanics from a different point of view.

A step further along this direction was made a few years ago by Feynman (1987), who has shown that, by dropping the assumption that the predictions of Quantum Mechanics can only be formulated by means of nonnegative probabilities, one can avoid the use of probability amplitudes, namely waves, in quantum mechanics. The physical meaning of Wigner's pseudoprobabilities is well clarified by Feynman's words: *"It is that a situation for which a negative probability is calculated is impossible, not in the sense that the chance for its happening is zero, but rather in the sense that the assumed conditions of preparation or verification are experimentally unattainable."*

The road is therefore open for a new reformulation of Quantum Mechanics, in which the concept of probability "waves" is *eliminated from the beginning.* After all, particles and waves *do not stand on the same footing* as far as their practical detection is concerned. We have remarked in the Introduction that the position of a particle assumes a sharp value as a consequence of a single interaction with a suitable detector, but we need a beam of particles to infer the sharp value of their common momentum. This means that *we never detect waves*: we only infer their existence by detecting a large number of particles.

This program has been recently carried on [9] by generalizing the formalism of classical statistical mechanics in phase space with the introduction of a single quantum postulate, which introduces mathematical constraints on the set of variables in terms of which any physical quantity can be expressed (usually denoted as characteristic variables). It turns out, however, that these constraints cannot be fulfilled by ordinary random numbers, but are satisfied by the mathematical entities called by Dirac *q-numbers.* It is not necessary, however, to provide an explicit representation of these mathematical objects, because the theory yields the equations (I will call them Wigner-Feynman equations) which allow the complete determination, for *any statistical ensemble defined by a sharp value of a given chosen variable*, of the average values of all the other physical variables of the system. It is also possible to reconstruct, for each ensemble, the correspondent Wigner function.

In this way, not only is the whole structure of Quantum Mechanics in phase space deduced from a single quantum postulate without ever introducing wave functions or probability amplitudes, but it is shown that the introduction of q-numbers in quantum theory is not a mathematical postulate coming out of the blue, *but is a consequence of a well defined physical requirement.*

The advantages of this approach are numerous. First of all, many paradoxes typical of the wave-particle duality disappear. On the one hand in fact, as already shown by Feynman, it becomes possible to express the correlations between two distant particles in terms of the product of two pseudoprobabilities independent from each other. All the speculations on the nature of an hypothetical superluminal signal between them becomes therefore meaningless. This is in complete agreement with our preceding discussion in §8. Similarly, the long time debated question about the meaning of the superposition of state vectors for macroscopic objects (§7) may also be set aside as equally baseless.

Secondly, this approach eliminates the conventional hybrid procedure of describing the dynamical evolution of a system, which consists of a first stage in which the theory provides a deterministic evolution of the wave function, followed by a hand made construction of the physically meaningful probability distributions. If the probabilistic nature of the microscopic phenomena is fundamental, and not simply due to our ignorance as in classical statistical mechanics, why should it be impossible to describe them in probabilistic terms from the very beginning?

The third advantage is connected with the possibility of dissipating the ambiguity of the conventional theory about two physically different aspects of the quantum uncertainties inherent to the Heisenberg inequality discussed in §7. In the present formulation of quantum theory, in fact, only the quantum ontological uncertainties are present, without any spurious statistical contribution. This is because the uncertainty principle in our theory is given by an *equality* involving only the minimum value of the Heisenberg *inequality*.

The elimination of waves from quantum theory (which, after all, is nothing else than a simple application of Occam's razor), might therefore play the same role as the elimination of aether in the theory of electromagnetism.

References

[1] BELL J., Physica **1** (1964), 194.

[2] BLANCHARD P., CINI M., SERVA M., *Ideas and methods in quantum and statistical physics. In memory of Raphael Hoegh-Krohn*, S. Albeverio et al. eds., Cambridge University Press 1992, 132.

[3] BUNGE M., in press.

[4] CINI M., DE MARIA M., MATTIOLI G., NICOLÒ F., Found. of Phys. **9** (1979), 479.

[5] CINI M., Nuovo Cim. **73B** (1983), 27; Nature **302**, 307.

[6] CINI M., LÉVY-LEBLOND J.M. (EDS), *Quantum theory without reduction*, A. Hilger, London 1991.

[7] CINI M., SERVA M., Found. of Phys. Lett. **3** (1990), 129.

[8] CINI M., SERVA M., Phys. Lett. A **167** (1992), 319.

[9] CINI M., Ann. of Phys. **273** (1999), 199.

[10] FEYNMAN R.P., in: "Quantum Implications", B.J. Hiley, F.D. Peats eds., Routledge & Kegan, London 1987, 285.

[11] GOTTFRIED K., *Quantum Mechanics*, J. Wiley, New York 1966, sect. 20.

[12] SCULLY M.O., WALTHER H., SCHLEICH W., Phys. Rev. A **49** (1994), 182.

[13] WIGNER E., Phys. Rev. **40** (1932), 479.

[14] WHEELER J., ZUREK W. (EDS), *Quantum Theory and measurement*, Princeton University Press 1983.

DECOHERENCE AND CLASSICAL BEHAVIOUR IN QUANTUM MECHANICS

Gianfausto Dell'Antonio
Dipartimento di Matematica,
Università di Roma "La Sapienza" - ITALY
and
Laboratorio Interdisciplinare SISSA-ISAS, TRIESTE - ITALY
gianfa@sissa.it

Rodolfo Figari
Dipartimento di Scienze Fisiche,
Università di Napoli "Federico II" - ITALY
figari@na.infn.it

Alessandro Teta
Dipartimento di Matematica Pura e Applicata,
Università di L'Aquila - ITALY
teta@univaq.it

Abstract We discuss the meaning of decoherence and the emergence of a classical behaviour of a quantum system in interaction with an environment. After a brief introduction to the formalism of Quantum Mechanics we present a simple two-particle model where the decoherence effect can be quantitatively evaluated.

Non relativistic Quantum Mechanics proved to be an accurate tool to describe and predict phenomena at the atomic level. Nevertheless the standard formulation provides a theory which is not ontologically complete inasmuch as it refers to a "macroscopic world" (e.g. measurement apparatus) governed by different physical laws (e.g. Classical Mechanics).

It is often argued that one can resolve the difficulty taking into account that macroscopic objects are made up of an enormous number of elementary components (obeying the laws of Quantum Mechanics). A Law of Large Numbers would "explain" the apparent "classical" behaviour of macroscopic objects which would then turn out to be a consequence of an averaging with negligible statistical fluctuations due to the large number of components.

Although this statement seems to be reasonable , it cannot be considered as an "explanation". On the other hand it is well known that there are macroscopic objects showing a quantum behaviour (e.g. superconductors) and, on the opposite side, microscopic systems whose properties can be well described in classical terms (e.g. molecules made up of a sufficiently large number of atoms).

Attempts to understand how microscopic systems can "fake" a classical behaviour, at least when their local properties are under observation, were made recently.

Superposition principle (see below for a short introduction to some basic aspects of Quantum Theory), allowing coherence and then possible interference between different parts of the wave packet describing a quantum system, has to be considered as the most peculiar and less classical feature of Quantum Mechanics. This justifies the use of the name "decoherence" to identify the mechanism producing (partial) suppression of the interference effects for certain quantum systems interacting with the environment.

Decoherence would be an explanation of the apparent classical behaviour of most macroscopic systems (see e.g. [7], [2] and references therein).

In this short contribution we want to discuss some theoretical aspects of this mechanism.

It should be noticed that in the theory of decoherence many results are accepted on a qualitative or even intuitive ground and only few of them are mathematically proved. Further analysis, in the direction of proving rigorously some key result, is required in this important subject which might become a flourishing research field in Mathematical Physics in the near future.

In this spirit we will present in the conclusive part of the paper a completely solvable model in which decoherence, i.e. reduction of interference, can be explicitly exhibited (see also [5] for results in a similar direction).

Let us start introducing briefly the formalism of Quantum Mechanics in such a way that we can go to a deeper level of analysis.

There are two equivalent formulations of Quantum Mechanics; their origin goes back to two founders of the theory : Schroedinger whose approach was guided by the analogy with classical wave propagation phenomena and Heisenberg whose starting point was the structure of the observed atomic emission spectra.
The equivalence of the two formulations and the correspondence between their formalised concepts was proved immediately after their appearance by Schroedinger, and, later on, by Dirac and von Neumann with greater precision.

In Schroedinger's formulation, where the wave aspect is more evident, the state of a system is represented by a complex valued function of the space variables of the particles forming the system. Its squared modulus is required to be integrable with respect to the Lebesgue measure. In symbols

$$\Psi \in L^2(\mathbb{R}^{3N}), \qquad \Psi : x \to \Psi(x) \qquad x \in \mathbb{R}^{3N} \tag{1}$$

where N is the total number of particle of the system.

If internal degrees of freedom (spin) have to be taken into account the representation space must be enlarged in a suitable way; for sake of simplicity we will limit ourselves to case (1).

In (1) x is a parameter allowing a concrete representation of the state Ψ. It acquires the meaning of "observable" as soon as one assumes that

$$\mathcal{P}(x \in \Omega \mid \Psi) \equiv \int_\Omega |\Psi(x)|^2 dx \tag{2}$$

represents the probability that the point x (which represents the position of the N particles) belongs to the set $\Omega \subseteq \mathbb{R}^{3N}$ when the system is in the state Ψ. This interpretation justifies the normalisation assumption:

$$\int_{\mathbb{R}^{3N}} |\Psi(x)|^2 dx = 1\,. \tag{3}$$

In this way $|\Psi(x)|^2 dx$ becomes a probability measure on $\mathbb{R}^{3N}$ induced by the function Ψ and

$$\int_{\mathbb{R}^{3N}} f(x)|\Psi(x)|^2 dx \tag{4}$$

gives the expectation value of a function f on $\mathbb{R}^{3N}$ assuming the value $f(x)$ in the configuration x.

It should be noticed that the knowledge of the value (4) corresponding to any function f does not uniquely determine the state of the system. In fact the substitution of $\Psi(x)$ with $\Psi(x)\exp(i\zeta(x))$ is not going to change the value (4). A different representation of the state Ψ is required.

This is provided by the identification, according to De Broglie, of a particle of momentum p with the plane wave $e^{i\frac{p}{\hbar}x}$, where $\hbar$ is the Planck's constant divided by 2π.

Let $\hat{\Psi}(k)$ be the Fourier transform of $\Psi(x)$.

Following the previous identification, the following interpretation will be given to $\hat{\Psi}(k)$: for any open set $\Sigma \subseteq \mathbb{R}^{3N}$

$$\mathcal{P}(p \in \hbar\Sigma \mid \Psi) \equiv \int_{\Sigma} |\hat{\Psi}(k)|^2 dk \tag{5}$$

represents the probability that the momenta of the N particles belong to the set $\hbar\Sigma$ when the system is in the state Ψ. In this framework for any function g on $\mathbb{R}^{3N}$ the expression

$$\int_{\mathbb{R}^{3N}} g(\hbar k)|\hat{\Psi}(k)|^2 dk \tag{6}$$

gives the expectation value of the function on $\mathbb{R}^{3N}$ assuming the value $g(p)$ when momentum is p.

Notice that, corresponding to a sequence of states $\Psi_n^{x_0}$ for which the support of $\Psi_n^{x_0}(x)$ is concentrating around the point $x_0 \in \mathbb{R}^{3N}$, the expected value of any function f is closer and closer to the value $f(x_0)$ with a dispersion becoming more and more negligible; analogously $g(p_0)$ will become the (almost) exact value of the function g for a sequence of states $\Psi_n^{p_0}$ for which the support of the functions $\hat{\Psi}_n^{p_0}(p)$ concentrates around momentum p_0.

Well known properties of Fourier transform imply that

$$\Delta_\Psi p_j \, \Delta_\Psi x_l \geq \frac{\hbar}{2}\delta_{jl}, \qquad j,l = 1,\ldots 3N \tag{7}$$

where $x = \{x_i\}_{i=1}^N$, $x_i \in \mathbb{R}^3 \ \forall i$, $p = \{p_i\}_{i=1}^N$, $p_i \in \mathbb{R}^3 \ \forall i$ and

$$(\Delta_\Psi p_j)^2 = \int_{\mathbb{R}^{3N}} \hbar^2 k_j^2 |\hat{\Psi}(k)|^2 dk - \left(\int_{\mathbb{R}^{3N}} \hbar k_j |\hat{\Psi}(k)|^2 dk\right)^2 \tag{8}$$

$$(\Delta_\Psi x_l)^2 = \int_{\mathbb{R}^{3N}} x_l^2 |\Psi(x)|^2 dx - \left(\int_{\mathbb{R}^{3N}} x_l |\Psi(x)|^2 dx\right)^2 \tag{9}$$

(Heisenberg uncertainty relations)

For the very same reasons we gave for the "x space" case, the distribution $|\hat{\Psi}(k)|^2$ does not determine completely the state Ψ. In order to have a complete knowledge of the state Ψ one should know the joint distribution of positions and momenta; Heisenberg uncertainty relations indicate that such a distribution cannot be a product measure.

It is easy to see that it is always possible to define a one-to-one (up to an irrelevant overall phase factor) correspondence between wave functions $\Psi \in L^2(\mathbb{R}^{3N})$ and measures H_Ψ (Husimi measures) defined on $\mathbb{R}^{3N} \times \mathbb{R}^{3N}$ in such a way that the marginals of each measure correspond respectively to the measures induced in $\mathbb{R}^{3N}$ by $\Psi(x)$ and by its Fourier transform $\hat{\Psi}(k)$, i.e.

$$|\Psi(x)|^2 = \int_{\mathbb{R}^{3N}} H_\Psi(x,k)\,dk\,, \qquad |\hat{\Psi}(k)|^2 = \int_{\mathbb{R}^{3N}} H_\Psi(x,k)\,dx \tag{10}$$

It turns out instead that it is more suitable to the framework of Schroedinger's picture of Quantum Mechanics to construct a correspondence between states and functions of x and p (Wigner functions). With this choice the role of the phase space of Classical Mechanics is particularly stressed. This is very useful to investigate the "semiclassical limit" of Quantum Mechanics, a key subject that we will not discuss here.

In the Schoedinger's formulation of Quantum Mechanics, whose essential features we are briefly outlining, "states" play a very important role.

"Observables" play somehow a minor role in this scheme at least as far as the formal apparatus is concerned. They have however a relevant part at the interpretative level as soon as one has to give a physical meaning to different representation for the state. For example when one states that in the "x representation" the parameter x must be interpreted as the global position vector of the particles forming the system. Analogously, to the parameter p appearing in the Fourier transform is given the meaning of momentum of the particles. In this way in the Hilbert space of the states of the system a major role is devoted to particular operators associated to physical properties of the system having a classical counterpart.

These operators, let us call them $F_{(x)}$ and $G_{(p)}$, act as multiplication by the functions F and G respectively in the representation of positions and momenta. In the x representation the action of $F_{(x)}$ is

$$(F_{(x)}\Psi)(x) = F(x)\,\Psi(x) \tag{11}$$

and the action of $G_{(p)}$ in the p representation is

$$(G_{(p)}\hat{\Psi})(p) = G(p)\,\hat{\Psi}(\hbar^{-1}p) \tag{12}$$

Commutation relations of $F_{(x)}$ and $G_{(p)}$ follows immediately by the properties of Fourier transform. In particular if $F_{(x)} = x_l$, $G_{(p)} = p_j$, for a dense set of vectors one has

$$[x_l, p_j] = i\hbar\delta_{lj} \tag{13}$$

(Canonical commutation relations).

With this notation the expectation value of the observable corresponding to $F_{(x)}$, when the state is Ψ, reads

$$\int_{\mathbb{R}^{3N}} F(x)|\Psi(x)|^2 dx = \langle \Psi, F_{(x)}\Psi \rangle \tag{14}$$

and the one of the observable corresponding to $G_{(p)}$, in the same state reads

$$\int_{\mathbb{R}^{3N}} G(\hbar k)|\hat{\Psi}(k)|^2 dk = \langle \Psi, G_{(p)}\Psi \rangle \tag{15}$$

where we denoted with $\langle \cdot, \cdot \rangle$ the scalar product in the Hilbert space $\mathcal{H} = L^2(\mathbb{R}^{3N})$.

Before heading toward a generalisation of the concept of quantum state we want to make some remarks on quantum dynamics and on the superposition principle.

The evolution of the state Ψ of a quantum system is modeled in terms of a unitary flow $U(t)$, whose generator is the selfadjoint operator H corresponding to the total energy observable

$$\Psi(t) = U(t)\Psi = e^{-i\frac{t}{\hbar}H}\Psi \tag{16}$$

It should be clear from the introduction to the x and p observables, which operator is associated to the energy, at least when the system has a classical analogue with hamiltonian given in the form $h(x,p) = f(x) + g(p)$ (we will not discuss here many important cases for which the quantization of energy or of other classical observable is more delicate). For example, in the simple case of one particle of mass m in $\mathbb{R}^d$ subject to an interaction potential $V(x)$ the quantum hamiltonian is the operator in $L^2(\mathbb{R}^d)$

$$H = -\frac{\hbar^2}{2m}\Delta_x + V(x) \tag{17}$$

By differentiation, the unitary flow (16) can be seen as the solution of a dynamical equation

$$i\hbar\frac{\partial\Psi(t)}{\partial t} = H\Psi(t)\,, \qquad \Psi(0) = \Psi \tag{18}$$

(Schroedinger equation).

To clarify peculiar features of the statistics of the outputs of an experiment on a quantum system we consider a particularly simple example.

Let Ψ_1, Ψ_2 be two states of a single particle system, given in the coordinate representation, such that $\operatorname{supp}\Psi_1 \bigcap \operatorname{supp}\Psi_2 = \emptyset$. Let us suppose that the particle is initially in the state

$$\Psi = c_1\Psi_1 + c_2\Psi_2 , \qquad |c_1|^2 + |c_2|^2 = 1 \tag{19}$$

Ψ_1 and Ψ_2 being orthogonal to each other, the particle has probability $|c_1|^2$ to be in $\operatorname{supp}\Psi_1$ and probability $|c_2|^2$ to be in $\operatorname{supp}\Psi_2$ at $t = 0$.

At any later time t the state of the system will be

$$\Psi(t) = c_1\Psi_1(t) + c_2\Psi_2(t) , \qquad \Psi_i(t) = U(t)\Psi_i , \quad i = 1, 2 . \tag{20}$$

Notice that $\Psi_1(t)$ is still orthogonal to $\Psi_2(t)$, due to the fact that the flow is unitary, but, in general, it will result $\operatorname{supp}\Psi_1(t) \bigcap \operatorname{supp}\Psi_2(t) \neq \emptyset$. Roughly speaking the sets of events defined by "a measure at time t will find the system in the state $\Psi_1(t)$ (resp. $\Psi_2(t)$)" are still nonoverlapping, with probabilities fixed at their initial value $|c_1|^2$ (resp. $|c_2|^2$). Classical intuition might suggest the erroneous conclusion that the probability of any event could be written as a linear combination of probabilities conditioned to the fact that the system is in the state $\Psi_1(t)$ (resp. $\Psi_2(t)$).

Taking as example the event "the particle is in a region Ω at time t", assumption (2) prescribes that its probability will result

$$\int_\Omega |c_1\Psi_1(t,x) + c_2\Psi_2(t,x)|^2 dx \tag{21}$$

rather than

$$|c_1|^2 \int_\Omega |\Psi_1(t,x)|^2 dx + |c_2|^2 \int_\Omega |\Psi_2(t,x)|^2 dx \tag{22}$$

In complete generality, the assumption that all linear combinations of states are possible states of the system (superposition principle) and the prescription that squared moduli of the states specify the probability densities of observables, makes probability in Quantum Mechanics highly non classical.

A different way of writing the expectation values (14),(15) is obtained by considering the operator π_Ψ projecting vectors of $\mathcal{H}$ along Ψ, i.e. $\pi_\Psi\Phi = \langle\Psi,\, \Phi\rangle\Psi$.

Then, for a given operator F one has

$$\langle\Psi, F\Psi\rangle = \operatorname{Tr}(\pi_\Psi F) \tag{23}$$

where Tr indicates the trace on the algebra $B(\mathcal{H})$ of bounded operators on $\mathcal{H}$.

A generalisation of the definition of state consists in considering a classical probabitity distribution on the set of pure states, analogously to what is done in Classical Statistical Mechanics.

Accordingly, for a suitable unitary bases Ψ_i in $\mathcal{H}$, the expectation value of an operator F is written as

$$\sigma(F) = \sum_i c_i \langle \Psi_i, F\Psi_i \rangle , \qquad \sum_i c_i = 1 , \qquad c_i \geq 0 \tag{24}$$

or in general terms

$$\sigma(F) = \mathrm{Tr}\,(\rho F) \tag{25}$$

where ρ is a positive trace class operator of trace one in $\mathcal{H}$ called density matrix. Such generalised states in Quantum Mechanics are called statistical mixtures (as opposed to pure states given by (1)).

Notice the profound difference between the linear combination (24), where coefficients c_i have an immediate interpretation in terms of classical probability, and the statistical properties in a pure state.

From (24) and (16) it is easy to argue that the evolution law for a density matrix is given by

$$\rho(t) = e^{-i\frac{t}{\hbar}H}\rho(0)e^{i\frac{t}{\hbar}H} \tag{26}$$

This generalisation of the notion of state allows the introduction of a key concept in the quantum description of a system S in interaction with an environment E, i.e. the notion of partial trace.

Let us consider the case in which the Hilbert space $\mathcal{H}$ is the tensor product $\mathcal{H}_S \otimes \mathcal{H}_E$ and the algebra of observables relative to S can be identified with $B(\mathcal{H}_S) \otimes I_E$. The expectation value of observables relative to S is then defined by the relation

$$\sigma_S(A) = \sigma(A \otimes I_E) \qquad \forall A \in B(\mathcal{H}_S) \tag{27}$$

and in turn this defines a density matrix ρ_{re} in $\mathcal{H}_S$.

The procedure giving ρ_{re} is referred to as partial trace on ρ and ρ_{re} as reduced density matrix.

All the information about expectation values for observables corresponding to operators in $B(\mathcal{H}_S) \otimes I_E$ is contained in ρ_{re}. In general no information on expected values, in the state Ψ, for any observable outside $B(\mathcal{H}_S) \otimes I_E$ can be extracted from the knowledge of ρ_{re}.

Notice that even in the case of a pure state of the entire system in general ρ_{re} is not a projection operator. A subsystem of a system in a pure state is in general described by a statistical mixture.

As we saw before it is impossible to interpret the statistical properties of the outputs of measurements on an isolated quantum system according to the rules of classical probability theory. It cannot be considered a remedy to divide the system in two parts, the observable and the observer, because even the observer should be considered and treated as a quantum system. In order to understand the observed loss of quantum coherence even in microscopic systems it is necessary to admit that the system is not isolated. The lack of information about the external system after the interaction should be seen as responsible of the disappearance, at the level of the subsystem, of quantum interference effects and to the onset of decoherence.

The time evolution of the reduced density matrix is induced by the dynamics (26) of the density matrix of the entire system.

In general $\rho_{re}(t)$ as a function of time does not satisfy a differential equation of the type

$$\frac{d\rho_{re}}{dt} = F(t, \rho_{re}) \tag{28}$$

for a suitable function $F(t, \cdot)$. A remarkable exception is the case in which the system under consideration is made of one particle interacting linearly with a large set of harmonic oscillators (a thermal bath). Only in the thermodynamic limit the function F does not depend on t and depends on the initial state of the bath only through the temperature. In [3], [9] this fact has been used in the analysis of decoherence and dissipation effects for the subsystem.

A totally different set-up for the study of the decoherence effect due to an external environment was described in [8], [6], [11]. These authors consider the system made of a heavy particle in interaction with a "light" particle, when the final state of the latter is not observed. This "conceptual experiment", could give some insight into the classical behaviour of macroscopic objects and suggestions on possible descriptions of a measurement apparatus in Quantum Mechanics.

The most interesting aspect of this approach is the intuition that even a weak interaction with an (unobserved) environment (the light particle), forces the state of the heavy particle to become a statistical mixture. The corresponding reduced density matrix, written in the coordinate representation, tends to "concentrate on the diagonal" (as compared to the one corresponding to the free motion of the heavy particle) in a way quantitatively proportional (only) to the scattering amplitude.

This fact can be taken as an indication that interactions subsequently experienced by the heavy particle with a large number of light particles contribute in reducing its density matrix more and more to diagonal form, making the state of the heavy particle indistinguishable from a statistical mixture of classical states.

The indications given in [8], [6], [11] are very interesting; still not all the approximations and deductions used are easily justified from first principles, i.e. from the evolution given by the Schroedinger equation.

The rest of this contribution is devoted to the discussion of a simple model of a heavy and a light particle in interaction where all the steps are rigorously worked out (in particular all the approximations are controlled).

Let us consider a system of two particles in one dimension with masses M, m, $M > m$, interacting via a repulsive delta-like potential. Denoting by R the position of the heavy particle and by r the position of the light one, the system is then described by the following hamiltonian in $L^2(\mathbb{R}^2)$

$$H = -\frac{\hbar^2}{2M}\Delta_R - \frac{\hbar^2}{2m}\Delta_r + \alpha\delta(R-r)\,, \qquad \alpha > 0 \tag{29}$$

Rigorously (29) defines a positive, symmetric, closed quadratic form. It is well known that a unique positive self-adjoint operator is associated with such a form (see e.g. [1]).

The interesting aspect of the hamiltonian (29) is that the corresponding Schroedinger equation can be explicitly solved. In fact, for any initial state Ψ_0 of the system one has ([10])

$$\begin{aligned}\Psi(t,r,R) &= \left(e^{-i\frac{t}{\hbar}H}\psi_0\right)(r,R)\\ &= \int dr'dR'\Psi_0(r',R')U_0^{\nu}\left(t,\frac{M}{\nu}(R-R')+\frac{\mu}{M}(r-r')\right)\\ &\quad\cdot\left[U_0^{\mu}(t,(r-R)-(r'-R'))\right.\\ &\quad\left.-\frac{\mu\alpha}{\hbar^2}\int du e^{-\frac{\mu\alpha}{\hbar^2}u}U_0^{\mu}(t,u+|r-R|+|r'-R'|)\right]\end{aligned} \tag{30}$$

where $\nu = m + M$ is the total mass, $\mu = \frac{mM}{m+M}$ is the reduced mass and

$$U_0^a(t,x,x') = \sqrt{\frac{a}{2\pi i\hbar t}}e^{i\frac{a|x-x'|^2}{2\hbar t}}, \qquad a > 0 \tag{31}$$

We are interested in the analysis of the motion of the heavy particle and then we consider the corresponding reduced density matrix

$$\rho_{re}(t,R,R') = \int dr\overline{\Psi}(t,r,R)\Psi(t,r,R') \tag{32}$$

The expression for (30) can be simplified if one considers the asymptotic regime

$$M \gg m \tag{33}$$

and chooses the initial condition

$$\Psi_0(r,R) = \frac{1}{\sqrt{2}}\left(f_\sigma^+(R) + f_\sigma^-(R)\right)\cdot g_\delta(r) \tag{34}$$

$$f_\sigma^\pm(R) = \frac{1}{\sqrt{\sigma}} f\left(\frac{R \pm R_0}{\sigma}\right) e^{\pm i\frac{P_0}{\hbar}R}\,, \qquad \sigma, R_0, P_0 > 0 \tag{35}$$

$$g_\delta(r) = \frac{1}{\sqrt{\delta}} g\left(\frac{r - r_0}{\delta}\right) e^{i\frac{q_0}{\hbar}r}\,, \qquad \delta > 0\,,\ r_0, q_0 \in \mathbb{R} \tag{36}$$

$$f, g \in C_0^\infty(-1,1)\,, \qquad \|f\| = \|g\| = 1 \tag{37}$$

where we assume

$$\begin{aligned} &\frac{\sigma}{R_0} \ll 1\,, \qquad r_0 \in (-R_0+\sigma, R_0-\sigma)\,, \\ &\frac{\delta}{|R_0 - |r_0||} \ll 1\,, \qquad \frac{q_0\delta}{\hbar} \ll 1\,, \qquad \frac{\sigma}{\delta} \ll 1 \end{aligned} \tag{38}$$

The initial condition (34) is in a product form, which means that no correlation is assumed between the two particles at $t = 0$.

The heavy particle is described by a superposition of the states f_σ^+ and f_σ^-, where f_σ^+ describes the particle well-localised in position around $-R_0$ with positive momentum P_0 and f_σ^- describes the particle well-localised around R_0 with average momentum $-P_0$.

Notice that, due to our assumptions (38), the probability distribution of the position of the heavy particle at $t = 0$ is given by

$$\rho_{re}(0,R,R) = \frac{1}{2\sigma}\left|f\left(\frac{R+R_0}{\sigma}\right)\right|^2 + \frac{1}{2\sigma}\left|f\left(\frac{R-R_0}{\sigma}\right)\right|^2 \tag{39}$$

i.e. no interference effect is present at initial time (see Fig. 1).

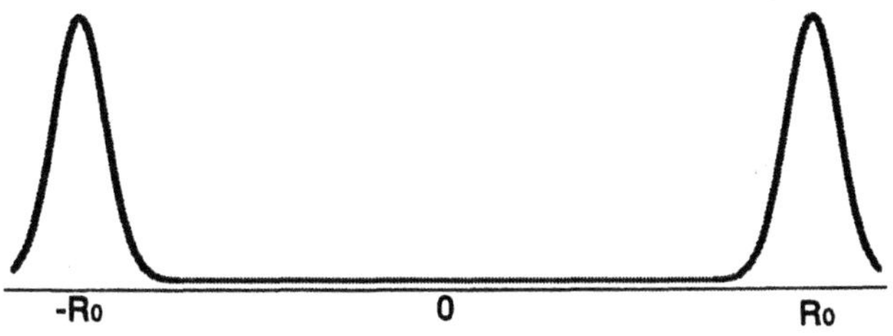

Figure 1. The probability density at time $t = 0$.

The light particle is localised in the region $(-R_0, R_0)$, well-separated by f_σ^+ and f_σ^- and has a zero average momentum. Moreover we require that the localisation in position of the heavy particle is much stronger than the one of the light particle.

As a matter of fact, in order to obtain a non trivial asymptotics, we need the further assumption that the "effective" strength of the interaction remains finite, i.e.

$$\beta \equiv \frac{m\alpha\delta}{\hbar^2} = O(1) \tag{40}$$

The key approximation result is the following ([4]).

Theorem 2. *Assume (33), (40) and $t > 0$. Then*

$$\|\rho_{re}(t) - \rho_{re}^a(t)\|_1 < C\,\frac{m}{M} \tag{41}$$

where $\|\cdot\|_1$ is the trace-class norm,

$$\rho_{re}^a(t) = e^{-i\frac{t}{\hbar}H_0}\rho_{re}^a(0)e^{i\frac{t}{\hbar}H_0}, \tag{42}$$

$$H_0 = -\frac{\hbar^2}{2M}\Delta_R, \tag{43}$$

$$\rho_{re}^a(0,y,z) = \frac{1}{2}\left(\overline{f_\sigma^+(z)} + \overline{f_\sigma^-(z)}\right)\left(f_\sigma^+(y) + f_\sigma^-(y)\right)\mathcal{I}(y,z), \tag{44}$$

$$\mathcal{I}(y,z) = \left((\Omega_+^{\beta,z})^{-1}g_\delta, (\Omega_+^{\beta,y})^{-1}g_\delta\right), \tag{45}$$

$\Omega_+^{\beta,\xi}$, $\xi \in \mathbb{R}$, is the wave operator of the one-particle hamiltonian

$$H_{\beta,\xi} = -\frac{1}{2}\Delta_x + \beta\delta(x-\xi), \tag{46}$$

and the constant C is a given function of t, β, $f_\sigma^\pm$, g_δ.

Theorem 2 shows that the motion of the heavy particle in presence of the light one is a free evolution of the modified initial state (44).

Notice that the kernel $\mathcal{I}(y,z)$ satisfies

$$|\mathcal{I}(y,z)| \le 1, \qquad \mathcal{I}(y,z) = 1 \quad \text{iff} \quad y = z \tag{47}$$

so that

$$\mathrm{Tr}\,\rho_{re}^a(0) = 1, \qquad \mathrm{Tr}\,(\rho_{re}^a(0))^2 < 1 \tag{48}$$

As a consequence, the heavy particle for any $t > 0$ and $\beta > 0$ is described by a mixture.

In order to see the decoherence effect in the position representation one needs a more detailed estimate of $\mathcal{I}(y,z)$ showing that $\mathcal{I}(y,z) \simeq 1$ for $|y-z|$ small and $\mathcal{I}(y,z) < 1$ for $|y-z|$ large.

More precisely, exploiting the explicit expression of the wave operator in terms of the generalised eigenfunctions and the assumptions (38), we obtain

Proposition 1. *If $y, z \in supp\, f_\sigma^-$ or $y, z \in supp\, f_\sigma^+$ then*

$$\mathcal{I}(y,z) = 1 + O\left(\frac{\sigma}{\delta}, \frac{\delta}{R_0 - |r_0|}\right) \tag{49}$$

If $y \in supp\, f_\sigma^-$, $z \in supp\, f_\sigma^+$ or $y \in supp\, f_\sigma^+$, $z \in supp\, f_\sigma^-$ then

$$\begin{aligned} \mathcal{I}(y,z) &= \Lambda_\beta + O\left(\frac{\delta}{R_0 - |r_0|}, \frac{q_0 \delta}{\hbar}\right), \\ \Lambda_\beta &\equiv \int dk |\hat{g}(k)|^2 \frac{k^2}{\beta^2 + k^2} \in (0,1) \end{aligned} \tag{50}$$

Notice that Λ_β is the fraction of transmitted wave of a particle in the state g subject to a point interaction of strength β; it reduces to one for $\beta = 0$ (absence of interaction) and to zero for $\beta \to \infty$ (limit of perfectly reflecting barrier).

In conclusion, under the assumptions (33), (34), (38), (40), the reduced density matrix of the heavy particle has the following asymptotic expression in the position representation

$$\begin{aligned} \rho_{re}(t,R,R') \simeq &\frac{1}{2}\Big[\left(e^{i\frac{t}{\hbar}H_0}\overline{f_\sigma^+}\right)(R)\left(e^{-i\frac{t}{\hbar}H_0}f_\sigma^+\right)(R') \\ &+ \left(e^{i\frac{t}{\hbar}H_0}\overline{f_\sigma^-}\right)(R)\left(e^{-i\frac{t}{\hbar}H_0}f_\sigma^-\right)(R')\Big] \\ &+ \frac{\Lambda_\beta}{2}\Big[\left(e^{i\frac{t}{\hbar}H_0}\overline{f_\sigma^+}\right)(R)\left(e^{-i\frac{t}{\hbar}H_0}f_\sigma^-\right)(R') \\ &+ \left(e^{i\frac{t}{\hbar}H_0}\overline{f_\sigma^-}\right)(R)\left(e^{-i\frac{t}{\hbar}H_0}f_\sigma^+\right)(R')\Big] \end{aligned} \tag{51}$$

The first two terms in (51) are sensibly different from zero in a neighbourhood of the diagonal $R = R'$ while the last two terms give the off-diagonal contributions, responsible for the interference pattern. The interaction with the light particle leaves unchanged the diagonal terms but reduces the interference terms by a factor Λ_β. In this way $1 - \Lambda_\beta$ turns out to be a quantitative measure of the decoherence effect on the state of the heavy particle due to the interaction with the light one.

Such decoherence effect is better visualised if one computes the probability distribution of the position at the time $t = \tau \equiv \frac{R_0 M}{P_0}$, which is the time of the maximal superposition of the two wave packets initially described by f_σ^+, f_σ^-.

Exploiting the further technical assumption

$$\frac{\sigma}{R_0} \ll \frac{\hbar}{\sigma P_0} \ll 1 \tag{52}$$

such distribution reads

$$\rho_{re}(t, R, R) \simeq \frac{\sigma P_0}{R_0 \hbar} \left| \tilde{f}\left(\frac{\sigma P_0 R}{R_0 \hbar}\right) \right|^2 \left(1 + \Lambda_\beta \cos \frac{2 P_0 R}{\hbar}\right) \tag{53}$$

It is evident from (53) that the typical oscillations due to the quantum

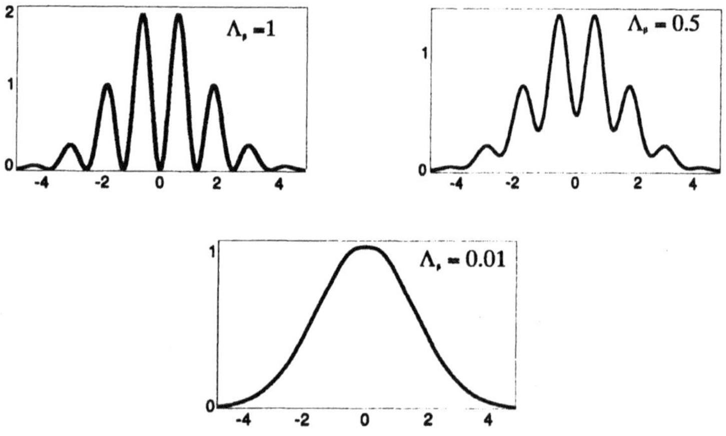

Figure 2. The probability density at time τ for different values of Λ.

interference of the two components of the heavy particle wave packet are reduced. When Λ_β is sufficiently small the probability density of the position of the heavy particle is indistinguishable from the one obtained as (incoherent) sum of the probability densities associated to the two parts of the wave packet (see Fig. 2).

We finally remark that our model of environment (one light particle) cannot be considered realistic. A more refined analysis is required to show decoherence induced by an environment modeled by a system of N (non-interacting in the simplest case) light particles.

It is reasonable to expect that the effect of N independent scattering events is cumulative so that the decoherence factor should be Λ_β^N, showing that the decoherence effect is exponentially increasing with the number of particles of the environment.

SCALING LAWS: MICROSCOPIC AND MACROSCOPIC BEHAVIOR

Raffaele Esposito
Dipartimento di Matematica Pura ed Applicata,
Università degli Studi dell'Aquila - ITALY
and
Centro di Ricerche Linceo Interdisciplinare "Beniamino Segre"
esposito@univaq.it

Abstract The relation between microscopic and macroscopic descriptions of many-particle systems is discussed in terms of scaling laws, following the Boltzmann original ideas. Models where a complete mathematical treatment is possible are outlined.

1. Introduction

The fluids are particularly important examples of complex systems. They in fact provide the simplest and most fundamental case of coexistence of two descriptions: the macroscopic theory, based on phenomenological arguments and the microscopic Newtonian dynamics, essentially free of assumptions, but for the nature of the molecular interaction.

One of the most challenging tasks of the Mathematical Physics is the understanding of the deep connection between the two descriptions. The interest of the problem was already recognized by Hilbert, who devoted part of his sixth question exactly to "*... the problem of developing mathematically the limiting processes ... which lead from the atomistic view to the laws of motion of continua...*" [1].

The limiting process Hilbert refers to is just the one we are going to discuss here, the *hydrodynamical limit.* In the last two decades significant contributions have been given by many Authors toward the solution of the sixth Hilbert question, although it is still open in realistic models.

The starting point of the theory is the Boltzmann [2] analysis of the thermodynamic systems, based on the enormous scale separation between the thermodynamic description and the microscopic Newton equations. The apparent contradiction between the microscopic reversibility and the macroscopic irreversibility was brilliantly solved by Boltzmann using the scale separation and his argument will be shortly presented below as a simple example of how scale arguments can fill the conceptual gaps between the two approaches.

The main achievements of the Mathematical Physics in the recent years are consequence of the introduction of simplified models where such difficult questions can be treated with adequate mathematical rigor, in the spirit of the Hilbert sixth question. A few models will be introduced and some ideas and recent results will be presented.

2. Microscopic description

A natural microscopic model for a gas is a system of N mass points with equal mass m. The role of the number N is crucial: indeed the same scheme holds both for the solar system and for a fluid, the only difference being that in the first case N is small ($N \approx 10$), while in the second N is large ($N \approx 10^{23}$).

The evolution of the system is, in principle, ruled by the Quantum Mechanics, but the approximation in terms of Classical Mechanics produces negligible errors on the macroscopic behavior, so we will assume classical equations for sake of simplicity. Therefore, the system is characterized by specifying at each time its coordinates and velocities q_i and v_i, $i = 1, \ldots, N$, with respect to a given inertial frame. A configuration of the system is a *phase point* $P = (q_1, v_1, \ldots, q_N, v_N)$. The set Γ_N of all such phase points is called *phase space*. The motion of the system is ruled by the Newton law

$$\frac{d}{dt} q_i = v_i \, , \qquad \frac{d}{dt} v_i = \sum_{j \neq i} F_{j \to j}(q_i - q_j) \tag{1}$$

where $F_{j \to i}(q_i - q_j) = -\nabla_{q_i} U(|q_i - q_j|)$ is the force of the particle j on i and U is the interaction potential. This system, regardless of the value of N is reversible in the following sense: suppose that the system starts at time $t = 0$ in a phase point P and evolves according to the Newton law for a time T. At this time all the velocities of the system are reversed and the system evolves for another time T. The final state is a phase point P' whose coordinates are the same of P and whose velocities are the opposite of the velocities in P.

The microscopic description, although fully deterministic, if N is very large, exhibits an erratic and apparently random behavior, as far as the motion of a test particle is concerned: in fact, due to the sensible dependence of the solutions to equation (1) on the initial conditions, the motion of a test particle is essentially unpredictable. Based on this,

Boltzmann assumed the point of view that only probabilistic statements on the system are relevant from the practical point of view. More specifically, Boltzmann assumed that, for N large, at the thermodynamic equilibrium, the system is described by a probability distribution on the phase space assigning equal weights to all the phases point with the same energy. The assumption is very convincing and effective in the construction of the modern Statistical Mechanics, but has not yet been proven to be true in realistic models.

Many Authors, including Zermelo and Poincaré, argued about the impossibility of describing macroscopic systems by means of the Newton law because of the contradiction between the *microscopic reversibility* discussed before and the *macroscopic irreversibility* which is part of the common experience. The (apparent) contradiction is illustrated by the following *ideal* experiment.

A system of N particles is enclosed in a region Ω of volume $2V$, which is divided in subregions Ω_1 and Ω_2 of volume V by a separating wall. At the initial time the N particles are all in Ω_1. The separation is removed and the system evolves according to the Newton laws. After some time T it invades the region Ω with uniform density. Because of the reversibility of the Newton law, one can show that there are configurations of the system with uniform density in Ω which, after a time T are all in Ω_1. This is "clearly" in disagreement with the irreversible behavior of a macroscopic system.

The argument Boltzmann used against this was that the value of N makes a big difference. The probability that a system of N particles returns in Ω_1 is proportional to the volume occupied by the phase points with all the coordinates in Ω_1 and it turns out to be of order 2^{-N}, which, in the case of a gas, means $2^{-10^{23}}$ so that the return in Ω_1 will never be seen for a gas.

From above argument we learn that, in order to get the macroscopic laws from the microscopic dynamics, we have to take the limit $N \to \infty$.

3. Macroscopic description

The macroscopic description of a fluid is given in terms of the density field, $\rho \geq 0$, the temperature field $T > 0$ and the velocity field $u \in \mathbb{R}^3$. They are smooth functions of the macroscopic space coordinates x and time t and satisfy a system of equations depending on the model of fluid considered. The most used models are the so called Euler and Navier-

Stokes fluid, which, for an ideal fluid, are described by the system

$$\partial_t \rho + \nabla \cdot (\rho u) = 0\,, \qquad \rho D_t u + \nabla P = \nabla \cdot \sigma\,, \tag{2}$$

$$\frac{3}{2}\rho D_t T + P \nabla \cdot u = \nabla \cdot (\kappa \nabla T) + \mathrm{Tr}\,(\sigma \nabla u)\,, \tag{3}$$

where $D_t := \partial_t + u \cdot \nabla$, $\sigma := \nu(\nabla u + \nabla u^\dagger - \frac{2}{3}\mathbf{I}\,\nabla \cdot u)$. P is the pressure, given by the equation of state of the perfect gases: $P = \rho T$; the *transport coefficients* ν and κ vanish in the Euler fluid while are strictly positive and represent the viscosity and thermal conduction coefficients in the Navier-Stokes fluid.

The practical advantage of above equations, compared with the microscopic description is evident. On the other hand, the parameters entering in the equations are only restricted by the agreement with experimental data. The validation of such equations in terms of the fundamental system (1) would fill an important conceptual gap.

4. Derivation of the Euler equations from the microscopic dynamics

The link between the microscopic, probabilistic, description and the macroscopic equation is provided by the *law of large numbers*, which holds because of the large size of the system ($N \approx 10^{23}$). It ensures that quantities depending on the behavior of many particles, assume with probability going to 1, the average values computed by means of the probability distribution of the system. The mathematical way to implement such an idea is the *scaling limit*. It is based on the introduction of a scaling parameter $\varepsilon > 0$ representing the ratio between the typical microscopic lengths (e.g. the size of the molecules) and the microscopic lengths (e.g. the size of the container of the fluid). This allows us to rewrite (1) in terms of macroscopic coordinates $x_i = \varepsilon q_i$, $i = 1 \ldots, N$. If one considers a small macroscopic region Δ, whose microscopic volume is $O(\varepsilon^{-3})$ in microscopic units, the number of particles it contains goes to infinity, when $\varepsilon \to 0$. Therefore, to such a small macroscopic region one can apply the Boltzmann considerations. Could we neglect the action of the particles outside of Δ, the system in Δ would be in equilibrium and hence described by a suitable equilibrium probability distribution. Of course such actions are not negligible, but they are so small that a long time, of order ε^{-1}, is necessary to change the state of the system in Δ. Since the equilibrium has strong stability properties, it is reasonable to expect that when the state of the system in Δ changes, at times ε^{-1}, it stays close to another equilibrium probability distribution with different values of the equilibrium parameters, i.e. density, temperature and mean velocity. This is roughly the *local equilibrium assumption*: the state of the system on the time scale ε^{-1} is characterized by a probability distribution which locally resembles an equilibrium distribution with parameters depending on the macroscopic space and time.

With this assumption it is not hard to derive *formally* the Euler equations from the microscopic dynamics and the first derivation does back to [3]. The derivation of the Navier-Stokes equations is more difficult also on the formal level because longer time scales are involved and, as a consequence, the transport coefficients, depending on the corrections to the local equilibrium, become relevant [4].

The rigorous derivation of the Euler equations is still lacking because there is no proof of the validity of the local equilibrium assumption. Given that, in [5] it has been proven (with some technical assumptions) that the Euler equations follow from the Newton law. Moreover, if some small stochastic noise is added to the deterministic dynamics, the local equilibrium assumption is also proven. The addition of stochastic noise to the dynamics may appear artificial, but the remark on the unpredictable character of the motion of a test particle indicates that the microscopic dynamics looks like a stochastic one, although we cannot give any proof of this statement at the moment for the original deterministic dynamics.

The one proposed in [5] is an example of the *models* on which the Mathematical Physics has worked in the last few decades, following the Hilbert advise, to obtain mathematically rigorous proofs of the macroscopic equations from the microscopic dynamics. Another model which has been quite fruitful in this respect is the well known Boltzmann equation which is the basis of the Kinetic Theory.

The Boltzmann equation is itself a kind of macroscopic equation, whose status is much more legitimate than the equations for a fluid, after the rigorous proof of its validity given by Lanford [6]. It describes the system on the scale of the mean free path, which is typically much smaller that the macroscopic one. Therefore, it makes sense, as already proposed by Hilbert in [1], to try to derive the macroscopic equations from it. The Hilbert expansion [7] is a technical tool to produce very easily the Euler equations, but its rigorous foundation was given only in 1980 in [8] where it was proved that the solution to the Boltzmann equation in the scaling limit can be approximated by suitable truncations of the Hilbert series. One should note that, after [8] the local equilibrium assumption, on the level of the Boltzmann equation, becomes a theorem.

5. Derivation of the Navier-Stokes equations

The Navier-Stokes equations can be derived from the Boltzmann equation, but on longer time scale than the Euler equation, namely times of order ε^{-2}. However, the lack of scaling invariance of the Navier-Stokes equations forces us to restrict the analysis to the low Mach number regime (i.e. for mean velocities sufficiently small compared with the sound speed), by using a truncation of a modified Hilbert expansion [9]. The equations thus obtained are the incompressible Navier-Stokes equations

$$\nabla \cdot u = 0\,, \qquad \partial_t u + u \cdot \nabla u = -\nabla p + \nu \Delta u\,. \tag{4}$$

As mentioned before, the corrections to the local equilibrium are necessary to compute the transport coefficient ν. On the level of the Boltzmann equation this can be nicely accomplished by using the expansion.

Starting from the Newton law, the problem is much more difficult because there is no way to give a mathematical sense to a power series expansion similar to the Hilbert expansion. A technique to compute the transport coefficient has been introduced in [10] in the framework of the stochastic particle systems on the lattice. For such models, many of the ideas driving to the macroscopic equations have been successfully tested. In [11] it has been possible to construct a model whose scaling limit is similar to the incompressible Navier-Stokes equations (4). The model is roughly the following: on the sites lattice $\mathbb{Z}^3$ there are particles which can jump from one site to a neighboring one, according to some transition probability depending on the velocity of the particles. The only restriction we have is an exclusion rule preventing particles with the same velocity to stay in the same site. The velocities of the particles (and hence the transition rules) change during collisions: two particles with velocities v_1 and v_2 sitting in the same site have a positive probability of having a collision which changes the velocities (v_1, v_2) into the new couple (v_1', v_2') with the condition that the total momentum is preserved:

$$v_1' + v_2' = v_1 + v_2\,.$$

The set of velocities $\mathcal{V}$ has finite cardinality and is not arbitrary, but has to be chosen properly to get the right limiting equations.

The method of [10] can be adapted to this model and it is possible to prove that the scaling limit does exist and corresponds to a slight modification of (4) due to the anisotropy of the lattice $\mathbb{Z}^3$. The transport coefficients can be characterized completely in terms of the microscopic dynamics and are shown to be given by the famous Green-Kubo formulas [12].

6. Conclusions

It is worth to stress again that the understanding of the relation between microscopic and macroscopic descriptions of complex systems has greatly improved after the use of scaling arguments.

We conclude by underlining once more the crucial role played by the models in the mathematical derivation of the macroscopic equations by means of scaling limits. It is only in the framework of specific models (as in [5] and [11]) that it has been possible to pass from heuristic considerations to proofs of mathematically precise statements. Of course just a few preliminary steps have been accomplished in this direction and most of the work is still to be done.

References

[1] HILBERT D., *Mathematical Problems*, Göttinger Nachroichten 1900, 253–297.

[2] BOLTZMANN L., *Über die Eigenshaften monzyklischer und anderer damit verwandter Systeme*, in: "Wissenshaftliche Abhandlungen", F.P. Hasenörl ed., vol. III, Chelsea, New York 1968.

[3] MORREY C.B., *On the derivation of the equations of hydrodynamics from statistical mechanics*, Commun. Pure Applicata. Math. **8** (1955), 279–326.

[4] ESPOSITO R., MARRA R., *On the derivation of the incompressible Navier-Stokes equation for Hamiltonian particle systems*, J. Stat. Phys. **74** (1993), 981–1004.

[5] OLLA S., VARADHAN S.R.S., YAU H.T., *Hydrodynamical limit for a Hamiltonian system with weak noise*, Commun. Math. Phys. **155** (1993), 523.

[6] LANFORD O.E., *Time Evolution of Large Classical systems*, in: "Lecture Notes in Physics" **38**, J. Moser ed., Springer, Berlin, Heidelberg, 1–111.

[7] HILBERT D., *Begründung der kinetischen Gastheorie*, Mathematische Annalen **72** (1916/17), 331–407.

[8] CAFLISCH R.E., *The fluid dynamical limit of the nonlinear Boltzmann equation*, Commun. Pure and Applicata. Math. **33** (1980), 651–666.

[9] DE MASI A., ESPOSITO R., LEBOWITZ J.L., *Incompressible Navier-Stokes and Euler Limits of the Boltzmann Equation*, Commun. Pure and Applicata. Math. **42** (1989), 1189–1214.

[10] VARADHAN S.R.S., *Nonlinear diffusion limit for a system with nearest neighbor interactions II*, in: "Asymptotic Problems in Probability Theory: Stochastic Models and Diffusion on Fractals", K.D. Elworthy and N. Ikeda eds., Pitman Research Notes in Math. **283**, J. Wiley & Sons, New York 1994, 75–128.

[11] ESPOSITO R., MARRA R., YAU H.T., *Navier-Stokes equations for stochastic particle systems on the lattice*, Commun. Math. Phys. **182** (1996), 395–456.

[12] GREEN M.S., *Markoff random processes and the statistical mechanics of time-dependent phenomena. II. Irreversible processes in fluids*, Jour. Chem. Phys. **22** (1954), 398–413.

[13] KUBO R., *Statistical-Mechanical theory of irreversible processes. I general theory and simple applications in magnetic and conduction problems*, Jour. Phys. Soc. Jap. **12** (1957), 570–586.

MEASURE OF DIFFUSION ENTROPY OF WEAK TURBULENCE IN SAMPLE OF NEMATIC LIQUID CRYSTAL

Laura Galeotti
Dipartimento di Fisica dell'Università di Pisa,
Via Buonarroti 2, 56127 PISA - ITALY
laura.galeotti@email.it

Giulia Menconi
C.I.S.S.C. Centro Interdisciplinare per lo Studio dei Sistemi Complessi,
Università di Pisa,
Via Bonanno 25/b, 56126 PISA - ITALY
menconi@mail.dm.unipi.it

Leone Fronzoni
Dipartimento di Fisica dell'Università di Pisa,
Via Buonarroti 2, 56127 PISA - ITALY
and C.I.S.S.C. Centro Interdisciplinare per lo Studio dei Sistemi Complessi,
Università di Pisa,
Via Bonanno 25/b, 56126 PISA - ITALY
and INFM,
Via Buonarroti 2, 56127 PISA - ITALY
fronzoni@df.unipi.it

Abstract The aim of this work is to characterize the complexity of nematic liquid crystal auto-organizing structures. The formation of these kinds of frames is due to the application of a particular electric field to the sample. Varying this control parameter, the structures evolve from very regular frames to very disorderly ones. For all these different cases we had acquired a string of data. We manage to classify these sequences, taken on the liquid crystal, according to their degree of complexity through an entropy method, the Diffusion Entropy Method.

1. Introduction

One of the most effective ways to describe and understand high-dimensional systems is the study of the diffusion function. The diffusion entropy method is a statistic method; it consists in an analysis of a signal with which a diffusive process is built using mobile windows. This method allows us to characterize the complexity of systems that show even turbulent behaviours. One of physical systems constituting a paradigm for the study of auto-organization phenomena and growth of complexity is the nematic liquid crystal. In the next paragraphs we'll show, first of all, how the auto-organizing structures of the crystal form, then what is the diffusion entropy method and finally our results obtained applying this method on particular strings of data taken on the surface of the crystal for different types of behaviours.

2. Nematic liquid crystal(NLC)

Liquid crystals are substances characterized by a quite orderly structure, but with physical proprieties typical of liquids like, for example, the capacity in assuming the form of a recipient in which they are contained. The characteristic of nematic liquid crystal is that the molecules orient themselves along a privileged direction, but they are free to move in all other directions.

For this kind of liquid crystals the formation of auto-organized structures is due to the anisotropy of conductivity. The Helfrich theoretical model describes this phenomenon that we are going to specify in a brief and schematic way. Let's consider the NLC contained between two conductive plates, with an omeotropic orientation, i.e. with the molecules' director of orientation set parallel to the walls [1].

Applying a difference of potential to the plates, we obtain an electric field perpendicular to the nematic's director. Initially there'll be no electrical current parallel to the walls, but, because of the anisotropy, just an eventually fluctuation of the orientation will cause an electric current of the type above mentioned. This current will produce spatial electric charges that will interact with the electrical field perpendicular to the walls. If the electric force will be greater than the viscosity force, it will be established an electro-convective motion with a periodicity defined by the sample's thickness.

This hydrodynamic instability assumes different geometrical properties on account of the variation of the threshold's values of the electric field. There will be a first threshold's value E_1 for which, if the electric field E is greater than E_1, we'll have cylindrical form convective motion with the axis parallel to the walls (Williams rolls) like those already described. For E greater of a second threshold's value, E_2, (with $E_2 > E_1$) the regularity of the structures is broken on account of defects' born that bring to a turbulent motion. When the electric field is between these two thresholds defects begin to form, with the result of the coexistence in the pattern of zones of order and disorder. More the E-value increases, more the NLC goes toward the disorder.

3. Diffusion entropy method (DEM)

In this paragraph we want to illustrate what is the diffusion entropy method. We report a brief description of the method taken from articles in references [2, 3, 4].

The diffusion entropy method rests on the evaluation of the entropy of the diffusion process generated by the time series. The details of the diffusion process depend on the walking rules, which determine the form and the time duration of the transition to the scaling regime, as well as the scaling parameter δ.

The main idea of this approach to scaling is remarkably simple. Let us consider a sequence of M numbers, $\xi_i(t)$, with $i = 1, \ldots, M$. The purpose of the DEM algorithm is to establish the possible existence of a scaling, either normal or anomalous, in the most efficient way as possible without altering the data with any form of detrending. Let us select first of all an integer number l, fitting the condition $1 \leq l \leq M$. We will refer to this integer number as "time". For any given time l we can find $M - l + 1$ sub-sequences defined by

$$\xi_i^{(s)}(t) = \xi_{i+s}(t)\,, \qquad s = 0, \ldots, M - l \tag{1}$$

For any of these sub-sequences we build up a diffusion trajectory, labelled with the index s, defined by the position

$$x^{(s)}(l) = \sum_{i=1}^{l} \xi_i^{(s)} = \sum_{i=1}^{l} \xi_{i+s} \tag{2}$$

Let us imagine this position as referring to a Brownian particle that at regular intervals of time has been jumping forward or backward according to the prescription of the corresponding sub-sequence of equation (1). This means that the particle before reaching the position that it holds at time l has been making l jumps. The jump made at the i-th step has the intensity $|\xi_i^{(s)}|$ and is forward or backward according to whether the number $\xi_i^{(s)}$ is positive or negative.

We are now ready to evaluate the entropy of this diffusion process. To do that we have to divide the x-axes into cells of size $\epsilon(l)$. When this partition is made we have to label the cells. We count how many particles are found in the same cell at a given time l. We denote this number by $N_i(l)$. Then we use this number to determine the probability that the particle can be found in the i-th cell at the time l, $p_i(l)$, by means of

$$\mu_i(l) = \frac{N_i(l)}{(M - l + 1)} \tag{3}$$

At this stage the entropy of the diffusion process at time l is determined and reads

$$S_d(l) = -\sum_i p_i(l) \ln[p_i(l)] \tag{4}$$

The easiest way to proceed with the choice of the cell size, $\epsilon(l)$, is to assume it to be independent of l and determined by a suitable fraction of the square root of the variance of the fluctuation ξ_i. In the case in which the numbers ξ_i are $+1$, 0 and -1, $\epsilon = 1$ is the natural choice.

Details on how to deal with a transition from the short-time regime, sensitive to the discrete nature of the process under study, to the long-time limit where both space and time can be perceived as continuous, are given in [3]. Here we make the simplifying assumption of considering large enough times as to make the continuous assumption valid. In this case, the trajectories, correspond to the following equation of motion:

$$\frac{dx}{dt} = \xi(t) \tag{5}$$

where $\xi(t)$ denotes the value that the time series under study gets at the l-th site of the sequence under study. This means that the time $t = l$ (with $l \gg 1$) is thought as a continuous and that the function $\xi(l)$ is a function of this continuous time. In this case the Shannon entropy reads

$$S(t) = -\int_{-\infty}^{\infty} dx\, p(x,t) \ln[p(x,t)] \tag{6}$$

We also assume that

$$p(x,t) = \frac{1}{t^{\delta(t)}} F\left(\frac{x}{t^{\delta(t)}}\right) \tag{7}$$

and that $F(y)$ maintains its form, namely that the statistics of the process is independent of time. Let us plug equation (7) into equation (6). Using a simple algebra, we get:

$$S(\tau) = A + \delta(\tau)\tau \tag{8}$$

where

$$A \equiv -\int_{-\infty}^{\infty} dy\, F(y) \ln[F(y)] \tag{9}$$

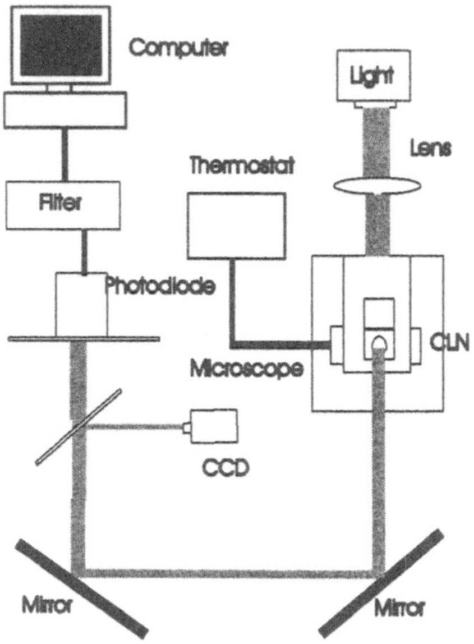

Figure 1. Scheme of the experimental apparatus for temporal signals acquisition.

and

$$\tau \equiv \ln(t)$$

The assumptions made to get the result of equation (8) are not correct during the transition process, and consequently the DE method can be used as a reliable way to detect scaling only in the long-time limit.

In the next paragraph we will show more practically how to calculate the diffusion entropy and the complexity index δ on our liquid crystals time series.

4. The experiment

First of all is necessary to illustrate what kind of sequence of the NLC we want to analyse and how they have been acquired.

Our time series are obtained taking with a photodiode the luminosity variations through the time in a point of the surface of the crystal. This luminous intensity variation is due to the auto-organizing structures' presence. In figure 1 there is a scheme of the experimental apparatus we needed to acquire these time sequences.

In order to reach the optimum for our analysis, we have to transform the time series in new sequences of only two symbols, 1 and 0, according to the asymmetric random walk rule. In this way we reduce the transition time. To transform a sequence taken on the NLC according with this rule we have to evaluate the difference between two data in succession, for example the i-th minus the $(i-1)$-th. If the difference is positive, i.e. the signal is growing, we put in the $(i-1)$-th place of the new string 1, otherwise we assign 0. We can see this thing as a walker that makes a jump forward every time the signal grows and stands if the signal remains constant or decreases. This criterium needs the choice of a sample time comparable with characteristic times of the dynamic in order to exclude, more than possible, the noise on the signal. We have previously said that the DEM is based on the analysis of a diffusive process. Therefore we have to build trajectories on our new time series in order to construct the diffusive process. The way to choose the trajectories is the following. First of all we decide a length L of a window for which we want to evaluate the diffusion entropy. Then, taking the new time series, we fix as the first trajectory the sequence that begins from the origin of the sequence and contains the first L data. The second trajectory could be simply determined moving the L-window of a place in the sequence: in this way we take data from the second to the $(L-1)$-th. Running the window on all the time series we obtain $N-L+1$ trajectories with N the series length.

Now we have to evaluate for every trajectory the effective walk made by the walker, i.e. the sum of all the symbols 1 and 0 included in the trajectory. Then, considering all the trajectories, we will establish the probability that the walker has made an effective walk of length i. Then we can finally calculate the diffusion entropy for the length L using equation (4).

At this point, to have the spectrum of how the diffusion entropy varies with the window length L, we have only to repeat this proceeding for other values of L. Then for every time series we evaluate the value of δ measuring the angular coefficient of the straight line of the plot of the diffusion entropy vs. the logarithm of the length of the window L. δ can assume values between $0,5$ and 1, where the first extreme is characteristic of chaotic systems while the second of regular ones.

5. Results and conclusions

We've acquired the time series for different values of the potential applied to the crystal in order to analyse different types of regimes of the sample. The result we've expected and obtained is to find a general decreasing trend of δ with respect to the values of this potential and to individuate three different zones of potential representing three different behaviours of the crystal. We have a first zone, at low voltage, where

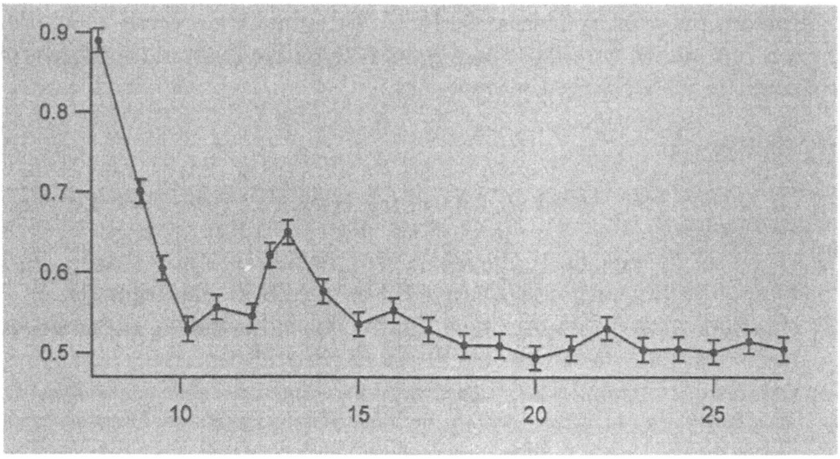

Figure 2. Complexity index δ as function of the RMS voltage applied to the sample.

δ initially assumes values near 1 and then decreases. This behaviour is the manifestation of the solitonic dynamic, i.e. on a very regular signal generated by Williams rolls, solitons appear as defects of the pattern and make their motion more complex more the potential grows. After this decreasing zone, there is a second zone where δ has a fluctuating behaviour. Initially we see a growing δ that means that solitons are organizing themselves in a more orderly way. After a peak δ has a new decreasing trend. In this part, in fact, the system goes toward a state of developed turbulence. This regime is the characteristic of the third zone in which, making growing the potential V, δ decreases going forward the asymptotic value $0,5$.

In figure 2 are reported the results for the analysis of the NLC time series with DEM.

These results are in agreement with the data obtained with an other method based on the variance measurements [3].

We suppose that the no-monotonic behaviour of the complexity as function of the control parameter is a more general property of systems that show a transition to the turbulence mediated by defects.

In order to verify most, we propose themselves to analyse Faraday's instabilities at the surface of an isotropic liquid in presence of gravity modulation.

Instead, for what concerns the NLC, we're making a study of spatial signals in order to compare the temporal dynamic (that we've shown in this article) with the spatial one.

References

[1] DE GENNES P.G., PROST J., *The physics of liquid crystals*, Clarendon Press, Oxford 1998.

[2] GRIGOLINI P., LEDDON D., SCAFETTA N., *Diffusion entropy and waiting time statistics of hard x-ray solar flares*, cond-mat/0108229 v1, (14 Aug 2001).

[3] SCAFETTA N., HAMILTON P., GRIGOLINI P., *The thermodynamics of social processes: the teen birth phenomenon*, Fractals **9** (2001), 193.

[4] GRIGOLINI P., PALATELLA L., RAFFAELLI G., *Asymmetric anomalous diffusion: an efficient way to detect memory in time series*, in press on Fractals, cond-mat/0104166.

[5] BRUGIONI S., Tesi di laurea, Dipartimento di Fisica.

COMPLEXITY IN PHYSICS OF AN ADHESIVE TAPE

Bruno Giorgini
Università di Bologna and INFN-sezione di Bologna - ITALY
bruno.giorgini@bo.infn.it

Enrico Lunedei
Università di Bologna and INFN-sezione di Bologna - ITALY
lunedei@bo.infn.it

Matteo Ciccotti
Università di Bologna and INFN-sezione di Bologna - ITALY
matteo@ibogeo.df.unibo.it

Abstract In this paper we show the emergency of a complex dynamics in the peeling of an adhesive tape. In particular we synthetically present some experimental results and theoretical interpretations of the stick-slip regime.

1. Introduction

The thin film peeling exhibits a rich and interesting fracture dynamics with stable and/or unstable regimes, depending on the control parameters value. In order to investigate this dynamics two different types of peel test are commonly used:

a) a strip of thin flexible film bonded to a flat rigid substrate is peeled apart to the underlying substrate (Fig. 1),

b) a thin strip of film bonded to a roller is peeled of and in this case generally the peel angle varies with time during the test [1].

In both the cases a stick-slip (or run-arrest) phenomenon is observed for certain ranges of peeling speed (the control parameter). This intermittent motion is created by a mechanism that generates cycles of crack-growth (or sliding) instability followed by subsequent arrest.

The stability and instability alternation in peeling is produced by the competition between the change in the driving force (or energy release rate) and the change in the crack-growth resistence.

In this paper we present some experimental results in the peeling of an adhesive tape (a viscoelastic system), i.e. in the b-case above described with a main interest for the stick-slip region. Moreover we will briefly discuss some theoretical models proposed to explain the observed dynamics.

At last we underline that a variety of phenomena as rock friction and earthquakes, or tearing of rubber and crack propagation in epoxy resine, etc. exhibit a stick-slip behaviour. More generally the stick-slip is a crackling external conditions through discrete, impulsive events spanning a broad ranges of sizes [2].

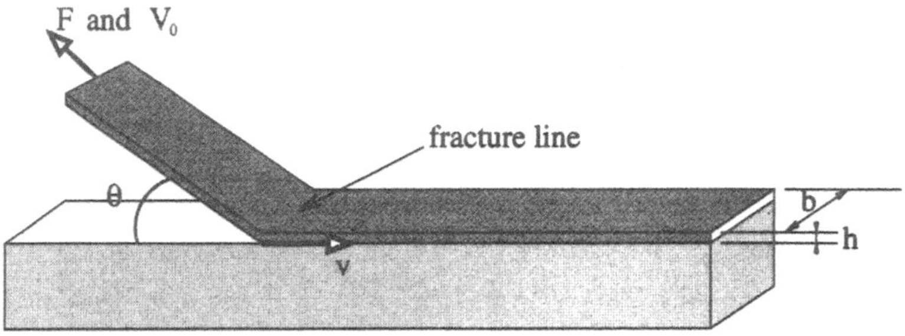

Figure 1. Peeling from a rigid substrate.

2. Stable and unstable regime

The every day experience shows that in a given range of temperatures and velocity, the peeling of an adhesive tape (scotch) is jerky with emission of a characteristic noise.

In more scientific way, the peeling was studied when a constant traction velocity V_0 was imposed onto the free end by the action of an electric motor (Fig. 2), where the limit between the adhesive tape ribbon and the free end may be assumed as a crack tip that propagates with speed v.

The experimental results [3] [4] are synthesized in the curve of figure 3 that relates the strain energy release rate G (see appendix) to the crack speed v.

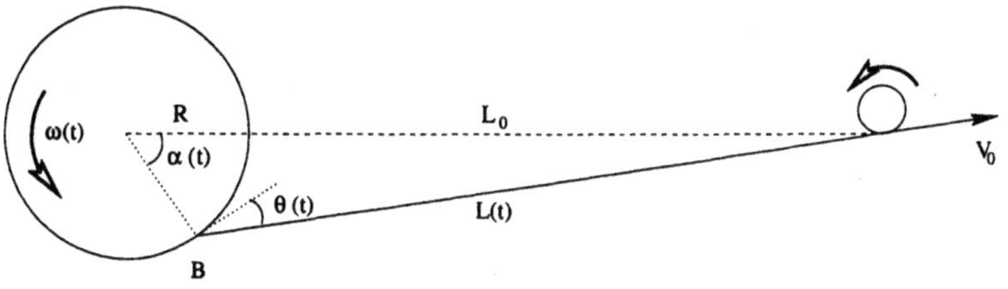

Figure 2. Scheme of unrolling apparatus.

The curve exhibits three peeling modes following the applied velocity V_0, that is the only control parameter of the system, once fixed the geometry of experimental apparatus. At slow V_0 the tape is peeled regularly, i.e. the fracture propagation appears stationary and stable with the crack propagation speed v equal to V_0 ($v = V_0$) and the peeling force F, that is proportional to G (see appendix), increases with V_0. At high traction speeds we find the same phenomenon, but with an important rise of F. Between these ranges, instead, the dynamics becomes strongly unstable and a highly irregular stick-slip phenomenon (self-sustained oscillations) emerges, i.e. the peeling is jerky with a typical noise emission.

Now the problem is to understand the physics underlying this observed behaviour of the system, or, at least, to set up a mathematical modelling which can fit with the experimental data.

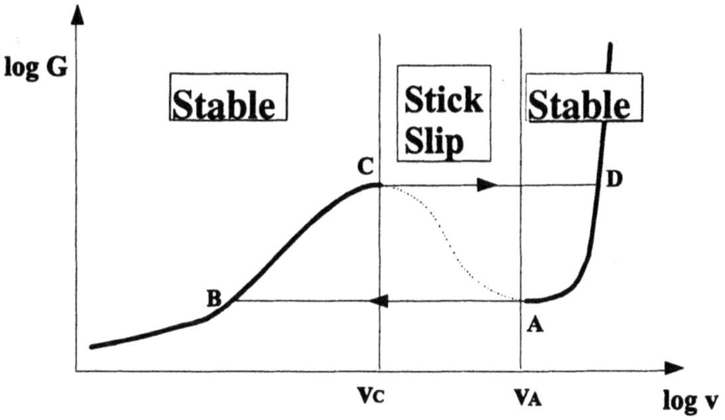

Figure 3. Schematic example of $G(v)$ in logarithmic scale.

3. The modelling

As a matter of fact we can observe that the macroscopic line of fracture is a combination of a huge number of microfractures in its neighbourhood, but at the moment we don't have a microscopic model for the adhesion force and for the fracture in a viscoelastic system. This is the main difficulty in order to give a deep physical interpretation; therefore we are constrained at the macroscopic level and some authors studied the system by means of dynamical equations.

Firstly a qualitative explanation of the empirical results can be given considering the curve of figure 3 as the superposition of the elastic solution with G increasing as v approaches the Rayleigh velocity and a broad peak due to viscoelastic losses. This interpretation is necessary for some following mathematical models, that needed a continuous $G(v)$.

The first model [3] takes into account only one degree of freedom (the elastic one) and explains the stable branches of the curve (G, v) but is completely unable in the stick-slip domain.

A second step [4] was to add another degree of freedom (the roller inertia), so a two variables model was obtained: it gives good results when applied to the initial part of the stick-slip region (slow speed), but its theoretical predictions fail when the traction speed is increased and the stick-slip oscillations become very irregular. More precisely Maugis and Barquins wrote the well known Liènard equation, which has Hopf bifurcations in the critical points A and C (see Fig. 3) where stable stationary equilibrium gives way to limit cycles on the branch with negative slope (unstable).

In the same optics, Hong and Yue [5] add a third degree of freedom (the peeling angle θ or the position α of the crack tip) obtaining the following system of dynamical equations:

$$\begin{cases} R \cdot \dot{\alpha} = \omega \cdot R - v \\ \dot{F} = -k \cdot [(v - V_0) + (\omega \cdot R - v) \cdot \cos\theta] \\ I \cdot \dot{\omega} = F(v) \cdot \cos\theta \\ F \cdot (1 + \alpha) = F_0(v) \end{cases} \tag{1}$$

where I is the inertial momentum of roller with respect its axis. The authors said that the numerical solutions are chaotic orbits in a very narrow range of V_0 (they found three positive Lyapounov exponents). So the problem of the irregular stick-slip regime seems to be solved as a chaotic deterministic configuration.

4. The criticism

A different experiment [6], where the tape was unrolled by the action of a load hooked to the free extremity (constant pull force), has shown new interesting results concerning the observation of speed jumps and the hysterical behaviour of the fracture advance. Moreover the branch of $G(v)$ for slow speed v in a neighbourhood of C was discovered as metastable, i.e. a very little perturbation in this region can prime the stick-slip.

From the theoretical point of view a further analysis [7] of the Hong-Yue equations shows that:

i) the equations system (1) is not a dynamical system, without to process them,

ii) the Hong-Yue solutions was obtained with forcing them, in order that there are jumps of speed v,

iii) processing (1) we obtained a 3-dimensional dynamical system with two singular points ($v = v_C$ and $v = v_A$).

Examining system (iii), we see that the numerical solutions admit a cycle only if they are forced by hand to avoid the singularities; without the system has not physical solutions.
So the deterministic chaos seems to be rather artificial and furthermore in the Hong-Yue system does not exist a well defined route to chaos.

This criticism has stimulated further theoretical and experimental studies [8]. The aim was to have a more accurate knowledge of the stick-slip, firstly empirical.

The highly non-linear phenomena, as the fracture dynamics, produce time evolutions which are really unpredictable. In order to extract useful information on the motion from experimental observations we must have sufficiently long time series (i.e. sequence of data representing the time evolution of one or more observables).

So we can use the statistical analysis and investigate the fine dynamical evolution to search the significant points (bifurcations and so on) and eventually to detect the emergency of hierarchical structures. For this we have set up a new experiment and the first provisional results show a stick-slip dynamics more complex of the expected one [9].

5. The new results

In our experiments we changed the philosophy with respect to the previous studies, taking into account a new observable, the time interval (period) Δt between two subsequent signals of the stick-slip characteristic noise, i.e. the temporal distance within a slip event and the following. The analysis of the collected time series of data has given the following main results [10]:

1) for V_0 (traction speed) $\leq 0.7\ mm/s$ the peeling is stable;

2) for $V_0 \geq 0.7\ mm/s$ the stick-slip begins with periodic or quasi-periodic cycles;

3) for $1.1 \leq V_0 \leq 1.3\ cm/s$ we observe the first bifurcation and sequences of multiplets appear with $\Delta t = 0.02, 0.04, 0.06\ s$; these structures remain stable until to $V_0 \simeq 3\ cm/s$ and for $V_0 > 3\ cm/s$ the $\Delta t = 0.02\ s$ becomes dominant;

4) at $V_0 = 6\ cm/s$ we have a sort of fine structure (sub-multiplets) with $\Delta t = 0.001, 0.002, 0.003\ s$; the structure evolves becoming more and more complicated as long as at $10\ cm/s$ the signal seems to be continuous; so we could conjecture that the chaos is installed, but really we cannot to be sure because other accurate experiments must be set up in order to have a more precise definition of the signal and a better power resolution;

5) at $V_0 = 3\ m/s$ the peeling becomes still regular and we have reached the second branch of the curve (G, v).

In figure 4 and 5 we show two examples to give the perception of the structures and multiplets emerging in the stick-slip dynamics.

We have also studied the series $\{\Delta t\}_i$ statistically finding an indication that two following events are not independent and that the process it is not a markovian one; but, at the moment, we are not able to describe in a coherent and mathematically consistent way our results.

Finally we have found that the mean period $\langle \Delta t \rangle$ of the stick-slip cycles is, at least in first approximation, inversely proportional to the traction velocity (Fig. 6).

6. Conclusions

As we have seen the main dynamical models constructed to explain the peeling evolution of an adhesive tape can predict correctly only the stationary behaviours and the periodic or quasi-periodic stick-slip cycles. But when the stick-slip becomes very irregular the proposed models are insufficient also if we increase the number of degrees of freedom. On the other hand the results of new experiments show that the stick-slip

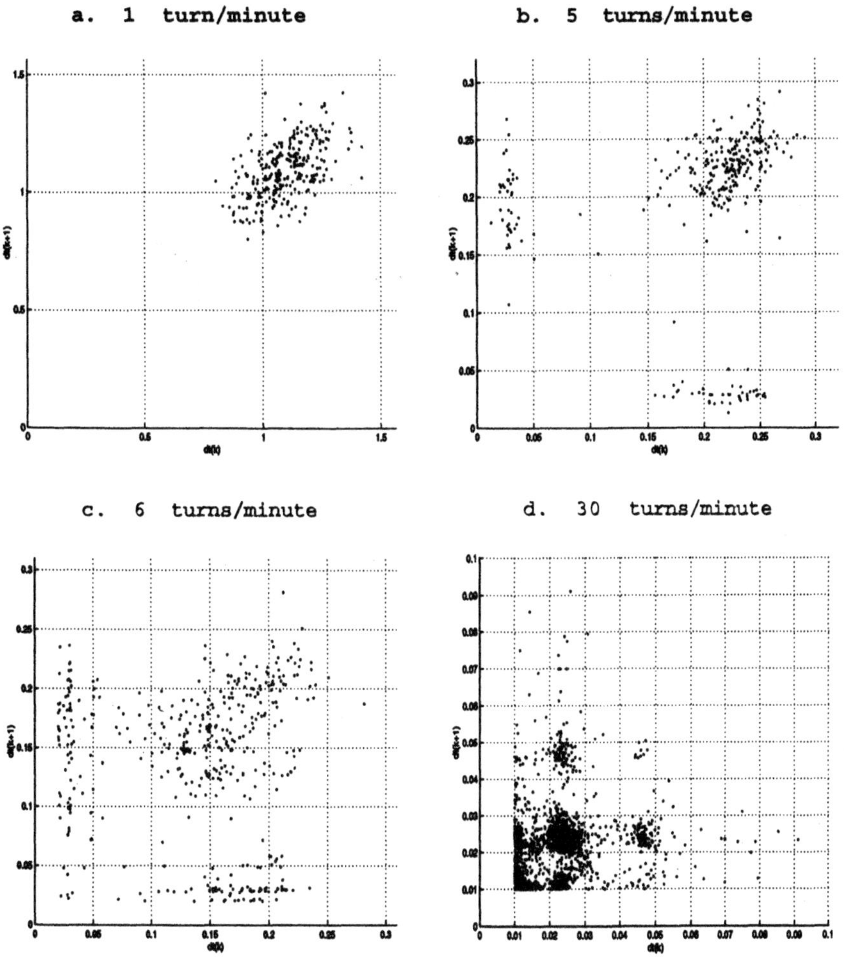

Figure 4. Example of bidimensional embedding of a $\{\Delta t\}_i$ serie.

dynamics is more rich and complicated of a simple bifurcation's route to chaos.

We observe hierarchical structures in a definite traction speed range that can suggest the emergency of complexity, at least in qualitative sense, when a fracture is produced and evolves in a viscoelastic system.

At this point many problems are open. Firstly the measure of the other observables as F, ω and, above all, the crack speed v. Furthermore, we must investigate the possible mathematical correlations of ours time

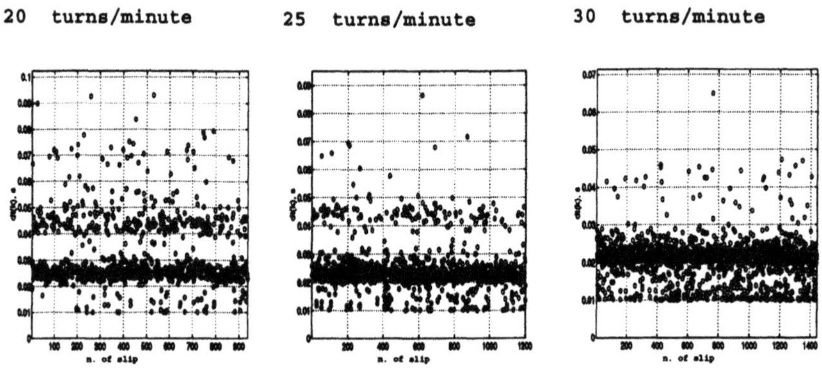

Figure 5. Example of multiplets emerging in a $\{\Delta t\}_i$ serie.

series of data. Finally, the goal would be to find a physical interpretation of the phenomenon or, more precisely, discovering the physical laws and equations that originate the observed fracture complex dynamics.

At last we want underline as a common scotch roller can be so enigmatic and full of mystery to justify the ancient saying "I know that I don't know". Perhaps this sentence could be the deep meaning of "complexity".

Acknowledgments

We thank a lot G. Turchetti for many useful discussions on the subject.

Appendix

In this appendix we give some simple definitions and formulas to understand the dynamical equations.

i) the quantity G, called the strain energy release rate, is defined as:

$$G = \frac{\partial U_E}{\partial A} \tag{A.1}$$

i.e. the rate between the mechanic energy ΔU_E absorbed by system when the fracture's surface advances of ΔA, for $\Delta A \to 0$;

ii) there exist a following relation:

$$G = \Phi(v) \tag{A.2}$$

where v is the crack speed, and Φ is a experimentally-obtained function;

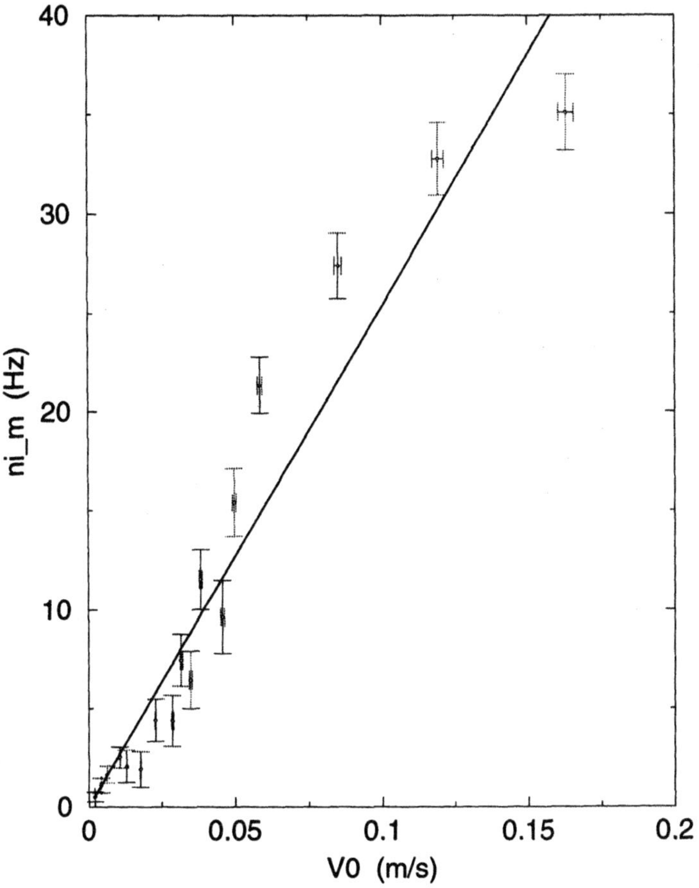

Figure 6. Fit of $\nu_m = \frac{1}{\langle \Delta t \rangle}$ vs V_0.

iii) when we unroll a tape of width b we define the adherence force:

$$F_0(v) \equiv G \cdot b = \Phi(v) \cdot b \tag{A.3}$$

iv) in 1975 Kendall has obtained the following relation:

$$G = \frac{F}{b} \cdot (1 - \cos\theta) + \frac{F^2}{b^2} \cdot \frac{1}{2Eh} \tag{A.4}$$

or:

$$F_0(v) = F \cdot (1 - \cos\theta) + \frac{F^2}{2ES} \tag{A.5}$$

where F = pull force, h = thickness of the film, E = Young's modulus of the film, $S = b \cdot h$, θ = the peel angle; for $\theta > 30°$:

$$G = \frac{F}{b} \cdot (1 - \cos\theta) \tag{A.6}$$

or:

$$F_0(v) = F \cdot (1 - \cos\theta) \tag{A.7}$$

References

[1] TSAI K.H., KIM K.S., Int. J. Solid Structures **30** (13) (1993), 1789.

[2] SETHNA J.P., DAHMEN K.A., MYERS C.R., Nature **410** (6825) (2001), 242.

[3] BARQUINS M., KHANDANI B., MAUGIS D., C. R. Acad. Sci. Paris **303** (1986), 1517.

[4] MAUGIS D., BARQUINS M., Adhesion **12** (Ed. Allen K.W.), Elsevier Applied Science, London 1988, 205.

[5] HONG D.C., YUE S., Phys. Rev. Lett. **74** (1995), 254.

[6] BARQUINS M., CICCOTTI M., Int. J. of Adhes. & Adhes. **17** (1997), 65.

[7] LUNEDEI E., Tesi di Laurea, Dip. di Fisica - Bologna 2001.

[8] CICCOTTI M., GIORGINI B., BARQUINS M., Int. J. of Adhes. & Adhes. **18** (1998), 35.

[9] VALLET D., CICCOTTI M., GIORGINI B., BARQUINS M., Recontre du non-linéaire 2000 - Paris onze Éditions, Bât. 338, Univ. de Paris-Sud, 91405 Orsay.

[10] CICCOTTI M., GIORGINI B., VALLET D., BARQUINS M., *New experimental results in the stick-slip dynamics of an adhesive tape*, preprint January 2002.

REFLECTIONS ABOUT THE TIME ARROW

Antonio Lepschy
Dipartimento di Elettronica ed Informatica,
Università di Padova,
Via Gradenigo 6A, 35135 PADOVA - ITALY
antonio.lepschy@unipd.it

Abstract After some preliminary remarks about causality, determinism and predictability, the paper discusses a paradoxical aspect of the discrete-time quantized dynamical systems. Precisely in such deterministic systems the future behaviour is uniquely predictable but the past cannot be uniquely reconstructed; correspondingly the reversed-time system associated with a system of the considered kind is not deterministic, because the future time-course of its output is not predictable (even if its past behaviour can be uniquely reconstructed).

1. Some preliminary remarks

1.1 Time arrow and causality

The question of the time arrow is usually faced from the point of view of the increase in entropy of the whole universe (or of an isolated system). Here, however, we will adopt a different point of view, referring to an open system and to the effects of causes acting on it from a given instant.

From this point of view we may recall that many scientific disciplines describe how a phenomenon evolves in time or is distributed in space, but there is, generally, a fundamental difference depending on whether the problem is studied from the point of view of time or from the point of view of space: in the case of time we can speak about an *arrow* which is oriented from the past towards the future, while in the case of space the arrow does not generally manifest itself.

Let us consider, for example, a metal bar to which a source of heat is applied on a specific point (but not at an extremity), and at a given instant of time. We are interested in the values of the temperature of the bar at each instant (that is to say, the time course of the phenomenon) and in each point (that is, the spatial distribution of the phenomenon). The instant when the heating action begins divides the time into two parts: the set formed by the instants preceding the one taken into consideration (the past in respect of that instant), and the set formed by successive instants (the future). Similarly the point to which the heat source is applied divides the bar's axis into two parts: supposing the bar is horizontal, the two parts are formed by the set of the points of the axis on the left of the heated section, and by the points on the right of that section. It is clear that variations in temperature due to the action applied occur only in instants successive to the one in which the heat source was applied (and not in preceding instants also), while they take place both on the right and on the left of the point of application of the heat source. In other words, if we introduce notions such as cause and effect, we could say that the time arrow corresponds to the view according to which causes precede effects (not only conceptually but even in time), while in the case of the phenomenon under scrutiny we cannot speak of a space arrow since the cause produces an effect both on the points situated to the left and to those situated to the right of the point in question[1].

As for the behaviour in respect of time, we can say that the 'arrow' is apparent in all phenomena of interest to physical sciences. In the field of life sciences some exceptions could be possible, for example in the case of goal-oriented actions in which a future event appears to have an effect on a preceding one. However, in many cases the situation is not too different from that of an artificial system, such as, for example, the automatic control system of an artillery gun which keeps the gun trained on the future position of a mobile target. In this case, however, the controlled action is manipulated using present and past data acquired by radar and extrapolated by the fire control so that an estimation of the future position can be made. It follows that we can say that the action exerted on the artillery piece at a given instant depends solely, (albeit not in an elementary way), upon data regarding the behaviour of the target up to that instant. In many cases the same kind of analysis can be applied to the interpretation of biological behaviour which is, in appearance at least, goal-oriented. The field of human and social sciences present a less

[1]There are, however, phenomena usually called 'transport' phenomena, in which an arrow occurs even in space, typically from top to bottom, for example, downstream in the case of a watercourse which is perturbed by the injection of extraneous matter at a given point: if the velocity of the watercourse is higher that that of the diffusion of the injected matter, this interests only the points which are downstream and not those which are upstream in relation to the point in which the perturbation occurs.

obvious scenario in which most behaviour appears to be geared towards the achievement of goals or the prediction of future events. It is generally not at all banal to interpret this prediction as an extrapolation from past events.

In several areas of physical sciences this dependency of an event exclusively from previous events or at most contemporary ones, and not from future events, is termed 'causality', in a sense of this word that is different from, but not in opposition to, the traditional use of the same word in philosophy, associated with the chain of causes and effects. To speak of 'cause' and of 'effects' has often been considered the residue of an inadequate terminology in a context in which it was held that a scientific argument, to be such, should free itself of all metaphysical layers and thus words like 'cause' and 'effect' were avoided. In some contexts, words like 'stimulus' and 'response' were preferred, in others, 'input' and 'output'. In the present note we will use the terms input and output and consider *open* dynamic systems on which inputs may act; the output will depend on the initial *state* of the system and on the actual input.

1.2 Determinism, predictability, and free will

In the sphere of the theory of dynamic systems the meaning of the word 'causality', with respect to the arrow of time, coexists with the meaning connected to the notion of 'determinism'.

One of the most famous formulations of determinism was expressed by Laplace [1] as follows: 'An intelligence that, for a given instant, were capable of knowing all the forces by which nature is animated and the respective situation of the beings that it is composed by, if it were vast enough to subject these data to analysis it would include in the same formula the movement of the largest bodies in the universe and that of the lightest atom; for it nothing would be uncertain: and the future and the past would be before its eyes'. Immediately afterwards, he adds though: 'The human spirit offers, in the perfection that it has been able to impart on astronomy, only an inkling of such an intelligence'.

In this formulation, however, one could point out two aspects: one implicit, connected to the tacit hypothesis of determinism, the other expressed, and linked to the possibility of forecasting the future (and also of backcasting the past). One could say, interpreting Laplace's words literally, that the future is certainly rigidly determined, but that it could be known only by a mind so vast as to embrace the state (in other words, the position and velocity, or better, the momentum), of every particle of the universe, no matter how small. In concrete terms then, the future could not fail to be seen as unpredictable and the past as something which could not be reconstructed, since any contemporary of Laplace (but today the situation is no different) would have been unable to exploit the enormous quantity of information needed for the task, (quite apart from the sheer volume of elaboration needed to make the transition, by means of the fundamental laws of physics, from the knowledge of that state to the prediction of the future state). Laplace himself, as appears from the above citation, was well aware of the fact that the human mind offers no more than an inkling of the 'intelligence' he had hypothesised as capable of embracing in the same formula the future and the past.

In this sense we could say that Laplace had already stated what is usually attributed to much more recent studies on deterministic chaos: that is, the unpredictability of the evolution of a phenomenon which is nonetheless described with a deterministic model.

Instead, in the current vulgate, Laplace is considered (and there is some truth in this perception), as the champion of a deterministic vision from which the predictability of the future, at least in principle, would result. In this he is seen in opposition to modern theories of deterministic chaos which underline the unpredictability of future evolution (and the impossibility of reconstructing past evolution), not only for phenomena one associates with a stochastic model, but also for at least some phenomena described with a deterministic model.

And actually, between Laplace's position and that of the theory of deterministic chaos, there is at least the substantial difference of the aspect which is given most emphasis. For Laplace such an aspect is undoubtedly that of determinism. In the theory of deterministic chaos, on the other hand, unpredictability is the central aspect which can be manifested even in relatively simple systems when there is a strong sensitivity to initial conditions, which can never be known 'exactly' but only with some approximation.

Another aspect of the compatibility of determinism with the impossibility of prediction regards open systems: one can know the laws that determine their evolution following a given input and therefore predict such an evolution, but very seldom is it possible to predict if and when the system will be subjected to the hypothesised input.

The point we have reached leads us to observe how the adoption of a deterministic model does not allow us to postulate predictability. One could, however, also consider another aspect of the problem (which is cited here, but not dealt with, since the writer is not an expert in the field). That aspect is the question whether deterministic laws in the physical field are compatible with the thesis of free will. The problem of the feasibility (or lack of it), of reconciling the free will of human being and the determinism of the physical world (including the human brain) is certainly formidable (even in the etymological sense of the word); we shall limit ourselves to pointing out the interest with which, in certain circles, first quantum theory and then chaos theory have been viewed as conceptual tools suitable to invalidate the perspective summed up, more or less appropriately, by Laplace in the text cited above.

As for quantum theory, the very basis of physical determinism is lost, but it would not appear that the idea of free will could profit from this, since the behaviour of the physical phenomena which characterise the functioning of the human brain were determined by a long chain of events with at bottom not a free choice of the mind, but an aleatory event at a subatomic level, unless one shares with Eccles (whose opinion, by the way, has hardly attracted a consensus) the view that, at the quantum level, the 'mind' can condition physical phenomena which take place in the atoms forming the molecules of the brain [2]. More immediately evident is the ambiguity implicit in invoking the theory of deterministic chaos which concerns only the possibility of an accurate forecast.

Another point of view could be the outlook of psychoanalysis, whereby the awareness of free will derives only from a kind of *a posteriori* rationalization of an event due to inputs operating below the threshold of consciousness.

There is no need to labour these points here. Instead, we shall limit ourselves to considering the dilemma of determinism versus possibility of forecasting or backcasting in the very simple case of an isolated system described by a difference equation and analysing a paradoxical aspect of the pertinent time arrow.

2. A paradoxical aspect connected to the prediction of the future and the reconstruction of the past in discrete-time dynamic systems

2.1 Continuoustime dynamic systems

Traditional mathematical physics considers time as a one-dimensional variable which assumes values in the set of real numbers; moreover, it considers physical quantities assuming values in the same set and presents them as functions of the independent time variable (and additionally, possibly of spatial variables).

It is assumed that the functions in question are differentiable and that the time behaviour of the isolated system, characterised for example by a quantity $y(t)$, be described by an ordinary homogeneous differential equation of the type:

$$f(y, y^{(1)}, \dots y^{(n)}; t) = 0 \tag{1}$$

in which the exponent in parenthesis indicates the differentiation order.

For the purposes of what follows, we can limit our attention to time invariant systems in which the function f does not depend explicitly on time t, and to linear systems, in which case f is a linear combination:

$$a_n \frac{d^n y}{dt^n} + a_{n-1} \frac{d^{n-1} y}{dt^{n-1}} + \cdots + a_1 \frac{dy}{dt} + a_0 y(t) = 0\,. \tag{2}$$

One can further narrow this down to the simplest case of $n = 1$, in which the equation becomes:

$$a_1 \frac{dy}{dt} + a_0 y(t) = 0 \tag{3}$$

and the solution is:

$$y(t) = y(0)e^{pt} = y(0)e^{-t/T} \tag{4}$$

with $p = -a_0/a_1$, $T = a_1/a_0$.

Equation (4), given the value of $y(0)$, enables one both to predict the value of y in any future instant ($t > 0$), and to reconstruct the value of y in any past instant ($t < 0$).

For higher order equations the $y(t)$ is the linear combination of *modes* analogous to that of the equation of the first order; in this case, however, the coefficient p of the exponent can be complex even if all the a_i coefficients are real. If the coefficients a_i are real, however, one can combine in pairs the terms in which the complex coefficients of the exponent are conjugate so as to obtain a real addendum corresponding to an oscillatory term multiplied by an exponential factor. In the case of p_i roots of the characteristic equation with multiplicity greater than 1, addenda will result in which the expressions hitherto considered are multiplied by a polynomial of the variable t.

Any way, once the values of y and of all its successive derivatives at $t = 0$ are known, up to that of the order $n-1$ (which uniquely determines the value in that instant of the derivative of the order n), the value of y and of its derivatives are also determined for any other future or past instant. This property follows from the theorem of existence and unicity of the solution of the differential equation (valid also for non linear differential equations, under appropriate hypotheses).

Instead of considering the evolution of only y as a function of time t, it is usual to refer to a one-dimensional trajectory which is parametered in t in the n-dimensional space, called *phase-space*, the coordinates of which are the values of y and of its derivatives up to that of the $n-1$ order. One and only one trajectory passes through each point in this space. In other words, in the hypotheses we have formulated, there can be no points of this space on which two trajectories until that time distinct converge at a finite value of t (thus making the reconstruction of the past impossible), nor can there be points from which two trajectories diverge (thus making the evaluation of the future impossible).

It is obvious that if the point at $t = 0$ of the considered phase-space is not sufficiently well known, (that is to say, we know only that it is internal to a 'small' volume of such space), prediction and reconstruction remain uncertain because the trajectories passing two points of that volume, no matter how close together, can diverge significantly (particularly in the presence of nonlinearities), with the outcome of an amplification of uncertainty. The phenomenon is typical of the so-called deterministic chaos (which characterises non-linear systems that are not too simple, even though 'quite' simple). Something similar, however, occurs in each case, even in equation (4); for if p is negative, the two evolutions beginning from values very close to $y(0)$, get closer and closer as t increases, making a prediction of the future acceptable even if $y(0)$ is not precisely known. But as t decreases they diverge, making the difference greater between two reconstructions of the value of y in an instant of the past, carried out on the basis of two estimations in close proximity of $y(0)$. The opposite occurs in the case of $p > 0$, in which the reconstruction of the past is acceptable, but the forecast of the future is increasingly uncertain. Clearly, in the case of differential equations of higher order, the solutions of which include converging and diverging *modes*, both the reconstruction of the past and the prediction of the future will be affected by growing uncertainties the further away from the initial instant one goes.

2.2 Discrete-time dynamic systems

If, in traditional mathematical physics, it is customary to resort to continuous-time models, in other fields of applied mathematics, for example, for the study of more or less artificial systems, like those of computer science or economics, discrete-time models are needed. In these the independent variable t assumes values in the set of the integers.

In this case the evolution of the quantities under scrutiny is made of sequences (rather than of functions which assign a value of y and of its derivatives to each *real* value of t) and the dynamics of the system is represented by difference equations rather than by differential equations.

In particular, the phase-variables, which in the previous case were y and its first $n-1$ successive derivatives, are now the values of y in n successive discrete instants.

In substance, however, the results mentioned earlier, (under the appropriate hypotheses), continue to hold true; that is to say, the existence and unicity of the solution of the difference equation: two distinct sequences which, from a certain time on, converge in the same succession, cannot exist. Nor is it possible that, from a given time onwards, a sequence of undriven evolution bifurcates into two different sequences. Similarly, the uncertainties about the phase-vector at $t = 0$ produces an increase of uncertainty in the prediction of the phase-vector in a future instant and/or in its reconstruction in a past instant (depending on whether the sequence contains only diverging *modes*, or converging *modes*, or *modes* of both types).

2.3 Discrete-time quantized dynamic systems

In the case of *digital* systems, the quantization of y is associated to the discretization of t and the system behaviour is described by difference equations with constant coefficients α_i multiplying the values of y corresponding to successive time instants $t - iT$:

$$Q[\alpha_n y(t-nT)]+Q[\alpha_{n-1}y(t-(n-1)T)]+\dots Q[\alpha_1 y(t-T)]+y(t) = 0 \quad (5)$$

where T indicates the sampling interval, t represents the generic instant, assumed as the basis of the distinction between present and future in the sequence $t+iT$, and where the symbol $Q(x)$ indicates the quantized value of x. In the last addendum the symbol of quantization has not been included, since $y(t)$, according to equation (5), is necessarily an integer, being the algebraic sum of integers (i.e., the quantized quantities).

It should be noted that quantization has been applied here to products of y times the coefficients α_i and this has a practical consequence (given that the values of y are only integers) if we also assume that the coefficients α_i are not whole numbers. We could allow that they may assume real values but, if this should seem too unrealistic, we can refer to the case in which the coefficients α_i are rational (or, more precisely, their quantization step is smaller than the quantization step of y).

In this case, if we refer back to a result already shown [3], it will be seen that there is no possibility of reconstructing the past univocally, since different sequences converging towards the same succession are simultaneously possible. Future forecasting, on the other hand, remains possible, since the computation procedure of each future value is deterministic.

This kind of behaviour is present even in the simplest of the considered systems; for example, in the one described by an equation of the first order, of the following type:

$$Q[\alpha_1 y(t-T)] + y(t) = 0 \tag{6}$$

According to this equation while each successor $y(t)$ is determined univocally by the relevant predecessor $y(t-T)$, several different predecessors could correspond to each successor.

If for example, $\alpha_1 = 0.2$, one can obtain $y(t) = 1$ with a $y(t-T)$ which could be worth 3 or 4 or 5 or even 7; because the product of the coefficient α_1 for each of these values varies from 0.6 to 1.4 and, with quantization corresponding to a round-off to the nearest integer, all these products are rounded off to 1; analogously, the predecessors of $y(t) = 2$ can be the integers from 8 to 12 and so on.

In this regard the following observations can be made:

- equation (6) can be considered the equation which describes an isolated system (because there are no inputs which could influence, by causality, only output values successive to their application);
- just as in the case of non quantized systems already discussed, the deterministic univocity of the future (that is of each successor) guarantees, at least in principle, the predictability of such a future;
- however, as opposed to the cases previously observed, the presence of more compatible predecessors with the same successor does not allow the reconstruction of the past.

This is undoubtedly a case of a time arrow, tied to a difference in situations concerning past and future, but the behaviour of the considered system appears to contrast with our sensation, on the psychological plane, of how the time arrow should act, i.e., with our belief that the past is unmodifiable while the future is open to various possibilities.

3. Reversed-time systems

A further reflection which can be made about the behaviour of systems described by equation (6) is connected to the viewpoint which suggested to J. Loschmidt the mental experiment that should have rendered problematic the so-called 'H-theorem' and the second principle of thermodynamics. In such an approach one considers what happens at molecular level in relation to the evolution in time of the volume, pressure, and temperature of a gas and therefore the effect produced by the movement of single particles on such variables, bearing in mind any related collisions. In this respect, Loschmidt noted that, if in a given instant the speed of all the molecules were inverted, the system would have undergone the same evolution it followed up to that instant and therefore the entropy would have decreased if it had increased in the preceding evolution. If the state the system had reached (consisting in a certain set of positions and speeds of the molecules), could be considered as probable as the one in which the vectors representing the speed of each molecule have the same absolute value but the opposite direction, then we should infer that the increase of entropy has the same probability as its decrease.

But, in this context, the aspect of Loschmidt's mental experiment we are interested in is the one relative to two physical systems which, finding themselves in a particular instant and certain conditions, have an (undriven) output that, for each of them, varies in the future in the same way in which the other varied in the past. To be more specific, if we assume that, at $t = 0$, the two systems are in an appropriate state, and call $y_1(t)$ the output of one of the two systems, and $y_2(t)$ the output of the other, we get:

$$y_1(t) = y_2(-t) \tag{7}$$

(note that in the mental experiment conceived by Loschmidt we consider a single system with two different initial states).

If we deal with systems described by differential equations of the type of equation (2), a situation of the considered type arises only if the a_i coefficients of the odd-order derivatives are equal to zero and the original system coincides with the associated reversed-time system. If, however, the coefficient a_i of at least one of the odd-order derivatives is not equal to zero, the system cannot produce an evolution in the future (starting from a suitable state) that coincides with an evolution which occurred in the past: equation (3) provides the simplest example. Nevertheless we may conceive two *different* systems with an equation of the type of equation (2) (e.g. those with the same value of a_0 and with opposite values of a_1 or viceversa), in which the evolution towards the future of the output of one coincides with the evolution towards the past of the other. Yet one can consider their mathematical models as being the schematization (at least for a certain time-interval of functioning) of equally possible physical systems (except that one is a passive system, that is to say, a system formed by components that only store or dissipate energy, but not including active elements such as power generators, while the other must necessarily incorporate a power generator to supply the energy for the increasing system output). The same is true for discrete-time systems represented by mathematical models in the form of difference equations.

The situation is no longer the same if quantization is introduced: equation (6) describes the behaviour of an easily set up physical system (typically a digital filter, or a digital calculator, however simple, which has been programmed to carry out the quantized difference equation we have taken into consideration).

A system which, placed in a given state, in the 'future' evolution of its output, plays backwards the succession of values assumed in the past by the output of the system described by equation (6), appears to contrast with some of the more common requirements associated with the physical feasibility of a deterministic system (such as the one described by equation (6)). Indeed, for the system under consideration several possible successions of the past could make it converge to the same state. So there is no way an isolated deterministic system could offer simultaneously possible multiple different futures, by means of which the different simultaneously possible pasts in the deterministic system described by equation (6) should be reproduced.

In this respect, we may note that it is not enough to introduce a (deterministic) procedure that makes it possible to compute all the values the output can have in the next instant (corresponding to those which, in the system described by equation (6), the output could have had in the preceding instant). To achieve this, it will be necessary to add a stochastic mechanism (or, more realistically, one based on producing a pseudo-causal sequence which is independent of the system equation), which chooses one of the simultaneously possible values (but not in a consistent way, as would occur, for example, if the smaller or the greater were always chosen).

A system that keeps producing the full ramification of simultaneously possible outputs, similar to just as many Schrödinger cats, simultaneously dead and alive, until one decides which one is the only one destined to be considered alive, could seem even less realistic as far as macroscopic physics is concerned. Similar considerations can be made with reference to the possibility of elements of ever increasing parallel universes, each one producing, at every phase, only one of the outputs we have mentioned, and an output which is different from all the others generated at that stage. And perhaps it is interesting to note that, to speak of a system presenting as its output the solution for negative values of t of the very simple quantized system described by equation (6) we have spoken about Schrödinger cats and parallel universes, which are hypotheses taken into account in quantum physics. An analogy could also be interestingly proposed with the theory of Fantappié.

References

[1] LAPLACE P.S., *Essai Philosophique sur les Probabilités*, Courcier, Paris 1814.

[2] ECCLES J.C., *Do mental events cause neural events analogously to the probability fields of quantum mechanics?*, Proceedings of The Royal Society of London **CCXXVII** (1986), 411–428.

[3] LEPSCHY A., MIAN G.A., VIARO U., *Un paradosso a proposito della freccia del tempo*, Atti e Memorie dell'Accademia Patavina di SS. LL. AA. **CII** (parte II) (1990), 59–69.

THE BIG COMPUTER. COMPLEXITY AND COMPUTABILITY IN PHYSICAL UNIVERSE

Ignazio Licata
Istituto di Cibernetica Non-Lineare per lo Studio dei Sistemi Complessi, Via Favorita 9, 91025 MARSALA (TP) - ITALY
licata@programmazione.it

Abstract Complexity permeates the space-time "thin" structure. We are going here to shortly outline the fundamental characteristics of a discrete space within which the phenomenology of the elementary particles can be described through a proper set of Cellular Automata (CA) as emerging process.

1. Introduction

During last years the problems linked to the Quantum Field Theory (QFT) and the quantum gravity developments have imposed to reconsider, in general terms, the computability characteristics of physical theories. As anyone knows the renormalizability in QFT and the singularities in General Relativity (GR) questions can be viewed as problems of intrinsic uncomputability.

In the sphere of quantum gravity a key role is played by a quantity defined observable theorically but whose value is given by the infinite sum of variable quantity, each one computable on a proper surface linked to the initial conditions. But it has been demonstrated the surfaces' enumeration is uncomputable, or it is an impossible calculation by a Turing machine [1]. Such kind of impasse suggests us to take into critic consideration the idea of a continuum space-time *texture*.

In 1936 Einstein expressed the idea that *"...perhaps the success of the Heisenberg method points to a purely algebraic method of description of nature, that is, to elimination of the continuous functions for Physics. Then, however, we must give up by principle, the space-time continuum..."* [2]. Really the idea of a discrete space-time had already been suggested frequently by Mach in 1908, and, later, in relation to the debate about Quantum Mechanics (QM) developments in the '20. Except for the approximation models the most serious attempts met with great difficulties and got a partial success because, in our opinion, they weren't sufficiently radical.

All these theories, such as Snyder operators' quantization, Heisenberg-Ivanenko pre-spinors, Finkelstein space-time code, Penrose twistors [3], used a method which, in informatics' slang, we can define *top-down*, that is trying to keep many peculiar aspects of the formulation on continuous space-time, only removing the problems. A typical example is the Lorentz invariance break [4].

Here our approach will be of the *bottom-up* kind: we are going to study the general characteristics of a quantized space-time on Planck scale by an essentially discrete language without being worried about the QFT "traditional" characteristics and introducing a **computational paradigm** to describe the *thin* structure of space-time.

2. Reticular Dynamics: Space-Time as Rubik Cube

Reticular Dynamics [5] replaces the usual continuous variety by a discrete structure of elementary space-time quanta, unitary little cells characterized by fundamental length Lp and time Tp.

$$Lp = \sqrt{\frac{hG}{c^3}}\,, \qquad Tp = \sqrt{\frac{hG}{c^5}}\,, \qquad \frac{Lp}{Tp} = c\,. \tag{1}$$

These are the well-known Stoney-Planck natural units, which, in such context, get a direct physical meaning in relation to the space-time structure.

Let's put the following fundamental assumptions:

a) To each space-time cell is associated a non-linear oscillator, subjected to natural vibrational limits owing to Lp and Tp.

b) The reticular space-time is at multiple vacuum states. Each cell is a topological mechanism of energetic confinement.

c) The non-linear interaction among oscillators of different cells, overcoming the vacuum state, produces particles which can be represented like-soliton object (*lumps*). Any particle with $M_0 \neq 0$ in absolute rest is a soliton with localized energy inside its Compton wavelength and has its own internal frequency (*like-clock particles*).

d) A remarkable consequence of Tp and Lp introduction is the existence of the superior limit for mass, energy and frequency in the cells:

$$M_{\max} = \frac{h}{L_p c} = \frac{h}{T_p c^2} = \frac{h V_{\max}}{c^2} = E_{\max}\,, \quad V_{\max} = \frac{1}{T_p}\,. \tag{2}$$

e) The mass of a particle depends on the number of the vibrant cells which form it.

These points suggest not only the discrete typology of such space-time but also how to introduce a dynamics. Just consider that there are several possibilities to study the non-linear oscillators interacting on a lattice. On the other hand, it is possible searching direct analytic solutions within the traditional language of differential equation, analogously to crystalline lattices. A further different way is to properly define the global quantity on averaged systems. Much more promising and of more direct physical significance in our context appears to be a third way: a rewriting in terms of second quantization formalism by the Bogoljubov transformations and diagonalization so as to get like-particle solutions as collective excitations of the reticule.

Using D. Bohm language [6], we are describing the activity of the sub-quantic medium in terms of the dynamic discrete action of a great amount of elementary cells. So we face a typical complexity problem, i.e. finding the correlations between the continuous language of the traditional physics *foreground* (Quantum Mechanics and General Relativity) and the one of the discrete space-time *background*. It is particularly interesting that the **Wigner-Feynman functions**, today used to get the QM formalism, are in perfect consonance with the suggestion of a discrete space-time (see M. Cini in this volume and in [7]), and that there is a close correlation between non-locality and quantization.

Consider an excitation propagating along a straight line formed by a succession of discrete segments. It is clear that the propagation from A segment to B segment will meet, at a certain point, a *crash down* of continuity and of every notion it connected: that is a situation in which the **excitation is either in A or in B, but not between A and B**. That's a binary situation to which it can't be applied the "time travelling" notion et similia. At the most we can say that the passing is instantaneous, or – but that's an undue extension of the classic language – that it occurs at infinite velocity. Such example allows us to point out the deep physical and topological relations between non-locality and discrete structures, moreover it allows us to look at super-luminality as an essentially classical language to express non-local quantum effects.

These considerations show us a close connection between the computational paradigm and the quantum description of the world, and also suggest us the most natural and mighty language to describe the reticular dynamics of space-time.

3. Cellular Automata: a language for computational complexity

A **Cellular Automata (CA)** is a n**-dimensional lattice** where: every cell can get a finite number of values or finite discrete states; each cell's behaviour is fixed by the value of the adjacent cells in relation to a set of transition rules which are applied simultaneously to all cells step by step by a discrete clock.

In the '80s, S. Wolfram [8] and C. Langton [9] set little order in the CA world. They identified four fundamental classes closely correlated among them at varying of a λ order parameter, which here, for plainness, we can identified with a generalized energy.

The Wolfram four classes are characterized by the transition rules and their development is independent of the initial conditions:

Class I: evolves to a homogeneous state (*limit points*);

Class II: evolves to simple periodic or quasi-periodic structures (*limit cycles*);

Class III: yields chaotic aperiodic patterns (*chaotic behaviour* associated with *strange attractors*);

Class IV: yields complex patterns of localized structure (with a very *long transient, for ex. Conway' "Life" Game*).

The C. Langton analysis showed that it was possible to order the classes in a sequence defined by λ order parameter variation:

$$I \to II \to IV \to III\,,$$

Such thing seemed suggesting a conceptual sequence of the following kind:

"rigid" order → periodicity → "complexity" → chaos.

From an essentially computational viewpoint we can wonder what cellular automata class could be really chosen to build up an universal computer in Turing sense. Now it is clear that the I and II classes are too "static" and the III class, on the contrary, is too chaotic. As for the first two cases it could be possible to memorize information, but it could be impossible to process it, while in the case of the III class every memory or signal would be overcome by noise. On the other hand, the class IV, using its strongly peculiar order-chaos *mixture*, gets all the characteristics to memorize and process information, so showing that for defining complexity and information efficiently both aspects have to be taken into consideration. Let's try now to look at the four fundamental classes in relation to **computation theory**. Within this ambit the difference between halting algorithms, i.e. which carry out their elaboration on the received data in a finite time, and the not-halting ones is fundamental. If we use the hint provided by the *material* analogy with the matter phases and the phase transitions, it is possible to say that I and II classes represent algorithms which can conclude their work in a **polynomial time**, *crystallizing* themselves in a definitive *output*. The III class automata keep on *bubbling* in continuous computational activity for an infinite non-polynomial time. On the contrary, the complex automata of the IV class at the *edge of chaos* [10] are a computational exemplification of one of the most stimulating mathematical outcome of any time, i.e. the existence of undecidable problems (Gödel) and unpredictable algorithms (Turing). In this case the whole initial data and rules are totally specifiable in a finite set of information, but during the computation process such information amplifies indefinitely and unpredictably so that, fixed a specified configuration, it's impossible going up again to the initial data and the rules which produced it. In other words it's impossible an **algorithm compression** of information in R. Solomonoff and G. Chaitin sense [11]. The universe of class IV automata, that is the universe of complexity, is uncompressible and undecidable, but computable at each step. The situation is similar to the ones of deterministic chaos (III class automata), since each point of the process is computable, but there's an essential difference which is, after all, the one we are looking for among order, chaos and complexity in relation to the information concept. Deterministic chaos is, theorically, algorithmically compressible, which thing often isn't clearly focused, so causing a big confusion. A phenomenon ruled by a set of non-linear equations characterized by one or two strange attractors **is totally defined** by the last ones, except for a *noise* term related to a finite quantity of uncertainty which can be valued [12]. The same doesn't go for the complex phenomena of the IV class, whose behaviours are considered by S. Wolfram and S. Kauffman [13], and us with them too, as particular peculiarities of the **complexity of discrete systems**.

So we can introduce another analogy, this time linked to the characteristics of the four classes in relation to computability:

I, II: halt in polynomial time→ IV: undecidable→III: don't halt.

4. Particles and Quantum Vacuum in Computational Paradigm

Since long time it's known that the CA IV class produces like-soliton configurations similar to particles. Now we get all the ingredients to develop our computational paradigm:

- **Any elementary particle is a coherence state of a CA;**
- **Any emission, absorption, decaying or scattering process can be described in terms of computational processes among CA;**
- **A stable particle is an information conserving system, an unstable one processes information in an exponential or polynomial time.**

Recently at Santa Fè Institute [14] had been made simulations which show a good concordance with the particle description of reticular dynamics. Since long time G. Chew [15] and H. Stapp [16], in the sphere of the S-Matrix Theory, have demonstrated that it is possible to get "finite" like-Feynman diagrams emerging from a complex *sea* of elementary events (*gentle events*). The last one select the observed states from the virtual ones. "*The problem of summing the gentle events series has taxed the ingenuity of two generations of particle theorists*" (G.F. Chew). In our case, the *gentle events* are represented by the vibrational activity of space-time cells. The problem of the number and of the type of families of the matter components can be brought back, in our computational paradigm, to the study of the initial conditions and to the kind of transition rules. In this case a T.O.E. (*Theory of Everything*) is the search for the smallest number of CA able to generate the known types of matter and energy.

The most natural language to describe the CA configuration is that of Sinai-Kolmogorov information dynamic theory [9]. In the continuous description it is necessary an infinite information to describe each single space-time point, but it isn't a problem in reticular dynamics. On the contrary, starting from Bekenstein formula [17], which correlate the information quantity I linked to a mass-energy E, it is possible describing "a natural history of information and energy".

If we admit a quantum description of vacuum, then Universe isn't an information polynomial amplifier (like in Fridman solutions of GR), but an exponential amplifier of information (like in the inflationary scenarios). Let's consider an initial state of our CA Universe. We have to consider (see point d) a maximum finite value of energy and, consequently, of the microscopic information within each cell. The vibrational activity is unstable by its nature (see point c). By means of what can be described as a phase transition of the second kind we get a symmetry breaking by which spread out an exponential growth of patterns of CA-particles. At the beginning we get a fluctuating vacuum at very high energy, where each cell gets a microscopic information maximum value. After the symmetry breaking, mass-energy structures are creating so getting macroscopic information over the Planck scale. Such passage from microscopic information to macroscopic one defines in a natural way the time arrow.

5. Conclusions

The computational paradigm is moving its first steps, but it promises to offer a telling solution to various classical problems, from the infinite quantity elimination in QFT to the nature of the global and local symmetries. We can state we live in a computable but unpredictable Universe, a very peculiar kind of computer: the only one able to simulate itself.

References

[1] GEROCH R., HARTLE J.B., Found. Of Phys. **16** (1986), 553.

[2] EINSTEIN A., *Physik und Realität*, Franklin Inst. Journ. **221** (1936), 313.

[3] IVANENKO DMITRI D., in: "Centenario di Einstein 1879-1979", Giunti-Barbera Ed., Firenze 1979, 131–220.

[4] ZENG XINCHUAN, Physics Essays **4** (1991), 344.

[5] LICATA I., Hadronic Journ. **14** (1991), 225.

[6] BOHM D., *Wholeness and the Implicate Order*, Routledge & Kegan Paul, London 1980.

[7] CINI M., Annals of Physics **273** (1999), 199.

[8] WOLFRAM S., Physica **D 10** (1984), 1.

[9] LANGTON CHRIS G., Physica **D 42** (1990), 12.

[10] BAK P., TANG C., WIESENFELD K., Phys. Rev. Lett. **59** (1987), 381.

[11] CHAITIN G., *Algorithmic Information Theory*, Cambridge Univ. Press, Cambridge 1987.

[12] TRAUB J.F., WASILKOWSKI G.W., WOŹNIAKOWSKI H., *Information-based Complexity*, Academic Press 1988.

[13] KAUFFMAN S., *At Home in the Universe*, Oxford Univ. Press 1995.

[14] HORDIJK W., CRUTCHFIELD J.P., MITCHELL M., in: "Parallel Problem Solving from Nature, Proc. Vth Workshop PPSN", 1998.

[15] CHEW G.F., in: "Quantum Implications Essays in Honour of David Bohm", Hiley B.J. and Peat D. eds., Routledge & Kegan Paul, London 1987.

[16] STAPP H.P., Phys. Rev. Lett. **50** (1983), 467.

[17] BEKENSTEIN J.D., Phys. Rev. Lett. **46** (1981), 623.

ON THE UNIQUENESS OR MULTIPLICITY OF PHYSICAL THEORIES: A NOTE ON THE MARGIN OF THE "SOKAL AFFAIR"

Claudio Pellegrini
Department of Physics and Astronomy, UCLA
pellegrini@physics.ucla.edu

Abstract In the light of the Sokal Hoax affair we reconsider the issues of the cultural meaning of science, and of the effect of culture, philosophy and religion on scientists and scientific theories. Following a comment to the Sokal affair by Steven Weinberg, we consider also the related questions of the objective validity and of the uniqueness or multiplicity of physical theories.

1. Introduction

About five years after the publication in Social Text of Alan Sokal's paper "Transgressing the Boundaries: Toward a Transformative Hermeneutics of Quantum Gravity" [4], the editors of Lingua Franca published a book, The Sokal Hoax, The Sham that Shook the Academy [1], with the original Sokal's paper and most of the comments that appeared at that time in many journals and magazines after Sokal admitted that his paper was an hoax. Social Text was a leading journal in the field of cultural studies, and Sokal's paper appeared in an issue dedicated to "The Science Wars".

One of the main claims of some of the scholars who contributed to the issue is that "Science is a culturally determined value system, and should be analyzed as such". Few practicing scientists would, in my opinion, agree with this statement.

As an example of the different viewpoints let me consider the first accurate measurement of the size of the solar system, made by Cassini and Richer, near the end of the XVII century, observing the parallax of Mars. A physicist or astronomer would discuss the history of this important scientific contribution by considering how it was made possible by:

a. The development of methods to measure astronomical distances going back to Aristarchus in the 3rd Century BC, in Hellenistic Alexandria;

b. The development of the telescope at the beginning of the XVII century, which made possible the discovery of Jupiter's satellites by Galileo;

c. The measure of their periods of rotation around Jupiter, and the painstaking preparation of tables with the relevant astronomical data on these satellites done by Galileo and improved by Cassini;

d. The improvements in the optical lenses of the telescopes made by Campani in Rome, and in the mechanical supports and pointing systems of telescopes made by other astronomers;

e. Finally the financial support given by Colbert to build the "Observatoire de Paris", to hire the best European astronomers, like Cassini and Roemer, and to fund the Richer's expedition to the Guyane.

Cassini's results, after a careful analysis of the possible measurement errors, would be considered to give an objective measurement of the distance between Mars and the Earth, to be confirmed and made more accurate by more measurements, possibly done by more accurate and improved instruments, and of universal validity. Using Kepler's analysis of the motion of the planets it was also possible to use Cassini's result to obtain the distance between the Earth and the Sun, and the Earth and the other planets. Aristarchus tried to measure directly the distance between the Earth and the Sun, and his results were quite different from those of Cassini, because he did not have the instruments needed to make an accurate measurement of the angles involved in the experiment. The differences can be traced back to errors in angular measurements.

A scholar in cultural studies of science would be more interested in seeing the connection of the same scientific work to the colonialist policy of France in Africa and America at that time, the consequent need of a strong and efficient navy that could easily navigate the oceans, and the resulting need of astronomical observations and tables to help the navigation.

While both points of view can contribute to our understanding of scientific work at the end of the XVII century, it could be interesting to ask another question: did Cassini, who followed the astronomical methods to measure distances that were established in the third century BC by Aristarchus, had the same motivation, to understand how large is the world in which we live that inspired also the great Alexandrine astronomer, even if the two were living in a totally different cultural and social environment? Was this motivation, in the end, simply the desire to better understand the world in which we live, using reason and observations? Would the result obtained by them be different if Aristarchus had been able to use better astronomical instruments?

2. Some comments on Sokal's paper

It is interesting to notice that Sokal's paper, or better Sokal's hoax, was the only one by a physicist in the Social Text issue on Science Wars. The paper was published even if it contained many historical and physical arguments, from Sokal himself, or quoted from other authors, that should have been easily seen as false or meaningless. In a note to the paper Sokal writes: "A careful analysis would show, I think, that the victory of cybernetics over quantum physics in the 1940's and 1950s can be explained, in large part, by the centrality of cybernetics to the ongoing capitalist drive for automation of industrial production, compared with the marginal industrial relevance of quantum mechanics."

The idea of a war between cybernetics and quantum physics would surprise many people, and the defeat of quantum mechanics would certainly surprise anyone who has been working on the development of the transistors, or lasers. In fact quantum mechanics is now taught in many engineering departments.

In another part of the paper Sokal writes: "A simple criterion for science to be post-modern is that it be free from any dependence on the concept of objective truth. By this criterion the complementarity interpretation of quantum physics due to Bohr and the Copenhagen School is seen as post modernist." Again quantum mechanics is a well defined mathematical theory, whose predictions have been verified experimentally to an extremely high level of precision, in experiments done all over the world.

Sokal also quotes Darrida: "The π of Euclid and the G of Newton, formerly thought to be constant and universal, are now perceived in their ineluctable historicity", or: "...the Einsteinian constant is not a constant, is not a center. It is the very concept of variability, it is, finally, the concept of the game." The Einsteinian constant is something new to any physicist, and in any case it would be just a number, exactly like π or G.

Sokal revealed that his paper was a hoax in a letter to Lingua Franca [5], which appeared almost at the same time as the Science Wars issue of Social text. What followed was a series of attacks and defenses of Sokal and his hoax. I will mainly be concerned with the contribution to the debate by Steven Weinberg, and the replies that followed. But before doing that let me show the depth of the problem of understanding what science is or is not, even after about four centuries of scientific development, by quoting a comment to Sokal by Marco d'Eramo [3]: "According to my senses the Sun moves around the Earth. If I state that the objective reality is different, and the Earth moves around the Sun, it is because I follow my theoretical description of reality (which by the way dramatically contradicts my eyesight). For millennia the objective reality was different ... That is why the same statement that inspires your (Sokal's) derision – "that scientific "knowledge" far from being objective reflects and encodes the dominant ideologies and power relations of the culture that produced it" – seems to me perfectly meaningful and sensible."

One clear problem with d'Eramo's comments is that he limits the possibility of observing natural phenomena to that allowed by direct use of our senses, refusing the help given by instruments like the telescopes to "extend our senses". His position is similar to that of the Aristotelian philosophers who refused to look through Galileo's telescope. But if one looks through the telescope, and observes, as Galileo did, the phases of Venus, one can immediately discard the Ptolemaic system and the notion that the Sun and the planets move around the earth. In fact the history of progress in physics is mainly, and with few exceptions, a history of "extending our senses" through the development of scientific tools, to see what we could not see before.

But perhaps d'Eramo never took the time to read Galileo, to understand the relativity of motion for different observers, and that the one has to account for the motion of all the planets in the solar system, not only the Earth. Perhaps he would approve Galileo's trial, and the sentence by the inquisition.

3. Weinberg's "Cultural meaning of Science"

Let us consider now Steven Weinberg's contribution to the discussion of Sokal's hoax [6]. In his paper Weinberg discusses the effect of science on culture and the effect of culture on science. He claims that, with a few exceptions, the results of science have no implications for culture, philosophy, and politics. He acknowledges that the technological applications of science do have an effect on society and thus on politics, but this kind of effect is not the subject of the "Science Wars" discussion.

The exceptions are:

1 The discovery, going back to Galileo and Newton, that nature is governed by impersonal mathematical laws. From this point of view the difference between Newtonian mechanics, Einstein relativity theory, and quantum mechanics, is inessential.

2 Space, time and matter are a subject for science, not anymore for philosophy. (In a restrictive sense we might consider matter as inanimate matter, but I would prefer to include in it biological systems.)

3 When and if we will learn about the origin of the Universe, or the final laws of nature it will have a tremendous cultural-philosophical impact. Presently, the cultural interpretation of quantum mechanics, and the related discussions on the mind-body problem, or on free will are simply wrong.

Weinberg recognizes that culture and philosophy are important in the process of "doing science", the choice of experiments and/or areas of research, but not in the final physical theory. The end result of scientific research, after having gone through the process of verification with additional experiments, is an objective representation of reality. One way to qualify the last statement is to say that if we would discover intelligent beings on another planet, and translate their scientific works, we would find that we on earth, and them on another part of the universe have discovered the same laws. On a more modest level, we can say that the laws governing electromagnetism, or the motion of bodies are the same for all people on Earth, with any ethnic and cultural background, religious or philosophical belief, women and men.

In his answer to Weinberg, Norton Wise [7] raises the issue of cultural influences on science, and the multiplicity of physical theories: "The issue of cultural relativism is not validity; it concerns multiplicity: the multiplicity of valid positions that have been available at any time, the many ways in which these positions have been embedded in the cultures from which they emerged, the divers processes through which they have crossed cultural boundaries and changed over time, with massive implications for both science and the rest of culture."

Wise denies the validity of the point that "as far as culture is concerned the difference between Newton, Einstein and Schrödinger is irrelevant". He mentions that Einstein never accepted Bohr's interpretation of quantum mechanics, and that the problem has been raised again by David Bohm with his deterministic and holistic reformulation of quantum mechanics. He then makes a long list of physicists from Kepler to

Maxwell, Oersted, and Jordan, who were motivated in their research by religious or philosophical beliefs.

The difference between Wise and Weinberg is that the first is interested in the process that leads to the discovery of physical laws, a process certainly influenced by the culture and the philosophy of the scientists involved. The other, Weinberg is interested in the end result, what remains after the discoveries and the resulting theories have been verified, purified of the cultural motivation, leaving us an "objective description of reality", valid for people in any culture, race or gender, hopefully also for the inhabitants of another planet.

This purification process, which has as a key element the comparison of theories with experimental data, with the possibility of their verification or "falsification", works in an almost miraculous way. During a discussion some time ago a philosopher asked me: is the standard model of elementary particles the only possible theory to describe all the known facts about protons, quarks and leptons? My answer was: yes. We can say (although we cannot prove this as a theorem) that if two physical theories, formulated in mathematical terms, can reproduce equally well the results of a large number of experiments, they must be mathematically equivalent.

This answer is based on the history of physics. The best example is perhaps that of quantum mechanics: Heisenberg discovered the matrix mechanics in July 1925; Schrödinger found the wave mechanics in January 1926; six weeks later Schrödinger submitted another paper proving that the theories are mathematically equivalent.

The same answer comes from the evolution of Newtonian mechanics into relativistic mechanics, and finally quantum mechanics. The later theories are an extension of the old one to cover new phenomena, but in the limit of a velocity small compared to that of light, or for a macroscopic system, they reduce to and give back to us Newton's theory. In a similar way quantum electrodynamics extends Maxwell's theory of electromagnetic phenomena, and reduces to it under the proper physical conditions mostly dominant in our everyday life.

Einstein dedicated much time to the problem of pluralities of theories. In a paper he presented at a symposium to celebrate Planck's 60th birthday in 1919 he said [2]: "The supreme task of the physicist is .... To search for those most general, elementary laws from which the world picture is to be obtained through pure deduction. No logical path leads to these elementary laws; it is indeed just the intuition ... In this state of methodological uncertainty one can think that arbitrary many, in themselves equally justified systems of theoretical principles were possible; and this opinion is, in principle, certainly correct."

Then he continues: "But the development of physics has shown that of all conceivable theoretical constructions a single one has, at any given time, proved itself unconditionally superior to all others. No one who has really gone deeply into the subject will deny that, in practice, the world of perceptions determines the theoretical system unambiguously, even though no logical path leads from the perceptions to the basic principle of the theory."

References

[1] *The Sokal Hoax, The Sham that Shook the Academy*, edited by the editors of Lingua Franca, University of Nebraska Press 2000.

[2] EINSTEIN A., *Induktion und Deduktion in der Physik*, Berliner Tageblat, 25 December 1919.

[3] D'ERAMO M., *Sfregio Accademico a Greenwich Square*, Il Manifesto, Rome 6 November 1996.

[4] SOKAL A., *Transgressing the Boundaries: Toward a Transformative Hermeneutics of Quantum Gravity*, Social Text **46/47** (1996), 217–252.

[5] SOKAL A., *Revelation: A Physicist experiments with Cultural studies*, Lingua Franca, May-June 1996.

[6] WEINBERG S., *Sokal's Hoax*, New York Review of Books, 8 August 1996.

[7] WISE N., *Letter to the Editors*, New York Review of Books, 3 October 1996.

AN INTERPLAY BETWEEN DETERMINISM AND ONE-PARAMETER SEMIGROUPS, AND APPLICATIONS

Silvia Romanelli
Dipartimento Interuniversitario di Matematica,
Università di Bari,
Via E. Orabona 4, 70125 BARI - ITALY
romansdm.uniba.it

Abstract We present a mathematical approach to motion and determinism in connection to the theory of one-parameter semigroups, according to Engel and Nagel [3]. Examples of generators of (C_o) semigroups which govern topical evolution equations occurring in Physics, Economics and Engineering are given.

1. Introduction

The general question concerning a relation of Mathematics and Reality, with particular reference to the connections between the theory of one-parameter semigroups (and evolution equations) and the philosophical concept of determinism, is well discussed in the last chapter of the recent book by K.J. Engel and R. Nagel [3, Epilogue, pp. 531–553]. In Section 2 we start from these ideas in order to present a possible approach to the theory of the (C_o) semigroups, which are a powerful tool for the investigation of many evolution problems arising in different fields, as Physics, Biology and Economics. A short introduction to (C_o) semigroups and to some meaningful classes of them is given in Section 3. In the last Section we consider a class of operators in divergence form, which are generators of (C_o) semigroups enjoying nice properties from both deterministic and probabilistic point of view.

2. A mathematical approach to motion and determinism

It is well-known that evolution equations describe some phenomena related to the *motion of systems* depending on time. Since the philosophical concept of *determinism* refers to the theory affirming that the course of all natural events is predetermined, it is essential to understand if mathematical models fit in this theory or some type of *freedom* is allowed.

Let us look at this problem in a fundamental case which often occurs in many problems of contemporary natural science: the motion of systems. Here by *motion* we will mean *all forms of temporal change.*

If we represent the time by the *additive group* $\mathbb{R}_+$ of all nonnegative real numbers, the system under examination can be characterized by the *state space* Z whose elements $z \in Z$ depend on the time $t \in \mathbb{R}_+$, according to some fixed laws (e.g. *planetary systems, ecosystems*).

The motion of the system is represented by means of the function

$$t \in \mathbb{R}_+ \mapsto z(t) \in Z\,.$$

Thus, if we restrict us to consider the single motion, it can already be viewed as deterministic.

The perspective changes if we require that the system can be restarted with any prescribed initial data, in the sense that

$$\begin{array}{l}\textit{for any} \quad t_0 \in \mathbb{R}_+ \quad \textit{and any initial state} \quad z_0 \in Z \quad \textit{there exists a} \\ \textit{unique motion} \quad z_{t_0,z_0} : \mathbb{R}_+ \to Z \quad \textit{such that} \quad z_{t_0,z_0}(t_0) = z_0\end{array} \tag{1}$$

In this way, the role of the observer is replaced by that one of an experimenter who can vary the initial value $z_0 \in Z$ at time t_0 and obtain a uniquely determined state $z_{t_0,z_0}(t_1)$ of the system at a target time t_1. Then, we can define

$$\Phi_{t_1,t_0} : Z \to Z$$

such that

$$\Phi_{t_1,t_0}(z_0) := z_{t_0,z_0}(t_1)\,. \tag{2}$$

Next, we take $z_1 := \Phi_{t_1,t_0}(z_0)$ as initial state and t_1 as initial time. By (1), at time t we obtain a unique state given by

$$\Phi_{t,t_1}(z_1) = \Phi_{t,t_1}(\Phi_{t_1,t_0}(z_0))$$

and unicity shows that

$$\Phi_{t,t_1}(\Phi_{t_1,t_0}(z_0)) = \Phi_{t,t_0}(z_0)$$

for all $z_0 \in Z$ and $t, t_1, t_0 \in \mathbb{R}_+$. Hence

$$\Phi_{t,s} \circ \Phi_{s,r} = \Phi_{t,r} \tag{3}$$

and

$$\Phi_{t,t} = I \tag{4}$$

provided that $t, s, r \in \mathbb{R}_+$ and I is the identity on Z. The system $(Z, \mathbb{R}_+, \Phi)$, where Z is the state space and $\Phi := \{\Phi_{t,s} : t, s \in \mathbb{R}_+\}$ is a family of mappings on Z which satisfies (3) and (4), is called *deterministic system*. Let us assume that the above system is *autonomous*, i.e.

the state $z_1 = \Phi_{t,s}(z_o)$ *at time* t *depends only on the initial state* z_o *at time* s *and the time difference* $\tau = t - s$. (5)

This implies

$$\Phi_{t,s} = \Phi_{r,u} \quad \textit{if and only if} \quad t - s = r - u\,.$$

Thus, the one-parameter family $(T(t))_{t \in \mathbb{R}_+}$ defined as follows

$$T(t) := \Phi_{r,r-t}\,, \quad t, r \in \mathbb{R}_+$$

consists of mappings acting on Z and satisfying

$$T(t+s) = T(s)T(t) \quad \textit{and} \quad T(0) = I \tag{6}$$

for all $t, s \in \mathbb{R}_+$. A family of mappings for which equations (6) hold is called *one-parameter semigroup* and this is our mathematical model of autonomous, deterministic motion of a system in time (this is the case when the governing physical laws do not change in the course of time and no external force acts on the system).

Therefore, we observe that in the framework of mathematical models described by one-parameter semigroups we must distinguish between the state space, which must include all properties we need, and the single state, which depends on the time. In order to obtain a good accuracy in the description of the system, we have to find the *right* state space, which contains all relevant properties and, at the same time, assures the deterministic motion.

3. (C_o) semigroups

In many applications the state space Z is a Banach space and the one-parameter semigroup $(T(t))_{t \in \mathbb{R}_+}$ consists of linear bounded operators $T(t)$ on Z which satisfy this additional *continuity property*

for any $z_o \in Z$, *the mapping* $t \in \mathbb{R}_+ \mapsto T(t)z_o$ *is continuous*. (7)

If this is the case, $(T(t))_{t\in\mathbb{R}_+}$ is called (C_o) semigroup on Z and a closed operator $(A, D(A))$ with domain $D(A)$ acting on Z is uniquely associated to $(T(t))_{t\in\mathbb{R}_+}$. It is defined as follows

$$\begin{aligned} D(A) &:= \left\{ z \in Z : \text{there exists } \lim_{t\to 0^+} \frac{T(t)z - z}{t} \in Z \right\} \\ Az &:= \lim_{t\to 0^+} \frac{T(t)z - z}{t} . \end{aligned}$$

In this way the semigroup $(T(t))_{t\geq 0}$ represents the aspects of permanent change in the considered system, while its generator stands for the timeless principle behind the system.

Among the properties of the (C_o) semigroups which allow to deduce useful information on the semigroup, starting from the knowledge of its generator, we mention:

(1) the invariance of the domain of A under the action of $(T(t))_{t\geq 0}$;

(2) the relations between the resolvent of A and the asymptotic behavior of $(T(t))_{t\geq 0}$.

Two classes of (C_o) semigroups, namely Feller semigroups and analytic semigroups, are worth to be considered here for the important role played in many settings.

A (C_o) semigroup $(T(t))_{t\geq 0}$ on a space $C(X)$ of real-valued continuous functions on a compact space X is called *Feller semigroup* if

$$f \in C(X), \quad 0 \leq f \leq 1 \quad \textit{implies} \quad 0 \leq T(t)f \leq 1 .$$

Feller semigroups can be identified by suitable Markov processes and generators of Feller semigroups are related to suitable martingales (see Ethier and Kurtz [4, Chapter 4]).

An *analytic semigroup* $(T(t))_{t\geq 0}$ on a Banach space Z (for its precise definition and characterizations see e.g. [3, Chapter II Section 4.a]), enjoys the property that the mapping defined in (7) can be extended analytically to a sector of the complex plane around axis x. In addition, it has smoothing effects on the initial data, more precisely $T(t)$ maps Z into $D(A)$ for any $t > 0$, and there exists a constant $C > 0$ such that $t\|AT(t)\| \leq C$ for any $t > 0$.

4. Applications

Let us consider a particular class of second order differential operators in divergence form of the type

$$A := \nabla \cdot (a\nabla) , \tag{8}$$

where a is a sufficiently smooth function defined in Ω, $a > 0$ in Ω, and Ω is an open subset of $\mathbb{R}^N$.

These operators often occur in several applications to Physics (see e.g. [13]), Economics (see e.g. [2], [12]) and Engineering (see e.g. [1]), sometimes in connection with inverse problems (see e.g. [5]).

We will present two examples of realizations of A which are generators of (C_o) semigroups.

Example 1. *Assume $N = 1$ and a continuous and strictly positive on $\Omega := (0, +\infty)$. Consider the linear equation*

$$\frac{\partial u}{\partial t} = \frac{\partial}{\partial x}\left[a(x)\left(\frac{\partial u}{\partial x}\right)\right], \quad t > 0\,,\ x > 0 \tag{9}$$

with initial condition

$$u(0, x) = u_o(x)\,, \quad x > 0 \tag{10}$$

and boundary conditions

$$\lim_{x \to 0^+, x \to +\infty} a(x)\frac{\partial u}{\partial x}(t, x) = 0\,, \quad t > 0\,. \tag{11}$$

This is a particular case of the linear Kompaneets equation *which arises in the context of plasma physics, where x is the normalized photon energy, and $u(t,x)$ represents the radiation density (which gives the total photon number). Notice that the coefficient a is allowed* to vanish *at the boundary. For the study of this equation from the point of view of (C_o) semigroups on various function spaces see e.g. [13] and [8]. Here we restrict us to describe the case $a(x) := x^2$ in the Banach space $C[0, +\infty] := \{u \in C[0, +\infty) : \lim_{x\to+\infty} u(x) < \infty\}$, equipped with the sup-norm. The problem (9)-(11) can be transformed into the* abstract Cauchy problem

$$\begin{cases} u'(t) = A_1 u(t)\,, & t > 0 \\ u(0) = u_0\,, & u_0 \in D(A_1) \end{cases}$$

where $A_1 u := (x^2 u')'$ and $D(A_1)$ must be suitably chosen according to the boundary conditions (11). As a consequence of the results proven in [8] and [9], the operator A_1 with domain

$$D(A_1) \;:=\; \Big\{u \in C[0, +\infty] \cap C^2(0, +\infty) : A_1 u \in C[0, +\infty)\,, \lim_{x \to 0^+, x \to +\infty} A_1 u(x) = 0\Big\}$$

generates an analytic Feller semigroup $\mathcal{T} = (T(t))_{t\geq 0}$ *on* $C[0,+\infty]$. *Thus* $\mathcal{T}$ *enjoys* all *properties described in Section 3. Recently, R. Mininni and the author [11] have also shown that the methods by Kessler and Sorensen [10] can be applied to* A_1, *in order to obtain optimal martingale estimating functions and inference for the drift coefficient of the corresponding diffusion process.*

Example 2. *Let us consider the heat equation*

$$\frac{\partial u}{\partial t} = Au\,, \quad \textit{for} \quad t > 0\,, \quad \textit{in} \quad \Omega$$

with the generalized Wentzell boundary condition

$$Au + \beta\frac{\partial u}{\partial n} + \gamma u = 0 \quad \textit{on} \quad \partial\Omega\,,$$

where Ω *is a bounded subset of* $\mathbb{R}^N$ *with a sufficiently smooth boundary* $\partial\Omega$, $a \in C^1(\overline{\Omega})$, *with* $a(x) > 0$ *in* Ω, $\beta, \gamma \in C^1(\partial\Omega)$ *with* $\beta(x) > 0$, $\gamma \geq 0$ *on* $\partial\Omega$, *and* $n(x)$ *is the unit outer normal at* x. *This type of boundary condition arises in a natural way in some problems of hydrauling fracture (see e.g. [1]). Assume that* $\Gamma := \{x \in \partial\Omega : a(x) \neq 0\} \neq \emptyset$. *Hence it is allowed that a may* degenerate *on some part of the boundary. Then, according to the results in [7], the* closure *of a realization of* A *on* $C(\overline{\Omega})$ *generates a Feller semigroup, while realizations of* A *on suitable* L^p *spaces,* $1 < p < \infty$, *generate analytic semigroups. In some cases, for instance when* $N = 1$ *and* $a \equiv 1$, *the operator* $A_2 u := u''$ *with domain*

$$D(A_2) := \{u \in C^2[0,1] : u''(j) + (-1)^{j+1}\beta_j u'(j) + \gamma_j u(j) = 0\,,\ j = 0,1\}\,,$$

where β_j, γ_j *are real numbers,* $\beta_j > 0$ *and* $\gamma_j \geq 0$, *is* closed *and hence it generates a Feller semigroup on* $C[0,1]$ *(see [6]).*

References

[1] APUSHKINSKAYA D.E., NAZAROV A.I., *The Venttsel' problem for nonlinear elliptic equations*, J. Math. Sc. **101** (2) (2000), 2861–2880.

[2] COLOMBO F., GIULI M., VESPRI V., *Generation of smoothing semigroups by degenerate elliptic operators arising in financial mathematics*, Quaderno n. 38 Dipartimento di Matematica "F. Enriques", Università di Milano 1997.

[3] ENGEL K.J., NAGEL R., *One-Parameter Semigroups for Linear Evolution Equations*, Graduate Texts in Mathematics **194**, Springer 2000.

[4] ETHIER S.N., KURTZ T.G., *Markov Processes*, J. Wiley & Sons, New York 1986.

[5] FAVARON A., LORENZI A., *Parabolic integro-differential identification problems related to radial memory kernels I*, Quaderno n. 18 Dipartimento di Matematica "F. Enriques", Università di Milano 2001.

[6] FAVINI A., GOLDSTEIN G.R., GOLDSTEIN J.A., ROMANELLI S., *C_o-semigroups generated by second order differential operators with general Wentzell boundary conditions*, Proc. Amer. Math. Soc. **128** (2000), 1981–1989.

[7] FAVINI A., GOLDSTEIN G.R., GOLDSTEIN J.A., ROMANELLI S., *The heat equation with generalized Wentzell boundary condition*, to appear in J. Evol. Eqns.

[8] FAVINI A., GOLDSTEIN J.A., ROMANELLI S., *Analytic semigroups on $L^p_w(0,1)$ and on $L^p(0,1)$ generated by some classes of second order differential operators*, Taiwanese J. Math. **3** (2) (1999), 181–210.

[9] FAVINI A., ROMANELLI S., *Degenerate second order operators as generators of analytic semigroups on $C[0,+\infty]$ or on $L^p_{\alpha-\frac{1}{2}}(0,+\infty)$* in: "Approximation and Optimization, Proceedings of International Conference on Approximation and Optimization (Romania) (ICAOR) Cluj-Napoca, July 29–August 1, 1996" (Eds. Stancu D., Coman G., Breckner W.W., Blaga P.) Vol. II, Transilvania Press 1997, 93–100.

[10] KESSLER M., SORENSEN M., *Estimating equations based on eigenfunctions for a discretely observed diffusion process*, Bernoulli **5** (2) (1999), 299–314.

[11] MININNI R., ROMANELLI S., in preparation.

[12] TOTARO S., *Explicit solutions of equations arising from economy problems*, in: "Atti IV Congresso Nazionale SIMAI, Giardini Naxos (Messina)" Vol. I, 1998, 230–233.

[13] WANG K., *The linear Kompaneets equation*, J. Math. Anal. Appl. **198** (1996), 552–570.

FROM DYNAMICAL SYSTEMS TO COMPLEX SYSTEMS

Giorgio Turchetti
Dipartimento di Fisica,
Via Irnerio 46, 40126 BOLOGNA - ITALY
turchetti@bo.infn.it

Abstract The framework of dynamical systems is appropriate to describe physical phenomena where regular, chaotic or intermittent behaviors are observed. The self organization with the emergence of cognitive and self replication properties typical of biological world requires the wider framework of complex systems.

1. Introduction

The rational description of nature is the root of civilization, having replaced the emotional phase, typical of primitive societies, where natural phenomena were considered as wonderful or terrific. The transition from magic to natural philosophy is a milestone in human evolution and is based on the cause-effect association of natural events. The search of simple laws underlying the variety of structures and events occurring in our world lead two thousands years ago to postulate the existence of elementary and invariable building blocks (atoms), which freely move and aggregate, The roots of modern physics are present in the pages of Lucretius poem *De rerum natura*: the search of elementary constituents and of simple laws which govern their interactions, the deterministic evolution and the presence of randomness, which prevents a detailed forecast of future in every day's life. Modern physics was born with celestial mechanics and the regular motions of celestial bodies were taken as a property of deterministic systems, which were assumed to be predictable. At the end of the nineteenth century, it became evident that it was not the lack of computing power, to prevent the trajectory a molecule to be followed for a long time, but rather an intrinsic instability rooted in non linear dynamical systems, the same which prevented the three body problem to be solved by quadratures [1, 2, 3]. On the other hand the statistical treatment of systems with many degrees of freedom allowed to recover the phenomenological laws of thermodynamics. In the past century the theory of dynamical systems, by coupling the deterministic evolution with a probability measure in phase space, allowed a unified treatment of deterministic systems integrable, chaotic or far from these well understood limits [4, 5].

Physical systems We consider a *microscopic description*, based on fundamental interactions and Hamilton's equations, a *macroscopic description*, based on phenomenological laws depending on few parameters, and a *mesoscopic description*, where a stochastic and damping force acts on each macroparticle to simulate the effect of collisions. Order and chaos are the extreme states. Order implies stability, long term predictability, low information flow. Chaos implies instability, short term predictability, high information flow. A chaotic system is *simple* for the statistical analysis. Is *complexity* [6] at the borderline? Coexistence of regular, and chaotic orbits, attractors with fractal boundaries, patterns formation, phase transitions can be used as complexity metaphors. We prefer to reserve the label *complex* to systems with cognitive [7] and replication properties.

Biological systems They are the result of an evolutionary process lasting 3 billions years. In this respect they have properties, that an organized physical system like a crystal does not possess. A living system has a structure (hardware) hierarchically organized on many space time scales. It has cognitive properties: receives, compresses, stores signals and learns. It undergoes a structural evolution process: mutates preserving the favorable changes and develops a short term and long term (phylogenetic) memory. It is a self replicating machine based on efficient information coding. A self organized dynamical system (hardware) and a replicating automaton with cognitive properties (information processing software) is the candidate to be a complex system.

2. Dynamical Systems

A dynamical system is specified by a phase space $\mathcal{M} \subset \mathbb{R}^d$, a deterministic flow S_t, a preserved probability measure μ. A state of the system is determined by a phase space vector $\mathbf{x} \in \mathcal{M}$, the change to $\mathbf{x} + d\mathbf{x}$ in the time interval dt, by a vector field Φ, depending on the state $\mathbf{x}$ at time t

$$\frac{d\mathbf{x}(t)}{dt} = \Phi(\mathbf{x}(t)) \tag{1}$$

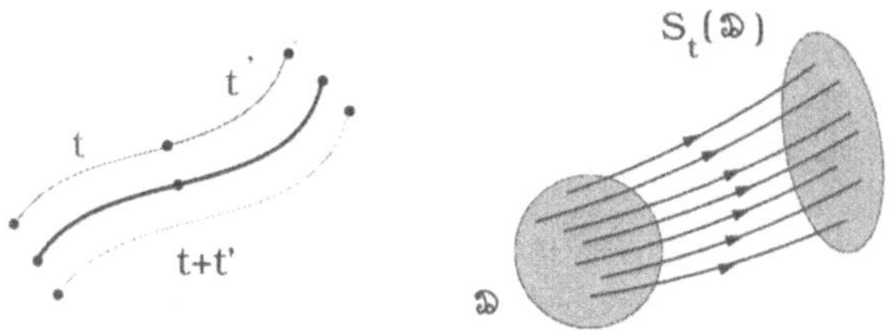

The evolution $\mathbf{x}(t) = S_t(\mathbf{x}_0)$ up to time t has the group property $S_t \circ S_{t'} = S_{t+t'}$ and the phase space density ρ satisfies the continuity equation

$$\frac{\partial \rho}{\partial t} + \operatorname{div}(\rho\, \Phi) = 0 \tag{2}$$

We interpret ρ as the density of a fluid in phase space, see figure 1. If the fluid is incompressible $\operatorname{div} \Phi = 0$, the volumes are preserved and ρ is constant along any trajectory, see figure 1. Systems with memory and damping have different properties[1].

2.1 Hamiltonian systems

Models of celestial bodies and microscopic systems are Hamiltonian. The flow preserves the volumes and other geometric (Poincaré) invariants. In this case the invariant measure is the normalized volume. Letting $(\mathbf{q},\, \mathbf{p}) \in \mathcal{M}$ where $\mathbf{q}$ are the space coordinates and $\mathbf{p}$ the momenta, the canonical form of Hamilton's equations is

$$\frac{d\mathbf{q}}{dt} = \frac{\partial H}{\partial \mathbf{p}} \qquad\qquad \frac{d\mathbf{p}}{dt} = -\frac{\partial H}{\partial \mathbf{q}} \tag{3}$$

The canonical transformations $\mathbf{q}, \mathbf{p} \to \mathbf{Q}, \mathbf{P}$ preserve the geometric invariants and the canonical form of the equations. A Hamiltonian system is integrable if there is a canonical choice of coordinates $\mathbf{P}, \mathbf{Q}$ such that $H = \tilde{H}(\mathbf{P})$ where Q_i are angular variables.
Integrable systems are a collection of uniform circular motions. They are simple and stable: an initial small error $\Delta\mathbf{q}$, $\Delta\mathbf{p}$ implies a small error in the normalized momentum $\Delta\mathbf{P}$ and a small change in the angular

[1]The systems with memory $\dot{\mathbf{x}} = \int_0^t K(t-s)\, \Phi(\mathbf{x}(s))ds$, are infinite dimensional and have no group property, nor a preserved measure. When dissipation is present $\dot{\mathbf{x}} = -\beta\mathbf{x} + \Phi(\mathbf{x})$ with $\operatorname{div} \Phi = 0$, the volumes contract. $V(t) = e^{-\beta t}\, V(0)$. The flow of $\mathbf{X} = e^{\beta t}\mathbf{x}$ preserves the volumes and satisfies $\dot{\mathbf{X}} = e^{\beta t}\Phi(e^{-\beta t}\mathbf{X})$.

speed $\omega = \partial H/\partial \mathbf{P}$. The system is equivalent to a collection of clocks, and initial errors change the angular speed of the arms of the clocks [8]. Small Hamiltonian perturbations change the long time behavior[2].

Uniformly chaotic systems. A classical example of chaotic system is the gas modeled by hard spheres colliding elastically. The relaxation to the equilibrium state having a Gibbs measure with density $\rho(\mathbf{x}) = \rho_0\,\delta(H - E)$, is insured by the mixing condition $\lim_{t\to\infty}\mu(A \cap S_t B) = \mu(A)\,\mu(B)$. The basic ingredients for mixing are compactness of phase space and hyperbolicity namely a sensitive dependence to the initial conditions.

Intermediate systems. The behavior of a system far from integrability and not yet fully chaotic is complicated. A variety of topological structures are present and many space-time scales coexist. A 2D map shows how intricate the intermediate region is: closed orbits are intercalated by chaotic regions at every scale. This fractal like structure can be directly visualized or detected by using dynamical indicators such as the Lyapounov exponents, return times, reversibility error, frequency map.

2.2 Information and entropy

Entropy is a measure of the amount of information required to specify the state of a system.

$$S = k \log N \qquad\qquad N = \#\text{ of states}$$

and k fixes the scale. In statistical physics N is proportional to the phase space volume Γ namely $N \propto \Gamma(E, V)$. Given a gas of M molecules in a piston of height $L = 2^n\,\ell$, we have $V \propto 2^{n\,M}$ and choosing $k^{-1} = \log 2$ we have $S = n\,M + S_0$. The entropy S is the number of bits needed to specify the state of the gas. More generally the entropy is defined for a sequence of strings[3].

---

[2]When $H = H_0(\mathbf{p}) + \epsilon H_1(\mathbf{q},\mathbf{p})$ it is still possible to find a coordinate change such that the perturbation is exponentially reduced $\exp(-1/\epsilon^\gamma)$ where $\gamma \simeq (d+1)^{-1}$ for $\mathbf{x} \in \mathbb{R}^{2d}$ (H analytic and periodic in t). The normalized momenta $\mathbf{P}$ are still almost exactly conserved for a time $T(\epsilon) \sim e^{1/\epsilon^\gamma}$ implying a remarkable scaling law $T(\epsilon/n) = [T(\epsilon)]^{n^\gamma}$.

[3]Consider an attractor whose points are specified by $x = 0.b_1 \cdots b_n \cdots$ where the digits range from 0 and $q-1$. Assume that only 0 and $q-1$ have non zero probability $p_1 = p$ and $p_2 = 1-p$. The number of strings $b_1 \cdots b_n$ with the same number n_1, $n_2 = n - n_1$ of digits equal to 0 and $q-1$ respectively is given by $N = \binom{n}{n_1}$. Denoting with $w_1 = n_1/n$ and $w_2 = n_2/n$ the corresponding frequencies, the large n behaviour is $N \simeq \left(w_1^{w_1} w_2^{w_2}\right)^{-n}$. The number of occurrences of the most probable string $w_1 = p_1$, $w_2 = p_2$ is given by $N \simeq \left(p_1^{p_1} p_2^{p_2}\right)^{-n} = q^{nD_1} = e^{nh}$ where h is the Kolmogorov entropy and D_1 is the information dimension. The entropy of the string is $S = \log N = nh$. The overall probability of strings with frequencies w_1, w_2 is given by $P = N\,p_1^{nw_1}\,p_2^{nw_2}$ whose large n behaviour is given by $P = [(p_1/w_1)^{w_1}(p_2/w_2)^{w_2}]^n$ and in the limit $n \to \infty$ we have $P = 1$ if $w_1 = p_1$, $w_2 = p_2$. Any other string has zero probability, but for n finite their probability density is a gaussian in $w_1 - p_1$, $w_2 - p_2$ with mean square deviation $(p_1/n)^{1/2}$ and $(p_2/n)^{1/2}$.

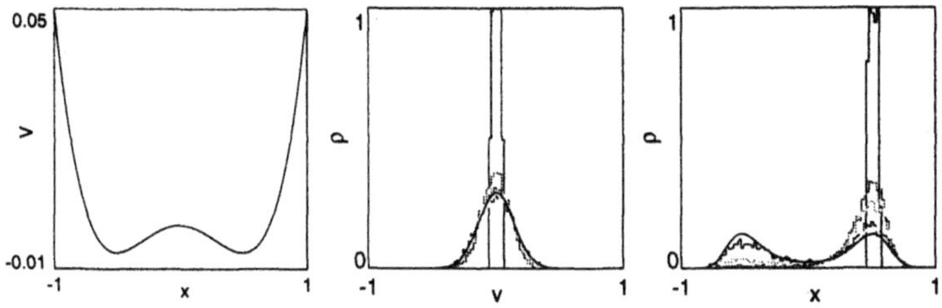

Figure 2. Equilibrium approach in a double well potential.

Information flow. A dynamical system evolves and the entropy is a measure of the information flow [5, 9]. The information produced at step n is the lost string. For a map $M(x) \bmod 1$ defined on the torus, letting $N(n)$ be the number of all distinct strings $[M^n(x)]$ for any initial x (by $[\,.\,]$ we denote the integer part), the flow rate is defined by

$$h = \lim_{n\to\infty} \frac{\log N(n)}{n}$$

For a chaotic map $M(x) = 2x$ we have $N(n) = 2^n$. For a regular map $M(x) = x + \omega$ we have $N(n) = n$ (from $M^{2^k}(0) = 2^k\omega$ setting $n = 2^k$). To summarize a regular map has a long forecast time, an information flow $\log n$ and zero entropy. A chaotic map has very short forecast time, linear information flow $n\,h$ and positive entropy $h > 0$.

2.3 Stochastic description

The chaotic motion of a particle (gas molecule) interacting with its environment, is phenomenologically described using a fluctuating force; this is much simpler than the dynamical description of a large number of molecules in a confining potential V. The Langevin equation for the one dimensional case reads

$$\frac{dx}{dt} = p \qquad \frac{dp}{dt} = -\beta p - V'(x) + \epsilon\xi(t)$$

where $\xi(t)$ is a *white* noise, $-\beta p$ a damping force. The phase space density satisfies the continuity equation, its average ρ the Fokker-Planck equation [10]

$$\frac{\partial \rho}{\partial t} + \frac{\partial}{\partial x}(p\rho) - \frac{\partial}{\partial p}[(\beta p + V'(x))\rho] = \frac{\epsilon^2}{2}\frac{\partial^2 \rho}{\partial p^2}$$

whose equilibrium solution $\rho = Z^{-1}\exp(-H/k_B T)$ is the Boltzmann distribution where $k_B T = \epsilon^2/(2\beta)$ and $H = p^2/2 + V(x)$. The tunneling effect due to noise allows to populate all the minima of V with relaxation times depending on β, see figure 2.

3. Complex systems

Complexity is a feature of living systems. A complex system can self organize, learn, reproduce itself and evolve. By evolution we mean random mutation with selection. Such a system is open and interacts with an environment, having exchange of matter, energy and information. As a consequence the order in the system can increase, the entropy decrease. A complex system has many different space-time scales. In a living system we have small molecules (amino-acid), large molecules (DNA, proteins), cells, organs, individuals, echo-systems whose sizes vary from 1 nm to 10^3 kM and whose time scales vary in a wide range. A complex system is itself a network of complex systems.
Models and simulations. In spite of the impressive achievements of molecular biology the reductionist program is far from being achieved: sectorial descriptions and modeling are needed. At the molecular level statistical analysis are used whereas at larger scales phenomenological models can be developed. Complex systems may be partially simulated by automata made of cells having physical properties and an information content.

3.1 The immune system

The IS has evolved since the appearance of life developing strategies adapted to the increasing complexity of the organisms [11]. In the vertebrates beyond the innate system we find a more sophisticated specific system, capable of learning from the antigenic experiences. The memory allows to react more promptly and efficiently to new attacks of previously experienced antigens (this is the reason of vaccines effectiveness). The system is based on various families of cells: B and T lymphocytes. The latter are split into helper and cytotoxic labeled by the surface markers, CD4+ and CD8+. Each sub-family has three states: virgin, effector and memory. In each state we find a large variety of cells related to the huge number of possible receptors (repertoire). The antigens are recognized by steric complementarity of surface molecules, and this process changes the virgin into memory or effector T cells. The effector cells attack the antigens and are short lived, the memory cells learn to recognize patterns and survive after proliferating during antigenic insults (clone expansion). The total number of T lymphocytes is basically constant. Aging corresponds to an increase of memory T cells, a decrease of virgin T cells and a squeezing of the repertoire.

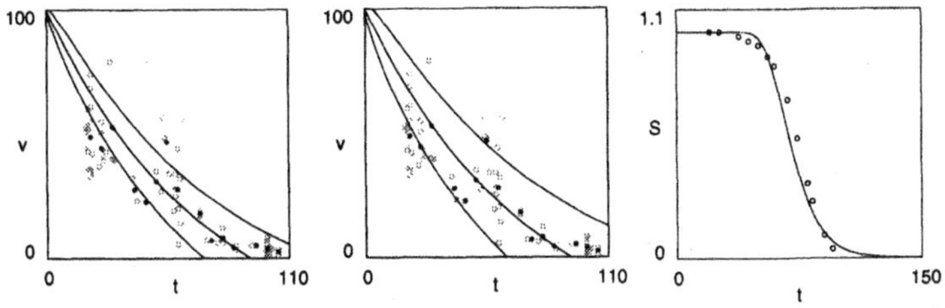

Figure 3. Average virgin T cells concentration $\langle v(t)\rangle$ and its deviation $\langle v(t)\rangle \pm 2\sigma(t)$ within two rms and comparison with experimental for two distinct noise amplitudes (left, center) and comparison of survival probability $S(t)$ with demographic data

Mesoscopic and macroscopic models of the IS have been proposed to describe the clonal expansion and the evolution of the antigen not experienced (ANE) and antigen experienced (AE) T lymphocytes along the whole life span [12]. Labeling by V and M the number of virgin (ANE) and memory plus effector (AE) T cells per unit volume, the simplest macroscopic model reads

$$\frac{dV}{dt} = -\alpha V - \beta M + \epsilon\xi(t) \qquad\qquad \frac{dM}{dt} = \alpha V + \beta M \tag{4}$$

where α is the conversion rate from ANE to AE compartment due to primary antigenic stimulus, β the increase of AE due to secondary restimulation. These factors take into account an average on the genetics and the average antigenic load whose fluctuations are described by a white noise $\xi(t)$. The thymic production, affecting the early stages of life, and the shrinkage of the total number of T lymphocytes are neglected. As a consequence the deterministic model has a conservation law, $V(t)+M(t) = V(0)+M(0)$. The concentrations $v = V/(V+M)$, $m = M/(V+M)$ satisfy the same deterministic equation and it has been shown that when the stochastic term is present the concentration v obtained from (4) is similar to the solution of

$$\frac{dv}{dt} = -\beta - (\alpha - \beta)v + \hat{\epsilon}\xi(t) \qquad\qquad \hat{\epsilon} = \frac{\epsilon}{V(0) + M(0)} \tag{5}$$

Assuming that the depletion of the virgin T cells compartment is a mortality marker we obtain a spread of the age at which this occurs with respect to t_* defined by $\langle v(t_*)\rangle = 0$, due to the fluctuations, which give to any individual its own immunological history. As a consequence one can compute a survival probability $S(t) = \frac{1}{2}\,\mathrm{Erfc}\,(\langle v(t)\rangle/\sqrt{2}\sigma(t))$ where $\sigma(t)$ is the variance of the process, see figure 3.

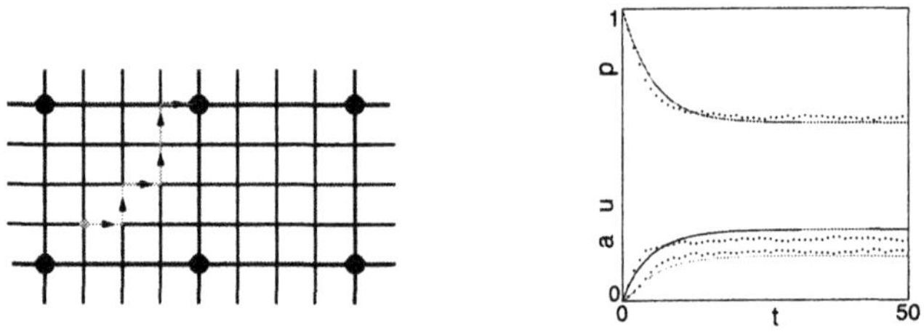

Figure 4. Pedestrians random walk (left). Comparison of the simulation and the mean free field (right) for pedestrians (p), train users (u), and station stoppers (a).

3.2 The city

The city is a frame (buildings, streets, public transport) where the citizens live and move. Analogies between the urban and the immune system can be found and similar modeling strategies can be adopted. We consider the citizens mobility in a network of streets and of bus or train lines [13]. We assume a lattice structure with N horizontal and vertical street, $n < N$ vertical and horizontal train lines, implying N^2 blocks and n^2 stations. The pedestrians mobility is a biased random motion, driven by some chronotopic attractors, areas where people converge for business, entertainment, shopping and other social activities. In the simplest approach, we might call *stupid mobility* the citizens move randomly taking the trains with assigned probabilities. The simulation of this random process is compared with mean field equations

$$p(t+1) = p(t) + P_{up}u - P_{pu}p \qquad u(t+1) = u(t) - P_{up}u + P_{pu}p$$

for the populations of pedestrians p and train users u, see figure 4. The mobility is defined by $\mu(t) = v_p\, p(t) + v_t\, u(t)$. At higher levels the complex structure of the city emerges. We may consider a *programmed mobility* providing each social category a fixed daily schedule: a string specifying the intensity of the attractors for any time interval. At the higher level the citizens record the daily mobility results and construct strategies to improve it so that an *intelligent mobility emerges.* At the top level the citizens interact and their number can vary; global objectives are fixed and the schedule of the citizens is determined accordingly. The city is no longer static but evolves according to some preselected objectives, can expand and eventually replicate, becoming a complex artificial life system.

4. Summary

The dynamical systems are the setting appropriate to describe physical systems, which exhibit regular and/or chaotic behavior. Complex systems have a physical self organizing hardware and a software with cognitive properties and self replication capabilities. Two examples, the immune system and the city, have been presented with a phenomenological description of their large scale behavior. At present an appropriate mathematical setting to treat complex systems does not exist and the only alternative to the phenomenological description is computer simulation. Having assumed that complexity is a distinctive feature of living systems, another basic question arises: how can a complex system emerge from a physical world governed by deterministic laws. The space-time invariance of physical laws and the randomness embedded in any non linear process are two basic ingredients. The emergence of coherent structures (macromolecules) is a starting point but their evolution to the self replicating stage is the result of a very long sequence of small random variations due to the nonlinear interaction with a rather stable environment. Some biologists claim that the appearance of life was extremely unlikely. On Earth the winning number came out from the lottery and this is a unique event in the Universe [14]. As a physicist I am more inclined to a Copernican rather than to a geocentric and mystic viewpoint. If the number of trials times the probability of an event is overwhelmingly close to 0 on the available time interval, then the event simply does not happen. unless you assume there are an infinity of parallel universes and the the event happened in our own, on the earth, because we are here to observe it. If this number is not very small so that we can imagine it came out from the box, then somewhere else in the universe, sometime this must have happened again. We know for sure since just one decade that most stars have planets and some of them, may be a very small fraction, will be earth like. Then the number of sites in the Universe where the lottery has taken place is huge. As a consequence why should we be alone and unique? Some presocratic philosophers came close to this thesis, which was openly defended at the price of life by Giordano Bruno[4]. The intellectual freedom, the strenuous defense of new ideas against the conservative attitude deeply rooted

[4]Giordano Bruno *La cena delle ceneri*. "Or questa distinzion di corpi ne la eterea reggione l'ha conosciuta Eraclito, Melisso,... Onde si vede che conobbero uno spacio infinito, selva infinita, capacità infinita di mondi innumerevoli simili a questo, i quali compiscono i loro circoli come la terra fa il suo, e però anticamente si chiamavano ethera cioè corridori, corrieri, ambasciatori della manificenza dell'unico Altissimo". In English: " This distinction of bodies in the ethereal region was known by Heraclitus, Melisso,... Hence we see how they knew an infinite space, an infinite region, an infinite forest, an infinite capacity of uncountable worlds

in any human (including the academic and scientific) community are the random seeds which allow the evolution of intelligent species.

References

[1] POINCARÉ H., *Methodes nouvelles de la mechanique celeste*, 1898; reprinted by D.L. Goroff, American Institute of Physics, New York 1993.

[2] BIRKHOFF G.D., *Dynamical systems*, American Institute of Physics, New York 1927.

[3] ARNOLD V.I., *Methodes Mathématiques de la Mécanique classique*, MIR, Moscou 1976.

[4] RUELLE D., *Hazard et chaos*, Odile Jacob, Paris 1991.

[5] BADII, POLITI A., *Complexity: Hierarchical Structures and Scaling in Physics*, Cambridge University Press, Cambridge 1997.

[6] KAUFMANN S., *The origin of order*, Oxford University Press, Oxford 1996.

[7] NICOLIS J.S., *Chaotic dynamics applied to biological information processing*, Akademie-Verlag, Berlin 1987.

[8] TURCHETTI G., *Meccanica classica dei sistemi fisici*, Zanichelli, Bologna 1998.

[9] BECK C., SCHLÖGEL F., *Thermodynamics of chaotic systems*, Cambridge University Press, Cambridge 1993.

[10] RISKEN H., *The Fokker-Planck equation*, Springer-Verlag, Berlin 1988.

[11] ABBAS A.K., *Basic Immunology*, Saunders, New York 2001.

[12] LUCIANI F. ET AL., VALENCIN S., VESCOVINI R., SANSONI P., FAGNONI F., FRANCESCHI C., BONAFÉ M., TURCHETTI G., *A stochastic model for CD^{*+} T cells dynamics in human immunosenescence: implications for survival and longevity*, Journal of Theoretical Biology **213** (2001), 587–597.

[13] BAZZANI A., GIORGINI B., SERVIZI G., TURCHETTI G., *A chronotopic model of mobility in urban spaces*, downloadable from http://www.physycom.unibo.it, to be published (2002).

[14] MONOD J., *Chance and necessity*, Vintage, New York 1971.

similar to this one, which describe their circles as Earth does. In the past they were called ethera namely runners, ambassadors, messengers of the magnificence of the unique Highest".

II

BIOLOGY

SHAPE AND SIZE IN BIOLOGY AND MEDICINE

Vincenzo Capasso
MIRIAM and Dipartimento di Matematica,
Università di Milano,
Via Saldini 50, 20133 MILANO - ITALY
vincenzo.capasso@mat.unimi.it

Abstract As D'Arcy Thompson pointed out in his pioneering book on *Growth and Form*, "THERE IS AN IMPORTANT RELATIONSHIP BETWEEN THE FORM OR SHAPE OF A BIOLOGICAL STRUCTURE AND HIS FUNCTION". Shape analysis deals with the statistical analysis of a family of "objects" in presence of stochastic fluctuations; stochastic geometry deals with the analysis of geometric aspects of "objects" subject to stochastic fluctuations. The scope of this presentation is to introduce relevant mathematical concepts and methods of shape analysis and of stochastic geometry, thus providing a guided tour in a selected bibliography. A relevant aspect of stochastic geometry is the analysis of the spatial structure of objects which are random in location and shape. In this case we may simply say that the mathematical interest is in spatial *occupation*. In various cases, described by specific examples (Birth-and-growth model; Boolean Model; a tumor growth model based on an inhomogeneous Boolean model), spatial occupation occurs via a random tessellation of the available space region. Hence a quantitative description of a random closed set can be obtained in terms of mean densities of volumes, surfaces, edges, vertices, etc., at the various Hausdorff dimensions.

1. Introduction

As D'Arcy Thompson pointed out in his pioneering book on "Growth and Form", [19]
"THERE IS AN IMPORTANT RELATIONSHIP BETWEEN THE FORM OR SHAPE OF A BIOLOGICAL STRUCTURE AND HIS FUNCTION".

The scope of this presentation is to introduce relevant nomenclature and mathematical tools for the analysis of shapes and forms, thus providing a guided tour in a selected bibliography.

First of all let us introduce the current terminology in this framework.

By *SHAPE* we mean the set of all geometrical informations of an object which are invariant under translations, rotations, and scaling.

By *FORM* we mean the set of all geometrical informations of an object which are invariant under translations, and rotations; i.e. $FORM = SHAPE + SIZE$.

The most important chapters of mathematical interest in this context are the following, for which we refer to the relevant literature.

MORPHOGENESIS (GENERATION OF PATTERNS) deals with the mathematical modelling of causal description of form (a direct problem) [20, 14].

IMAGE ANALYSIS deals with the filtering of images from noisy data (an inverse problem) [15].

SHAPE ANALYSIS deals with the statistical analysis of a family of "objects" in presence of stochastic fluctuations (a classification problem) [2, 9].

STOCHASTIC GEOMETRY deals with the analysis of geometric aspects of "objects" subject to stochastic fluctuations (direct and inverse problems) [18, 1].

2. Shape Analysis

In shape analysis the relevant problem is the geometrical "aspect" of an object.

2.1 Landmarks

For an obvious mathematical/statistical simplification it is much more convenient to reduce a *geometric object* belonging to an infinite dimensional space to an element of a finite dimensional space. In order to do this the concept of *landmark* has been introduced [2, 9] as a point in a finite dimensional space. In this way any element of a sample of objects is identified by its specific values of the landmarks.

Anatomical landmarks are chosen among points having a biological relevance (angle of an eye, a special point of the skull, etc.), while *mathematical* landmarks are chosen among points having a mathematical relevance (point of maximum curvature, etc.). Figure 1 shows an example of six mathematical landmarks (+) along the contour of a T2 vertebra of a mouse.

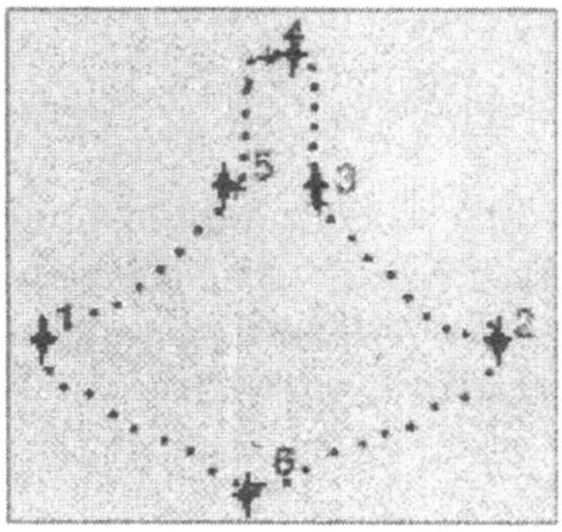

Figure 1. 6 mathematical landmarks (+) along the contour of a T2 vertebra of a mouse.

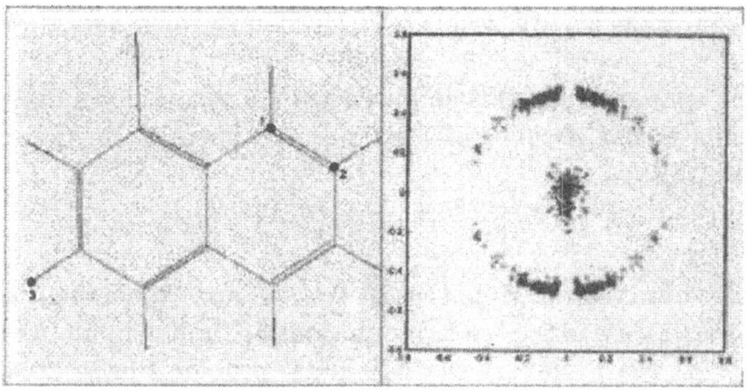

Figure 2. Left: 3 landmarks on a molecule of an Oxygenated compound, identified as the vertices of the triangle having maximum area with vertices 1 and 2 fixed and the third moving on the centers of the atoms. Right: clusters of 5 landmarks obtained from a sample of simulated crystals of a temperature-dependent crystallization of polymers: every crystal is identified by the 4 vertices and the center of the circumscribed rectangle.

A specific object is represented by an element of the space of *configurations* $(\mathbb{R}^m)^k$, if m is the dimension of the Euclidean space to which each landmark belongs. So that a sample of objects singled out of a population of similar objects is represented by a cluster of points in such km dimensional space of configurations (see Fig. 2).

In presence of randomness we have a *random configuration* which is a random vector

$$X = (\Omega, \mathcal{F}) \to \left((\mathbb{R}^m)^k, \mathcal{F}_{(\mathbb{R}^m)^k}\right)$$

where $(\Omega, \mathcal{F}, P)$ is a suitable underlying probability space.

Given a space of configurations Σ, two elements of Σ are $\sim$-equivalent if one of them is obtained from the other one by application of any (one or more) of the following transformations

- scaling $\beta \in \mathbb{R}_+$
- rotation $\Gamma \in SO(m)$
- translation $\gamma \in \mathbb{R}^m$

The *SPACE of SHAPES* is given by the quotient space $\widetilde{\Sigma} = \Sigma / \sim$.

Any element of $\widetilde{\Sigma}$ is called an *icon*.

The *SPACE of FORMS* is the quotient space $S\Sigma / \sim'$ where now $\sim'$ is the equivalence with respect to the family of all rotations and translations.

When the space of shapes is endowed with a metric d, we may define the *mean* or *expected element* μ of a random element X as any element of $\widetilde{\Sigma}$ such that

$$\mathbf{E}\left(d(X,\mu)^2\right) = \inf_{a \in \widetilde{\Sigma}} \mathbf{E}\left(d(X,a)^2\right).$$

Problems of statistical Hypotheses Testing may arise such as comparing the mean elements of two populations of similar random objects. Technically the most relevant mathematical problems arise due to the fact that the space of shapes is not an Euclidean space but a general manifold, so that reference to the tangent space and corresponding linearization around a typical shape (a *pole*) is needed when dealing with normal distribution of the stochastic fluctuations [17]. A further example is the dependence (nonlinear regression analysis) of the yield of bioconversions upon the shape-and-size of molecules of some oxygenated compounds [3].

2.2 Radius-Vector

A different approach for analyzing randomness for shape of planar objects is based on the statistical analysis of the *radius-vector* of a random object X, $\{r_X(\phi), \phi \in [0, 2\pi]\}$ (see Fig. 3). In this case the *mean element* is defined as

$$r(\phi) = \mathbf{E}[r_X(\phi)], \quad \phi \in [0, 2\pi],$$

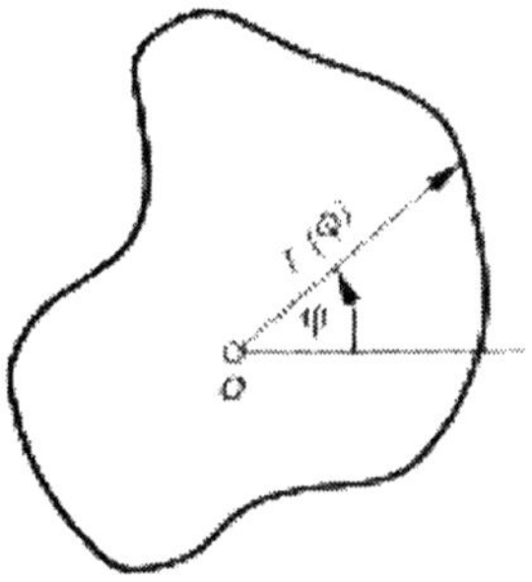

Figure 3. Radius-vector function.

and the *radius-vector variance* as

$$V_X(\phi) = \mathbf{E}\left[(r_X(\phi) - r(\phi))^2\right], \quad \phi \in [0, 2\pi].$$

An estimator of the mean radius-vector may be obtained based on a random sample of i.i.d. random objects $X_1, \ldots, X_n$ all having the same distribution of X. The sample mean object $\overline{X}$ is the one having radius-vector function $\overline{r}$ given by

$$\overline{r}(\phi) = \frac{1}{n}\sum_{j=1}^{n} r_{X_j}(\phi), \quad \phi \in [0, 2\pi].$$

Correspondingly an estimator of the radius-vector variance of X is given by

$$\widehat{V_X(\phi)} = \frac{1}{n}\sum_{j=1}^{n}(r_{X_j}(\phi) - \overline{r}(\phi))^2, \quad \phi \in [0, 2\pi].$$

If, for any fixed ϕ, the real random variables $r_{X_j}(\phi)$, $j = 1, \ldots, n$ make an i.i.d. sample for $r_X(\phi)$, we may apply the Central Limit Theorem to compute a confidence region for $\mathbf{E}[r_X(\phi)]$ as follows

$$\left[\overline{r}(\phi) - k_\alpha\sqrt{\frac{\widehat{V_X(\phi)}}{n}}, \overline{r}(\phi) + k_\alpha\sqrt{\frac{\widehat{V_X(\phi)}}{n}}\right], \quad \phi \in [0, 2\pi],$$

where k_α is determined once confidence level $1 - \alpha \in [0, 1]$ is given (see Fig. 4).

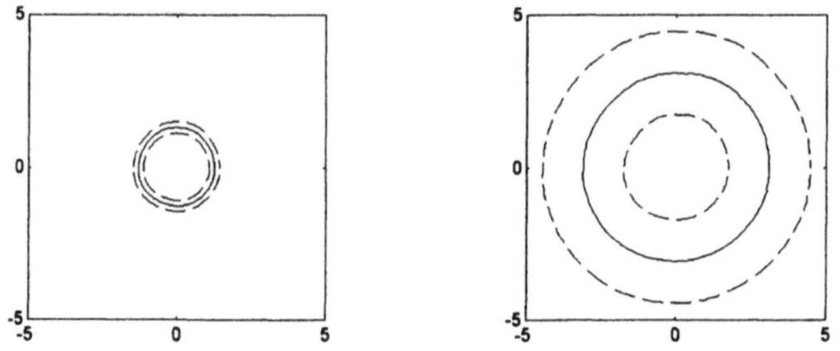

Figure 4. Examples of means and confidence regions for radius-vector functions.

Other methods for the analysis of the radius-vector function have been proposed based on Fourier series expansions [8].

3. Stochastic geometry

A relevant aspect of stochastic geometry is the analysis of the spatial structure of objects which are random in location and shape. In this case we may simply say that the mathematical interest is in spatial *occupation*, so that geometric measure theory is involved in presence of stochastic fluctuations.

Given a random object $\Sigma \in \mathbb{R}^d$, a first quantity of interest is for example the probability that a point x belongs to Σ, or else the probability that a compact set K intersects Σ.

In presence of stochastic fluctuations, we need to refer to an underlying probability space $(\Omega, \mathcal{A}, P)$. The theory of Choquet-Matheron [12], shows that it is possible to assign a unique probability law associated with a *RACS* (*random closed set*) $\Sigma \in \mathbb{R}^d$ on the measurable space $(\mathcal{F}, \sigma_{\mathcal{F}})$ of the family of closed sets in $\mathbb{R}^d$ endowed with the σ-algebra generated by the hit-or-miss topology, by assigning its *hitting function* T_Σ.

3.1 Hit or miss topology

Let E be a locally compact, separable, Hausdorff space (i.e. $\forall x \in E$ an open neighbourhood of x exists which includes a compact neighbourhood of x; $\forall x, y \in E$ two open and disjoint neighbourhoods A and B exist such that $x \in A$, $y \in B$; the topology on E admits a countable basis).

We will always assume $E \subseteq \mathbb{R}^d$, $d = 1, 2, 3$.

Let

$\mathcal{F}$ = the family of the closed subsets of E

$\mathcal{G}$ = the family of the open subsets of E
$\mathcal{K}$ = the family of the compact subsets of E.
$\forall A \subset E$ let us denote by

$$\begin{aligned}\mathcal{F}_A &= \{F \in \mathcal{F}: F \cap A \neq \emptyset\} \\ \mathcal{F}^A &= \{F \in \mathcal{F}: F \cap A = \emptyset\}\end{aligned}$$

For any compact $K \in \mathcal{K}$, for all $n \in \mathbb{N}$ and for any n-tuple $G_1, \ldots G_n$ of open sets, define

$$\mathcal{F}^{\mathcal{K}}_{G_1,\ldots,G_n} = \mathcal{F}^K \cap \mathcal{F}_{G_1} \cap \cdots \cap \mathcal{F}_{G_n}$$

The family $\{\mathcal{F}^{\mathcal{K}}_{G_1,\ldots,G_n}\}$ is a basis for a topology in E called *hit-or-miss topology*. By $\sigma_{\mathcal{F}}$ we shall denote the σ-algebra generated by the hit-or-miss topology.

3.2 Random Closed Sets (RACS's)

A RACS Σ is a random object

$$\Sigma : (\Omega, \mathcal{A}, P) \longrightarrow (\mathcal{F}, \sigma_{\mathcal{F}}) .$$

Let us denote by $\mathcal{K}$ the family of compact sets in $\mathbb{R}^d$.
The hitting function of Σ is defined as

$$T_\Sigma : K \in \mathcal{K} \longmapsto P(\Sigma \cap K \neq \emptyset) .$$

Actually we may consider the restriction of T_Σ to the family of closed balls $\{B_\varepsilon(x);\ x \in \mathbb{R}^d,\ \varepsilon \in \mathbb{R}_+ - \{0\}\}$.

3.2.1 Example: Homogeneous Boolean Model [18, 1]. Let $N = \{X_1, X_2, \ldots, X_n, \ldots\}$ be an homogeneous spatial Poisson process in $\mathbb{R}^2$, with intensity $\lambda > 0$:

$$\forall A \in \mathcal{B}_{\mathbb{R}^2} : P[N(A) = n] = \exp(-\lambda\nu^2(A))\frac{\lambda\nu^2(A)}{n!}, \quad n \in \mathbb{N}.$$

Let $\{\Sigma_1, \Sigma_2, \ldots, \Sigma_n, \ldots\}$ be a sequence of i.i.d. RACS all having the same distribution as a *primary grain* Σ_0, (e.g. a ball with a random radius R_0)

$$\Sigma := \bigcup_n \{X_n \oplus \Sigma_n\} .$$

In this case the hitting function is given by

$$T_\Sigma(K) = 1 - \exp\left\{-\lambda \mathbf{E}\left[\nu^2\left(\Sigma_0 \oplus \check{K}\right)\right]\right\}, \quad K \in \mathcal{K},$$

where

$$\check{K} = \{-x \mid x \in K\},$$

and

$$A \oplus B = \{x + y \mid x \in A, \; y \in B\}.$$

3.2.2 Example: Birth–and–growth model; an inhomogeneous Boolean Model [4, 6]. Consider the marked point process (MPP) N defined as a random measure given by

$$N = \sum_{n=1}^{\infty} \epsilon_{T_n, X_n}$$

where

- T_n is an $\mathbb{R}_+$-valued random variable representing the time of birth of the n-th nucleus,
- X_n is an E-valued random variable representing the spatial location of the nucleus born at time T_n,
- $\epsilon_{t,x}$ is the Dirac measure on $\mathcal{B}_{\mathbb{R}_+} \times \mathcal{E}$ such that for any $t_1 < t_2$ and $B \in \mathcal{E}$,

$$\epsilon_{t,x}([t_1, t_2] \times B) = \begin{cases} 1 & \text{if } t \in [t_1, t_2], \; x \in B, \\ 0 & \text{otherwise}. \end{cases}$$

We have

$$N(A \times B) = \sharp\{T_n \in A, \; X_n \in B\}, \quad A \in \mathcal{B}_{\mathbb{R}_+}, \; B \in \mathcal{E}$$

is the (random) number of nuclei born during A, in the region B.

Let $\Theta^t_{T_n}(X_n)$ be the RACS obtained as the evolution up to time $t > T_n$ of the nucleus born at time T_n in X_n, according to some growth model. The *germ-grain* model is given by

$$\Theta^t = \bigcup_{T_n < t} \left[\Theta^t_{T_n}(X_n) \oplus X_n\right].$$

In this case, if the nucleation rate in the free space is $\lambda(t)$, $t \in \mathbb{R}_+$, let S be the r.v. in $\mathbb{R}_+$ with probability density

$$f(s) = \frac{\lambda(s)}{\lambda(0,t)} I_{(0,t)}(s), \quad s \in \mathbb{R}_+,$$

with $\lambda(0,t) = \int_0^t \lambda(s)\, ds$ and let $\Theta^t_S = \Theta^t_{\bullet} \circ S$. The hitting function of Θ^t is now given by

$$T_{\Theta^t}(K) = 1 - \exp\left\{-\lambda(0,t)\mathbf{E}_S\left[\nu^d\left(\Theta^t_S \oplus \check{K}\right)\right]\right\}, \quad K \in \mathcal{K}.$$

3.2.3 Example: a tumor growth model based on an inhomogeneous Boolean model [8]. A model for tumor growth has been proposed in [8] based on a discrete time birth-and-growth model in $\mathbb{R}^d$, $d = 2, 3$. Given a RACS X_i representing the region occupied by the tumor mass at time $i \in \mathbb{N}$, a spatial point process $\Phi_{i+1} = \{s_k\}_{k \in \mathbb{N}}$ is generated with an intensity

$$\lambda_{i+1}(s) = \begin{cases} \lambda & \text{if } s \in X_i \\ 0 & \text{if } s \notin X_i \end{cases}$$

where λ is a given positive real number.

The region occupied by the tumor mass at time $i + 1$ is modelled as a Boolean model

$$X_{i+1} = \bigcup_k \{Z_{i+1}(s_k) \mid s_k \in \Phi_{i+1} \cap X_i\}$$

where the $Z_{i+1}(s_k)$ are i.i.d. balls (disks) with a random radius R_{i+1} centered at s_k (see Fig. 5).

The hitting function of the model is now

$$T_{X_{i+1}}(K) = 1 - \exp\left\{-\lambda \mathbf{E}\left[\nu^d((Z_{i+1} \oplus K) \cap X_i)\right]\right\}, \quad K \in \mathcal{K},\ i \in \mathbb{N}.$$

Based on this model, estimation procedure have been proposed for the estimation of the relevant parameters of birth λ and growth R_i [8].

3.3 Stochastic geometric measures

Consider the measure spaces $(\mathbb{R}^d, \mathcal{B}_{\mathbb{R}^d}, \nu^d)$, and $(\mathcal{F}, \sigma_\mathcal{F}, P_\Xi)$ where $\mathcal{F}$ is the family of closed subsets of $\mathbb{R}^d$, $\sigma_\mathcal{F}$ is the σ-algebra generated in $\mathcal{F}$ by the hit-or-miss topology [12], and P_Ξ is the probability measure induced by a RACS Ξ on $(\mathcal{F}, \sigma_\mathcal{F})$. Correspondingly we shall denote by E_Ξ expected values computed with respect to this law.

A quantitative description of the RACS Ξ can be obtained in terms of mean densities of volumes, surfaces, edges, vertices, etc., at the various Hausdorff dimensions, in the following way.

Definition 1. *We call* **volume measure** *of the random closed set Ξ the measure V_Ξ defined on $(\mathbb{R}^d, \mathcal{B}_{\mathbb{R}^d})$ by*

$$V_\Xi(B) := \mathbf{E}_\Xi\left[\nu^d(\Xi \cap B)\right], \qquad B \in \mathcal{B}_{\mathbb{R}^d} \tag{1}$$

where ν^d denotes the d-dimensional Lebesgue measure.

Since obviously $V_\Xi(\cdot) \ll \nu^d(\cdot)$, by the Radon-Nikodym Theorem V_Ξ has a density $V_V(x) \in L^1(\nu^d)$ such that for all Borel sets B,

$$V_\Xi(B) = \int_B V_V(x)\nu^d(dx)\,. \tag{2}$$

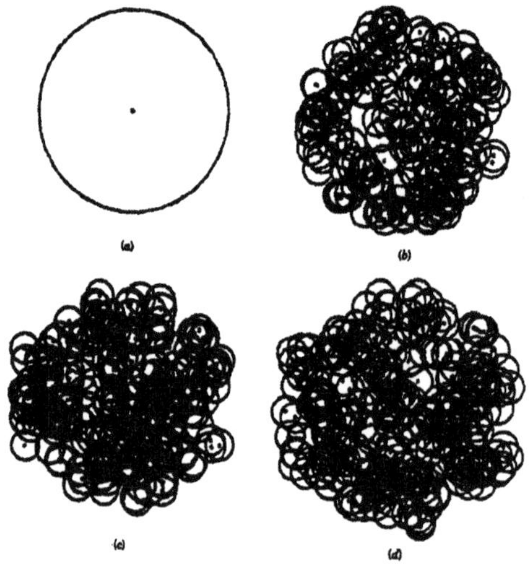

Figure 5. Simulation of the tumor growth via a Boolean model.

Definition 2. *The Radon-Nikodym derivative* $V_V(x) = \dfrac{dV_\Xi}{d\nu^d}(x)$ *is called* **mean local volume density** *or* **mean volume fraction** *of the region occupied by* Ξ *(see Fig. 7, 8).*

In a similar way we may define also the surface measure of a RACS Ξ:

Definition 3. *Let* Ξ *be a RACS having a boundary of Hausdorff dimension equal to* $d-1$*. Then we define the surface measure of* Ξ *on* $(\mathbb{R}^d, \mathcal{B}_{\mathbb{R}^d})$ *by*

$$S_\Xi(B) = \mathbf{E}_\Xi\left[\nu^{d-1}(\partial\Xi \cap B)\right], \qquad B \in \mathcal{B}_{\mathbb{R}^d}. \tag{3}$$

Suppose that (this is not always true, see e.g. [5] for any Borel set B with $\nu^d(B) = 0$ we have

$$S_\Xi(B) = \mathbf{E}_\Xi\left[\nu^{d-1}(\partial\Xi \cap B)\right] = 0,$$

so that $S_\Xi \ll \nu^d$; again by the Radon-Nikodym Theorem a density $S_V \in L^1(\nu^d)$ (Radon-Nikodym derivative) exists such that, for any Borel set B,

$$S_\Xi(B) = \int_B S_V(x)\nu^d(dx).$$

Definition 4. *Under the above assumptions, the Radon-Nikodym derivative*

$$S_V(x) = \frac{dS_\Xi}{d\nu^d}(x)$$

is called **mean local surface density** *of the RACS Ξ (see Fig. 7, 8).*

The question is raised on how to compute these densities. The following holds [16]

Proposition 1. *Let μ be a measure defined on $\mathbb{R}^d$, absolutely continuous with respect to the Lebesgue measure ν^d. The density of μ with respect to ν^d is given by*

$$\frac{d\mu}{d\nu^d}(x) = \lim_{r\to 0} \frac{\mu(B(x,r))}{\nu^d(B(x,r))}$$

(where $B(x,r)$ is a ball centered at x of radius r) provided that the limit exists and is ν^d-a.e. finite.

As a consequence for our densities we have

Proposition 2. *Let Ξ be a d-dimensional RACS in $\mathbb{R}^d$, having boundary of Hausdorff dimension $d-1$. The* **mean local volume density** *and* **mean local surface density**, *respectively, of the RACS Ξ at point $x \in \mathbb{R}^d$ are given by*

$$\begin{aligned} V_V(x) &= \lim_{r\to 0} \frac{\mathbf{E}_\Xi\left[\nu^d(\Xi \cap B(x,r))\right]}{\nu^d(B(x,r))} \\ S_V(x) &= \lim_{r\to 0} \frac{\mathbf{E}_\Xi\left[\nu^{d-1}(\partial\Xi \cap B(x,r))\right]}{\nu^d(B(x,r))}, \end{aligned}$$

provided that the limits exist and are a.e. finite.

It can be shown [11, 5] that

$$V_V(x) = P(x \in \Xi).$$

3.4 Stochastic tessellations

In the birth-and-growth process described above we may include an additional feature known as impingement, by assuming that at points of contact of their growth front grains stop growing. In this case the spatial region in $\mathbb{R}^d$ in which the process occurs is divided into cells (random Johnson-Mehl tessellation [10, 13]), and interfaces (n-facets, $n = 0, 1, 2, \ldots, d$) at different Hausdorff dimensions (cells, faces, edges,

Figure 6. **A copper-tungsten alloy.**

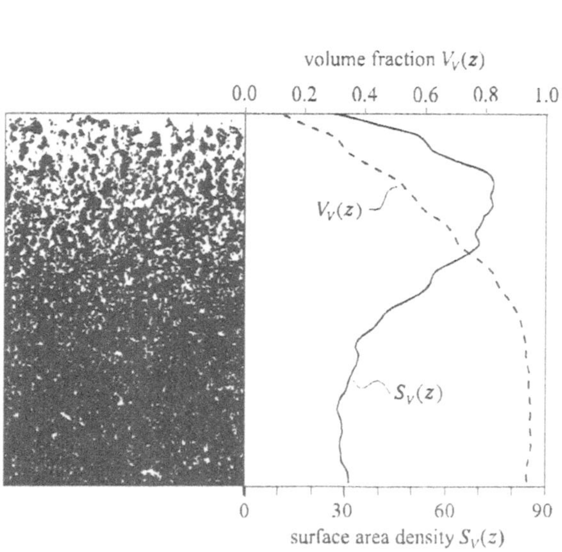

Figure 7. **An estimate of S_V and V_V on the previous picture.**

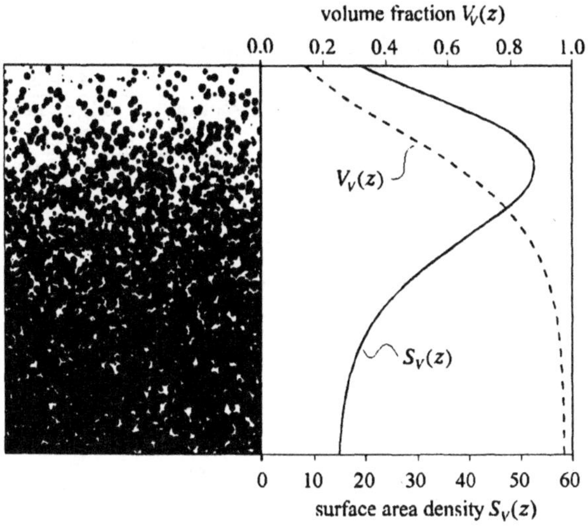

Figure 8. A Boolean model approximating the real process of figure 7.

vertices) appear. As above, we may describe quantitatively the tessellation by means of mean densities of the n-facets with respect to the d-dimensional Lebesgue measure [7].

By referring to the example in Section 3.2.2 we introduce the following definitions [7].

Let us denote by $\tau_i(y)$ the random time at which a point $y \in E$ is reached by a grain freely grown (disregarding impingement) from a germ $a_i = (X_i, T_i)$, where X_i is its random space location and T_i its random time of birth. We admit $\tau_i(y) = \infty$, which corresponds to the case in which point y is never covered by the i-th grain.

Definition 5. *The* **cell** $C_i(t)$ *of the (incomplete) tessellation, generated by the nucleus* $a_i = (X_i, T_i)$*, at the time of observation* t *is the non-empty set*

$$C_i(t) = \{y \in E \mid \tau_i(y) \leq t \text{ and } \tau_i(y) \leq \tau_j(y)\ \forall j \neq i\}.$$

We will denote by $C_e(t)$ the **empty region** at time t, i.e.

$$C_e(t) := \{y \in E \mid \tau_i(y) > t,\ \forall i \in \mathbb{N}\}.$$

Let us now introduce a rigorous concept of "interface" at different Hausdorff dimensions.

	$d = 1$	$d = 2$	$d = 3$
$n = 0$	**vertex**=intersection of 2 cells	**vertex**=intersection of 3 cells	**vertex**=intersection of 4 cells
$n = 1$	**cell**	**edge**=intersection of 2 cells	**edge**=intersection of 3 cells
$n = 2$		**cell**	**face**=intersection of 2 cells
$n = 3$			**cell**

Table 1. Specific meaning of "n-facets" in spaces of dimension d.

Definition 6. *An* **n-facet** *at time t ($0 \leq n \leq d$) is the non-empty intersection between $m + 1$ cells, i.e.*

$$F_n(t, k_0, \ldots, k_m) = C_{k_0}(t) \cap C_{k_1}(t) \cap \cdots \cap C_{k_m}(t)\,,$$

with $m = d - n$ and $k_0, \ldots, k_m \in \mathbb{N}$.

Note that in the previous definition

- $d =$ dimension of the space in which the tessellation takes place
- $n =$ Hausdorff dimension of the interface under consideration
- $m + 1 =$ number of cells that form such an interface

(see also Table 1, and Fig. 9).

Consider now the set of all n-facets at time t

$$\Xi_n(t) = \bigcup_{k_0,\ldots,k_m \in \mathbb{N}}^{\neq} F_n(t, k_0, \ldots, k_m) \quad \text{for } n = 0, \ldots, d\,;$$

the symbol $\neq$ denotes that the indices $k_0, \ldots, k_m$ are distinct (and so also the corresponding cells).

For any Borel set B in $\mathbb{R}^d$ one can define the *mean n-facet content* of B at time t as the measure

$$\mathcal{M}_{d,n}(t, B) = E\left[\lambda_n(B \cap \Xi_n(t))\right] \tag{4}$$

where λ_n is the n-dimensional Hausdorff measure (coinciding with the n-dimensional Lebesgue measure ν_n for integer and positive values of n). Note that, with the previous definitions, $\Xi_d(t) \equiv \Theta^t$, so that $\mathcal{M}_{d,d}(t, B)$ is the d-dimensional volume of the portion of the set B occupied by cells at time t.

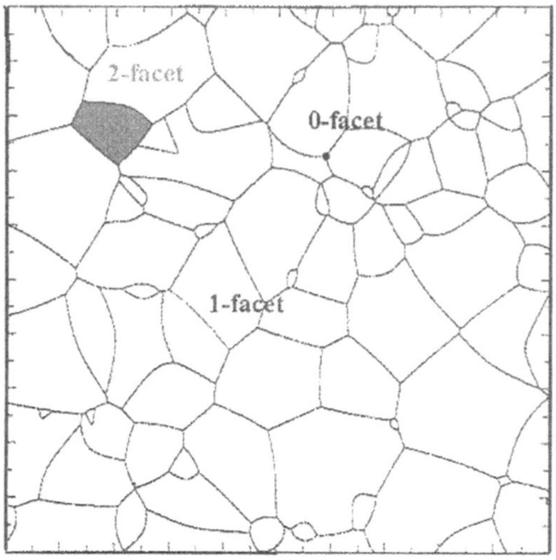

Figure 9. n-facets for a tessellation in $\mathbb{R}^2$.

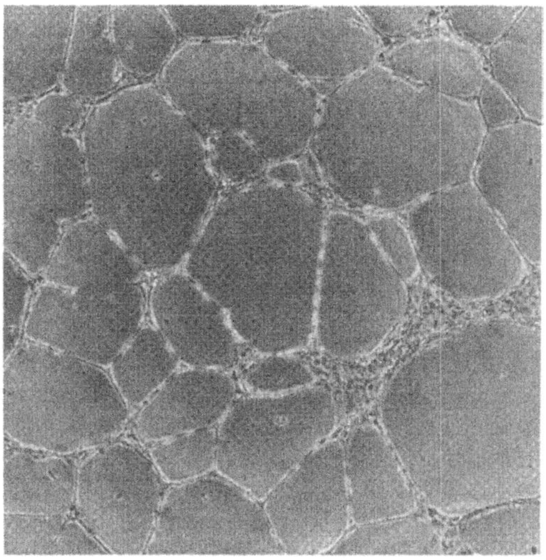

Figure 10. A real picture showing a spatial tessellation due to vascularization of a biological tissue: endothelial cells form a vessel network.

If the probability distribution of the germs a_i which generate the tessellation is non-atomic, then $\mathcal{M}_{d,n}(t,\cdot) \ll \nu_d$ where ν_d is the d-dimensional Lebesgue measure [5]. In this case a density $\mu_{d,n}(t,x)$ exists such that for all Borel sets B in $\mathbb{R}^d$

$$\mathcal{M}_{d,n}(t,B) = \int_B \mu_{d,n}(t,x)\,dx\,. \tag{5}$$

Definition 7. *The function $\mu_{d,n}(t,x)$ defined by (5) is called local mean n-facet density of the (incomplete) tessellation at time t.*

In particular $\mu_{d,d}(t,x)$ is the mean local volume density of the occupied region at time t, and $\mu_{d,d-1}(t,x)$ is the surface density of the cells.

Acknowledgments

It is a pleasure to acknowledge the important contribution of A. Micheletti in the development of joint research projects relevant for this presentation.

References

[1] BARNDORFF-NIELSEN O.E., KENDALL W.S., VAN LIESHOUT M.N.M. EDS., *Stochastic Geometry. Likelihood and Computation*, Chapman & Hall-CRC, Boca Raton 1999.

[2] BOOKSTEIN F.L., *The Measurement of Biological Shape and Shape Change*, Lecture Notes Biomath **24**, Springer-Verlag, Heidelberg 1978.

[3] BRAGHIERI E., *Teoria Matematica della Shape Analysis* (in Italian), Thesis, Mathematics, University of Milano 1999-2000.

[4] CAPASSO V. ED., *Mathematical Modelling for Polymer Processing. Polymerization, Crystallization, Manufacturing*, Springer-Verlag, Heidelberg, in press.

[5] CAPASSO V., MICHELETTI A., *Local spherical contact distribution function and local mean densities for inhomogeneous random sets*, Stochastics and Stoch. Rep. **71** (2000), 51–67.

[6] CAPASSO V., MICHELETTI A., *Stochastic geometry of spatially structured birth-and-growth processes. Application to crystallization processes*, in: "Spatial Stochastic Processes", E. Merzbach ed., Lecture Notes in Mathematics - CIME Subseries, Springer-Verlag, Heidelberg, in press.

[7] CAPASSO V., MICHELETTI A., BURGER M., *Densities of n-facets of incomplete Johnson-Mehl tessellations generated by inhomogeneous birth-and-growth processes*, preprint 2001.

[8] CRESSIE N., *Statistics for Spatial Data*, Wiley, New York 1993.

[9] DRYDEN I.L., MARDIA K.V., *Statistical Shape Analysis*, Wiley, New York 1998.

[10] JOHNSON W.A., MEHL R.F., *Reaction Kinetics in processes of nucleation and growth*, Trans. A.I.M.M.E. **135** (1939), 416–458.

[11] KOLMOGOROV A.N., *Foundations of the Theory of Probability*, Chelsea Pub. Co., New York 1956.

[12] MATHERON G., *Random Sets and Integral Geometry*, Wiley, New York 1975.

[13] MØLLER J., *Random Johnson-Mehl tessellations*, Adv. Appl. Prob. **24** (1992), 814–844.

[14] MURRAY J.D., *Mathematical Biology*, Springer-Verlag, Heidelberg 1989.

[15] SERRA J., *Image Analysis and Mathematical Morphology*, Academic Press, London 1982.

[16] SHILOV G.E., GUREVICH B.L., *Integral, Measure and Derivative: a Unified Approach*, Dover Pub., New York 1977.

[17] SMALL C.G., *The Statistical Theory of Shapes*, Springer, New York 1996.

[18] STOYAN D., KENDALL W.S., MECKE J., *Stochastic Geometry and its Application*, John Wiley & Sons, New York 1995.

[19] THOMPSON D.W., *On Growth and Form (1917)*, Cambridge University Press, Cambridge 1970.

[20] TURING A.M., *The mechanical basis of morphogenesis*, Phil. Trans. Roy. Soc. Lond. **237** (1952), 37–72.

ASSESSMENT OF THE QUALITY OF WATERS AND THE ENVIRONMENT BY MEANS OF BIOLOGICAL INDICES (MACROINVERTEBRATES, DIATOMS, FLUVIAL FUNCTIONALITY): THE RIVER FRIGIDO

Nicola Ceccopieri
Dipartimento di etologia, ecologia ed evoluzione,
Via A. Volta 4, 56126 PISA - ITALY

Rosalba Banchetti
Dipartimento di etologia, ecologia ed evoluzione,
Via A. Volta 4, 56126 PISA - ITALY
rbanchetti@deee.unipi.it

Abstract During spring-summer 2000, a monitoring campaign with Extended Biotic Index (IBE) and two with Diatomic Eutrophication-Pollution Index (EPI-D) were carried out on nine stations of the Frigido river. In autumn 2000, the survey of Index of Fluvial Functionality (IFF), published recently in 2000, was carried out along the whole flow of the river. The results of IBE show a biological fragility, connected with the oligotrophy of the waters, in the spring area, a good quality in the tributaries, a slight deterioration (I-II and II Class of Quality, C.Q.) in the middle reach of the river and a sensible worse situation (III and IV C.Q.) in the flat reach. The comparison with the campaigns carried out between 1980-2000 shows that the reclamation of the chemical-hygienical quality of the waters (got principally by the interruption of the waste waters coming from marble sawing) is a necessary but not sufficient condition to recover the biological quality and the colonies of macro-invertebrates. As the negative effects of the waste waters continued (even years after their interruption) in 1997 the bed of the Frigido was reclaimed, removing about ten thousand tons of marble dust. The IBE campaign in 2000 put in evidence a remarkable reclamation of the biological quality, so confirming also the peculiar ways of environmental impact of sawing waste waters and marble dust, explained not only by the deterioration of waters but also by the destruction of the varieties of micro-habitats in the substratum of the river. The index EPI-D seems to point out a lower sensitivity of IBE to the intermediate levels of pollution and a quite superior sensitivity as regards the two extreme situations, of extremely clear or polluted waters. The index IFF points out a marked critical state in the area in Massa and in the flat reach, connected principally with the compromise of the strips of vegetation on the banks, due to the works to put the same banks in a steady state. From the elements of the environmental diagnosis drawn from the survey, suggestions for environmental reclamation and restoration have been developed.

1. Introduction and methods

In the years of post-war reconstruction and the economic miracle, the protection of the environment and the landscape remained beyond the cultural horizons of most sections of society and the environmental consequences were an unprecedented urbanisation of the coasts and alluvial plains and the pollution of waters, of the air and of the ground.

The extent of the environmental damage led to the enactment of the Merli law, which centred on the protection of waters from pollution. This law regulated the activity of pollution monitoring for more than twenty years. The main limitation of the Merli law was that it took into consideration only one part of the environment: water. This was because the law's objectives were essentially utilitarian in nature, rather than considering the characteristics of the ecosystem, i.e. a cleaning of the rivers aimed not so much at the restoration of their ecological functionality, but rather at guaranteeing the availability of a resource of adequate quality for human use (productive, energetic, for irrigation, for drinking).

The cultural leap occurred with the introduction of the Extended Biotic Index [8], based on macroinvertebrates, which made it possible to analyse water not only in relation to human use, but above all in relation to the organisms that naturally inhabit that environment.

The Extended Biotic Index (IBE) is based on the study of the communities of bentonic macroinvertebrates (*Insecta, Crustacea, Mollusca, Oligochaeta, Hirudinaea, Platelminta*) composed of organisms captured with a handled fishing net. These organisms, because of their varying sensitivity to pollution, represent an extremely sophisticated network of informants. An occasional toxic discharge might elude chemical monitoring, but will leave a clear scar in the community of organisms which, being permanently immersed in these waters and present in every square metre of the river, constitute the most sophisticated monitoring network.

The study of the structure of the community makes it possible to express a judgement regarding the quality of the river environment.

The Diatomic Eutrophication-Pollution Index [2] is based on the study of the communities of algal Protists, particularly the Diatoms (or *Bacillariophyceae*), collected by scraping off the patina of periphiton present on the surface of completely submerged stones. The theoretical basis of the Diatomic Eutrophication-Pollution Index (EPI-D) relies on the outstanding sensitivity of many diatoms to eutrophication and pollution. Through microscopic analysis it is possible to formulate a judgement regarding the quality of the water [2].

Attention given to the wet river-bed (rather than to the water only) soon broadened its horizons. Thus the close functional correlations between the river and the immediate surrounding territory were identified and the central importance of the bands of riparian vegetation was recognised. From examination of the single drop of water, research moved on to consider the entire river environment.

The particularly appropriate technique to satisfy these new requirements, is represented by the Index of Fluvial Functionality (IFF). Centred on the observation of morphological and biological characteristics of ecological importance (e.g. extension and continuity of the bands of riparian vegetation, scrapings, holes, meanders, substrate, periphytic and macrophytic cover, organic detritus, macrobentonic communities), the IFF provides not only synthetic assessments of fluvial functionality and information regarding the causes of its deterioration, but also precise indications for planning clean-up operations and evaluating their general effectiveness. The application of the IFF [1] requires the filling in of a record for each homogeneous section of the river surveyed, proceeding from the mouth to the source. The description of the morphological, structural and biotic parameters of the ecosystem, interpreted in the light of the principles of fluvial ecology, makes it possible to formulate a judgement of fluvial functionality.

2. Results and discussion

The study was carried out on the River Frigido, Massa Carrara (Tuscany), one of the main waterways that flow down from the Apuan Alps. For the biological surveys (EPI-D and IBE), analyses from 9 sampling stations were used, two from the mouths of the main tributaries (Torrente Renara and Fosso di Antona) and the remaining ones from along the main river course, positioned in such a way as to absorb the impact of settlements and their discharges. For the IFF the whole course of the river from the mouth to the source was covered on foot.

The results of the biological mapping with the macroinvertebrates show, for the first time after twenty years, a significant improvement. This is attributable above all to the removal from the river bed of the deposits of marble accumulated over the years (Fig. 1). Although the discharges of the stone-extracting companies have ceased since the early 1990's, and there has been a clear improvement in the quality of the waters (finally limpid once again), the biological quality remained very low.

This confirms that the mechanisms of the polluting action of marble dust, as hypothesised in the early 1980's, was not due so much to the turbidity of the waters as to the occlusion by the dust of the interstices between the pebbles, and resulting destruction of the variety of microenvironments. For this reason the need to remove the deposits of marble from the bed of the Frigido was recognised. Therefore today, after long years of commitment to halting the discharges of marble dust and after its removal from the river bed (about 60,000 tons in the period 1997-98), a clear improvement in the quality of the river can be observed.

The diatomic index also shows an acceptable situation in most of the Frigido, with the exception of its terminal section, still heavily compromised by the discharge, of low quality, of the Querce depurator (Fig. 2).

It should be remembered, however, that the fluvial ecosystem is not limited solely to the flowing water or to the wet river bed: the bands of riparian vegetation and the interactions with the surrounding territory are also an integral part of it. Awareness of this, combined with the obvious risks deriving from waterway regulation works, has led to the application of the IFF. This is essentially a carefully developed census of the environmental characteristics of ecological importance.

The results, summarised in figure 3, show considerable damage to the Frigido from Massa to the mouth. The restorative proposals illustrated schematically in figure 4 are derived from this analysis: demolition of walls in concrete, widening of the river bed, renaturalisation of the river banks, delocalisation of local firms, creation of a river reserve with naturalistic and recreational functions, diversification of the substrate, restoration of the vegetation, radical improvement of the Querce depurator and pumping of its discharge as far as Massa (to impede the summer drying up of the Frigido and to exploit its self-depurating capacity).

3. Conclusions

In the 1990's the IBE showed that the restoration of the chemical and hygienic quality of the waters was a necessary, but not sufficient, condition for the recovery of the biological quality of the macrobentonic communities. It indicated the need to remove from the bed of the River Frigido the sediments of marble dust that had accumulated over the years.

Today, now that the cleaning operation is concluded and the biological restoration has started, the IFF shows that further operations are necessary, above all regarding the qualitative and quantitative restoration of the bands of riparian vegetation. These observations, together with those concerning the depuration of the waste discharges and the renaturalisation of the fluvial and perifluvial zones, are intended to be a contribution to the growth of a more mature awareness and to the beginning of a new campaign of operations to restore the river environment.

References

[1] ANPA, *I.F.F. Indice di funzionalità fluviale*, Siligardi *et al.*, Manuale ANPA, Roma 2000.

[2] DELL'UOMO A., *Popolamenti vegetali delle acque correnti*, in: "Atti del Seminario di aggiornamento *Ecologia dell'Ambiente Fluviale*", Manzini P. and Spaggiari R. eds., Reggio Emilia 31 maggio 1° giugno 1985, Ammin. Prov. Reggio Emilia, USL n. 9, CISBA (1996), 87–106.

[3] GHETTI P.F., *I macroinvertebrati nell'analisi di qualità dei corsi d'acqua. Manuale di applicazione - Indice Biotico E.B.I. modificato*, Provincia Autonoma di Trento 1986.

[4] GHETTI P.F., *Manuale di Applicazione: Indice Biotico Esteso - I macroinvertebrati nel controllo della qualità degli ambienti di acque correnti*, Provincia Autonoma di Trento, Servizio Protezione Ambiente 1997.

[5] GHETTI P.F., BONAZZI G., *I macroinvertebrati nella sorveglianza ecologica dei corsi d'acqua*, Collana del Progetto Finalizzato "Promozione della Qualità dell'Ambiente", CNR AQ/1/127, 1981.

[6] SANSONI G., SACCHETTI P., BARABOTTI P.L., *Corsi d'acqua del litorale apuano. Effetti inquinanti della polvere di marmo*, Comunità Montana delle Apuane, Massa 1983, 95.

[7] SANSONI G., *La comunità dei viventi di un fiume quali soggetti dell'autodepurazione*, in: "Ambiente '91, Atti del Conv. Naz. *L'acqua. Storia, Cultura, Natura*", Terme di Comano (TN), 4–5 ott. 1990, Prov. Aut. di Trento 1991, 191–199.

[8] WOODIWISS F.S., *The biological system of stream classification used by the Trent River Board*, Chemistry and Industry **14** (1964), 443–447.

[9] WOODIWISS F.S., *Comparative study of biological-ecological water quality assessment methods. Second practical demonstration*, Summary Report 1978, Commission of the European Communities.

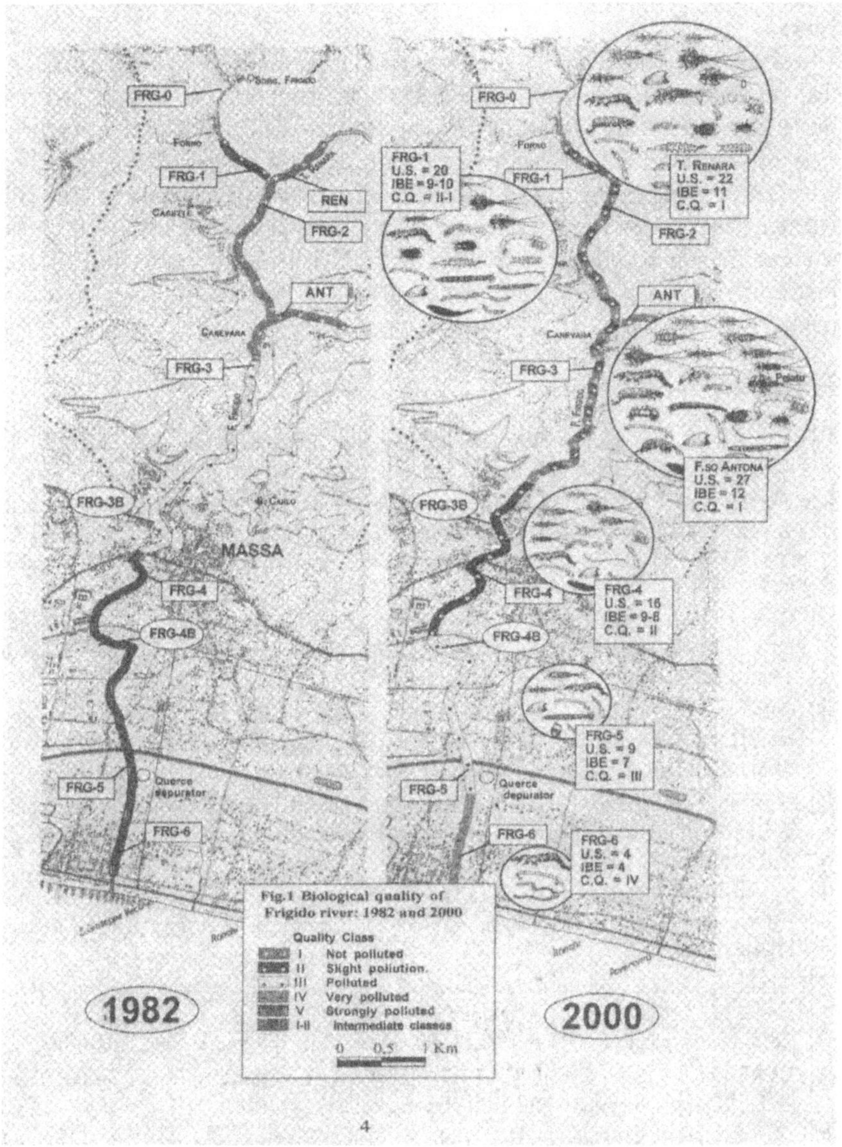

Figure 1. Biological quality of Frigido river: 1982 and 2000.

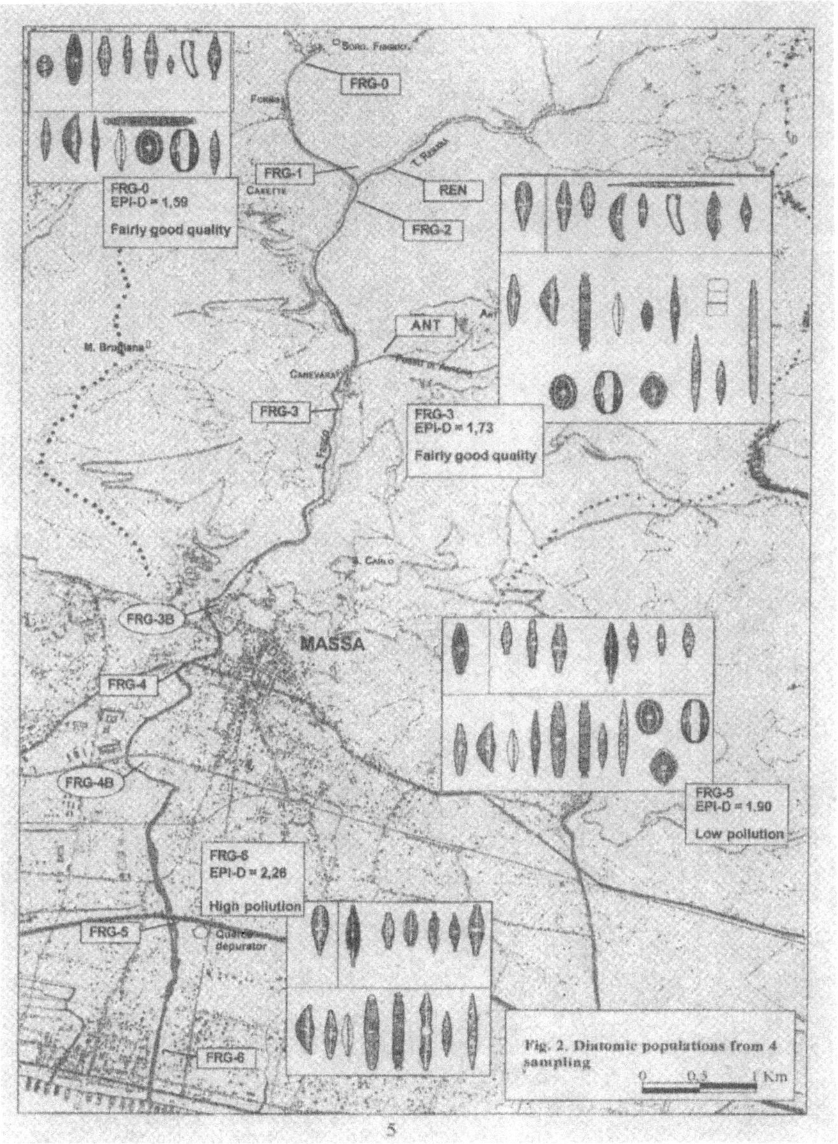

Figure 2. Diatomic populations from 4 sampling.

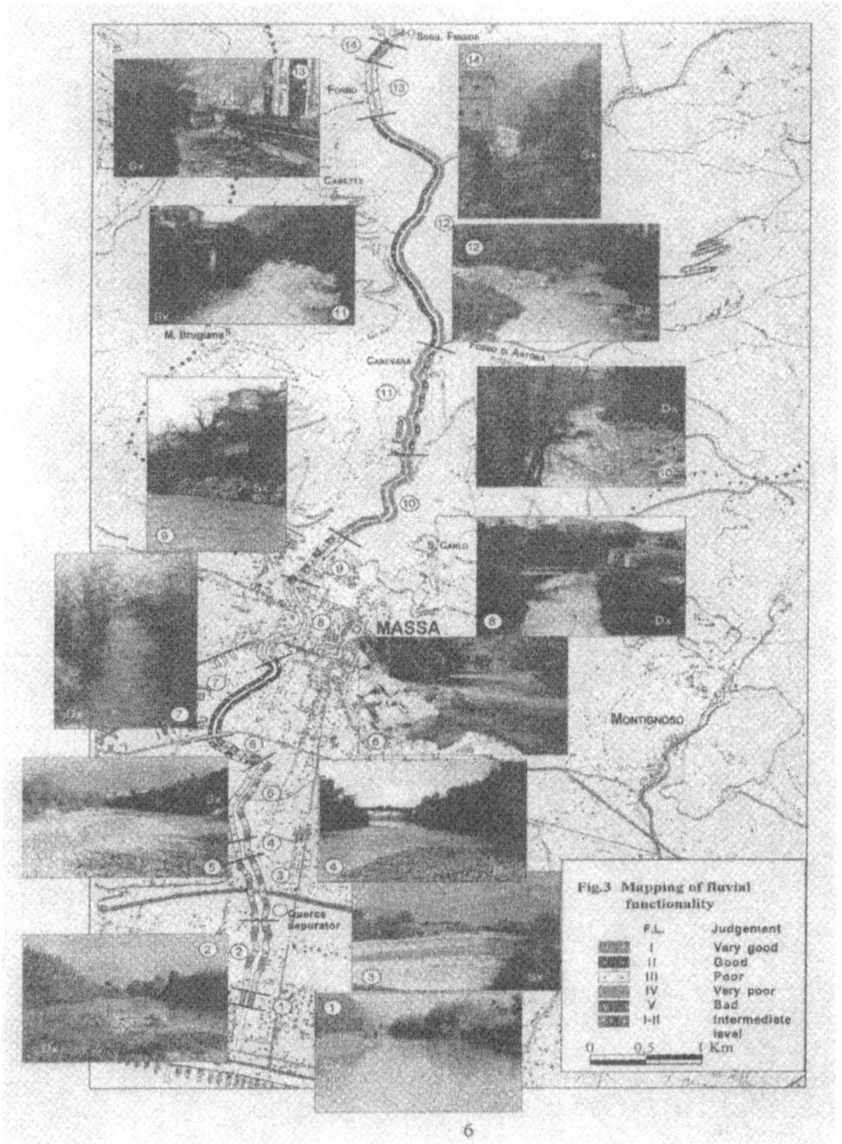

Figure 3. Mapping of fluvial functionality.

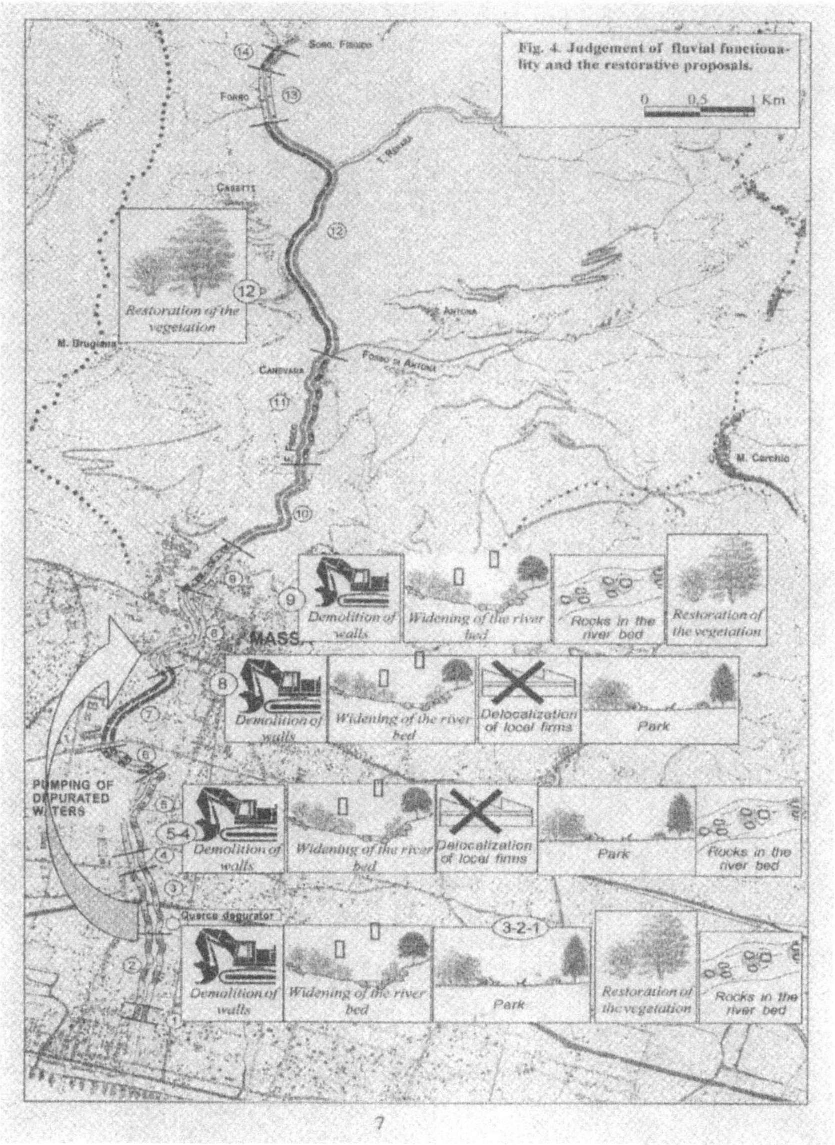

Figure 4. Judgement of fluvial functionality and the restorative proposals.

SYNCHRONIZATION OF NEOCORTICAL INTERNEURONS

Santi Chillemi
Istituto di Biofisica CNR,
Via G. Moruzzi 1, 56124 PISA - ITALY
chillemi@ib.pi.cnr.it

Michele Barbi
Istituto di Biofisica CNR,
Via G. Moruzzi 1, 56124 PISA - ITALY
barbi@ib.pi.cnr.it

Angelo Di Garbo
Istituto di Biofisica CNR,
Via G. Moruzzi 1, 56124 PISA - ITALY
and
Istituto Nazionale di Ottica Applicata,
Largo E. Fermi 6, FIRENZE - ITALY
digarbo@ib.pi.cnr.it

Abstract As shown by recent experiments, the Fast-Spiking inhibitory interneurons of the neocortex interact also by electrical synapses (gap-junctions). Moreover, their synchronous discharge enhances their inhibitory control of pyramidal neurons. By using a biophysical model of these interneurons the synchronization properties of a network of two synaptically coupled units are investigated: in the presence only of inhibitory coupling, well defined regions exist in the space of strength and duration of the synaptic current, where synchronous regimes occur. When electrical synapses are included, the region of synchronous discharge becomes larger. In both cases the coherent states are characterized by discharge frequencies in the gamma range. Lastly, the effects of heterogeneity obtained by using either different stimulation currents or unidirectional inhibitory coupling, are studied.

1. Introduction

Spectral analysis of EEG signals shows that different kinds of rhythms characterize the ongoing brain electrical activity. The fact that these power spectra exhibit well defined frequency bands implies that the electrical activities of many nearest neurons have some degree of coherence. Recently there has been an increasing interest in the study of synchronization phenomena among neurons for their possible functional role during information processing tasks. In particular it was suggested that the formation of neural assemblies (perhaps of thousand of neurons), whose units discharge in a coherent fashion, could be used to carry a representation of the physical world much richer than the simple rate codes of individual neurons [8]. There is experimental evidence that synchronization of cortical neurons, occurring in the gamma frequency range (30 – 80 Hz), could be the basic mechanism underlying the features binding process [6]. In the neocortex there are different kinds of neurons, ranging from the regular spiking ones, like pyramidal cells, to different classes of interneurons [7]. Pyramidal neurons interact by means of excitatory chemical synapses, while receiving inhibitory synaptic inputs from interneurons. Although all inhibitory interneurons use GABA as their primary transmitter, they differ in the gross morphology, neurochemistry, the inputs they receive and in their targets. An important group of interneurons are those belonging to the class of basket cells which innervate the soma and proximal dendrites of pyramidal cells, often exhibiting a fast-spiking behavior with little frequency adaptation [7]. Experimental findings suggest that networks of GABAergic inhibitory interneuron contribute to generate rhythms in neocortex by synchronizing their own firing and, consequently, that of many principal cells [4]. Moreover, experimental and theoretical works ascribe the synchronous discharge of a population of inhibitory interneurons to some features of their coupling, like the duration and intensity of the synaptic current [10, 9, 1]. The novelty, during the last two years, was the discovery that two groups of inhibitory neocortical interneurons, the Fast-Spiking (FS) and the Low-Threshold-Spiking (LTS) cells, make electrical synapses too [3, 5]. The FS interneurons exhibit non-accommodating firing activity, while the LTS cells are accommodating [5]. A peculiarity of FS cells concerns their electrical and chemical synaptic coupling: while any FS cell makes inhibitory connections with other FS units and pyramidal neurons, gap-junctions are prevalent among FS interneurons. At present it is not well understood how electrical coupling affects and promotes synchrony among inhibitory interneurons. In this paper we study the synchronization properties of a pair of FS interneurons interacting by inhibitory and electrical synapses. In particular, by using a biophysical model of FS cell, we will investigate how the coupling features – like synaptic current duration, chemical and electrical coupling intensities, and heterogeneity – influence and determine the emergence of a coherent activity in the network.

2. Interneuron biophysical model

FS interneurons do not exhibit adaptation: the average firing frequency after 600 msec is $\sim$ 97% of the initial one. The other basic properties of FS cells are: a) their action potentials are shorter than those of pyramidal neurons and exhibit afterhyperpolarization; b) they are GABAergic and the decay time constant of synaptic currents averages $\sim$ 10.5 *msec*; c) the inhibitory synapses between them can be either unidirectional or bidirectional, while the electrical coupling is always bidirectional [3, 5]. We model each FS cell by the following set of HH-like equations [2]:

$$\begin{aligned} C\frac{dV}{dt} =& I_E - g_{Na}m^3h(V - V_{Na}) - g_K n^4(V - V_K) \\ & - g_L(V - V_L) - I_{Sy} - I_{El} \\ \frac{dm}{dt} =& (m_\infty - m)/\tau_m; \; m_\infty = \dot{a}_m/(\dot{a}_m + \beta_m),\; \tau_m = 1/(\dot{a}_m + \beta_m) \\ \frac{dh}{dt} =& (h_\infty - h)/\tau_h; \quad h_\infty = \dot{a}_h/(\dot{a}_h + \beta_h),\; \tau_h = 1/(\dot{a}_h + \beta_h) \\ \frac{dn}{dt} =& (n_\infty - n)/\tau_n; \quad n_\infty = \dot{a}_n/(\dot{a}_n + \beta_n),\; \tau_n = 1/(\dot{a}_n + \beta_n) \end{aligned} \tag{1}$$

where $C = 1\mu F/cm^2$ and I_E is the external stimulation current, and

$$\begin{aligned} &\dot{a}_m = 4.2\exp[(V + 34.5)/11.57] && \beta_m = 4.2\exp[-(V + 34.5)/27] \\ &\dot{a}_h = 0.09\exp[-(V + 45)/33] && \beta_h = 0.09\exp[(V + 45)/12.2] \\ &\dot{a}_n = 0.3\exp[(V + 35)/10.67] && \beta_n = 0.3\exp[-(V + 35)/42.68]\,. \end{aligned}$$

The maximal specific conductances and the reversal potentials are, respectively: $g_{Na} = 100\ mS/cm^2$, $g_K = 40\ mS/cm^2$, $g_L = 0.1\ mS/cm^2$ and $V_{Na} = 55\ mV$, $V_K = -90\ mV$ and $V_L = -68\ mV$. The GABAergic synapse is modeled as: $I_{Sy} = g_{Sy}s_{Pre}(t)\,(V_{Post} - V_{Rev})$ where g_{Sy} is the maximal specific conductance for the inhibitory coupling, $V_{Rev} = -75\ mV$ is the reversal potential for the neurotransmitter and V_{Post} is the voltage of the postsynaptic interneuron. The gating variable $s_{Pre}(t)$ represents the fraction of receptors in the open state and obeys the first order kinetics $s_{Pre} = \alpha T(V_{Pre})[1 - s_{Pre}] - s_{Pre}/\tau$, where α is the channel opening rate, $T(V_{Pre}) = 1/\left(1 + e^{-V_{Pre}}\right)$ represents the dependence of the transmitter concentration on presynaptic voltage, V_{Pre}, τ

is the synaptic decay time constant. The electrical synapse is modeled as: $I_{El} = g_{El}\,(V_{Post} - V_{Pre})$ where g_{El} is the maximal conductance. In the following we will use physiological g_{Sy} values within the interval $[0.001, 1\ mS/cm^2]$ and g_{El} values within $[0, 0.3\ mS/cm^2]$. All numerical simulations were performed by using a 4-*th* order Runge–Kutta method with an integration step of 0.002 *msec*; moreover all quantities were computed after transient times ranging from 1 to 6 *sec*.

3. Synchronization phenomena

We firstly study the occurrence of synchronous regimes when the two units interacts only by chemical synapses; then the electrical coupling is set on to investigate its effects on the network coherence.

3.1 Chemical coupling alone

Let us firstly investigate the synchronization phenomena occurring in the network of two FS cells interacting by inhibitory synapses alone. The degree of coherence is quantified by the normalized cross-correlation $C = \langle x(j)y(j)\rangle$. The total number of bins, each 1 *msec* of duration, is n_{bin}; moreover, for each spike train, the variable $x(j)$ (or $y(j)$) is defined to be 1 or 0 according as almost a spike is fired or not in the j-*th* bin. To integrate equation (1) the initial value of V was randomly chosen within a neighborhood of the resting potential of the unperturbed neuron. Also, to get firing in both cells, they were stimulated with currents having the same intensity $I_E > 0$. The results of numerical simulations using different intensities are reported in figure 1.

For a fixed I_E value the two cells synchronize in a well defined region of the parameter space (τ, g_{Sy}), the so-called synchronization manifold (SM). The transition from asynchronous to synchronous state and the reverse one occur fastly as those parameters are varied, so the boundaries of SM are quite sharp.

Moreover, the plots in figure 1 show that the network synchronizes for values of τ greater than ~ 5 *msec*. As the strength of the inhibitory synapses decreases, the lower border of SM shifts to the right along the τ axis. No synchronization occurs, for any τ, as the intensity of the inhibitory synapses gets values lower than $\sim 0.046\ mS/cm^2$. The average frequency of the network increases as I_E gets higher values: in panel a) the frequency values are within the beta range (12–30 *Hz*), while they are in the gamma range (30–80 *Hz*) in the remaining ones (data not shown). We also investigated the dependence of the average network frequency on τ keeping g_{Sy} constant. Moving from the left, firstly the transition from asynchronous to synchronous firing and then the reverse one are met, both marked by a positive jump of the average frequency. Within SM the frequency exhibits a slow decrease (data not shown). The contour lines, delimiting each SM, suggest they can be fitted by exponential functions of the form $G_{Sy}(\tau) = g_0 + g_1 \exp[-(\tau - \tau_0)/\gamma]$. The corresponding results for the data of figure 1 suggest that the relationship $\gamma_{Low}/\gamma_{Up} \cong RI_E$ holds, where γ_{Up} and γ_{Low} characterize the decay of the exponential functions, respectively, for the upper and lower border of SM, while R (~ 1) is some proportionality constant.

3.2 Chemical and electrical coupling

The experimental evidence of electrical coupling among FS cells urged us to investigate its effects on the synchronization behaviour of our model network. So we performed numerical simulations of two mutually inhibitory FS interneurons interacting via electrical synapses too. The results show that the main effect of the electrical coupling is to widen the SM in the (τ, g_{Sy}) space (data not shown). Moreover, even in this case, the average firing frequencies in the synchronous regime fall in the beta and gamma bands. We also performed some numerical simulations to understand the effects of heterogeneity in the network. Starting with the parameter values corresponding to the case reported in panel c) of figure 1, we first introduced heterogeneity by using two different intensities ($I_E(1)$, $I_E(2)$) of the stimulation currents, and then the electrical coupling.

As shown in panel b) of figure 2, a moderate heterogeneity realised, by adopting different values of I_E for the two cells, breaks the synchronous discharge of the cells. When electrical coupling is set up, the network coherence deteriorates even more if g_{El} is low (see panel c), but synchronization increases again for higher g_{El} (see panel d). The conclusion can be drawn that in the presence of heterogeneity the main effect of electrical coupling is to promote higher levels of coherence in the system. This findings could explain why in neocortex FS interneurons make electrical synapses among them; in fact two real neurons are never identical and heterogeneity is an inherent feature of a neural population.

There is experimental evidence that some pairs of FS interneurons in neocortex possess unidirectional chemical coupling. So this new kind of heterogeneity was introduced in the network by using a bidirectional electrical coupling together with a unidirectional inhibitory synapse. The corresponding results are reported in figure 3 (both FS cells receiving the same stimulation current). For all panels the intensity of chemical coupling is $g_{Sy} = 0.28\ mS/cm^2$ while the duration of synaptic currents is $\tau = 11\ msec$. The first result is that the network coherence increases with the strength of the electrical coupling, while it deteriorates with increasing I_E. This last behaviour can be explained by the low-pass filtering properties of electrical synapses.

For the left and middle panels there are intensity values of electrical coupling for which suppression of firing occurs (of the inhibited neuron), but this phenomenon disappears as the stimulation current is further increased (right panel). In such case before the synchronization regime is achieved, irregular spike trains characterize the dynamic behaviour of FS cells. The transition from aperiodic to synchronous firing is marked by a bifurcation. The above findings suggest that the presence of electrical coupling promotes the synchronization even if the chemical coupling is unidirectional. Moreover, the frequencies corresponding to these coherent states are either within the beta or the gamma range, depending on the I_E values.

4. Conclusions

Experimental and theoretical results suggest that the synchronization phenomena occurring in the brain are important to understand the processes underlying sensory information coding. Moreover, it was recently emphasized that the emergent activity of a population of interneurons could be critical for the onset of the synchronous discharge in principal cells. We studied the behavior of a network of two FS interneurons interacting by chemical and electrical synapses. We found that, when the coupling is only inhibitory, a well defined SM exists in the parameter space (τ, g_{Sy}). For moderate values of I_E each border of the SM can be fitted by an exponential function in the plane (τ, g_{Sy}). Also, the two decay constants (corresponding to the lower and upper borders of SM) satisfy the approximate relation $\gamma_{Up}/\gamma_{Low} \sim I_E$. The introduction of the electrical coupling leads typically to widen the SM region and prevents the decoherence effects arising from heterogeneity.

References

[1] CHILLEMI S., BARBI M., DI GARBO A., *Synchronization mechanisms in neuronal networks*, Lecture Notes in Computer Science **2084** (2001), 87.

[2] DURSTEWITZ D., SEAMANS J.K., SEJNOWSKI T.J., *Dopamine-mediated stabilization of delay-period activity in a network model of prefrontal cortex*, J. Neurophysiol. **83** (2000), 1733.

[3] GALARRETA M., HESTRIN S., *A network of fast-spiking cells in the cortex connected by electrical synapses*, Nature **402** (1999), 72.

[4] GALARRETA M., HESTRIN S., *Electrical synapses between GABA-releasing interneurons*, Nature Neuroscience **2** (2001), 425.

[5] GIBSON J.R., BELERLEIN M., CONNORS B.W., *Two networks of electrically coupled inhibitory neurons in neocortex*, Nature **402** (1999), 75.

[6] GRAY G.M., SINGER W., *Stimulus-specific neuronal oscillations in orientation columns of cat visual cortex*, Proc. Natl. Acad. Sci. USA **86** (1989), 1698.

[7] GUPTA A., WANG Y., MARKRAM H., *Organizing principles for a diversity of GABAergic interneurons and synapses in the neocortex*, Science **287** (2000), 273.

[8] SINGER W., *Search for coherence: a basic principle of cortical self-organization*, Concepts in Neurosci. **1** (1990), 1.

[9] WANG X.J., RINZEL J., *Alternating and synchronous rhythms in reciprocally inhibitory model neurons*, Neural Comput. **4** (1992), 84.

[10] WHITTINGTON M.A., TRAUB R.D., JEFFERYS J.G.R., *Synchronised oscillations in interneuron networks driven by metabotropic glutamate receptor activation*, Nature **373** (1995), 612.

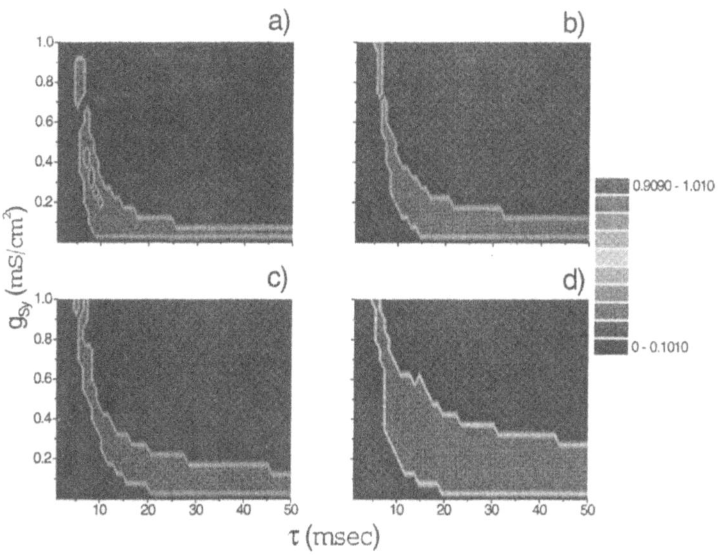

Figure 1. Coherence measure C of the spike trains for two mutually inhibiting FS cells when $g_{El} = 0$: a) $I_E = 0.5\ \mu A/cm^2$; b) $I_E = 1\ \mu A/cm^2$; c) $I_E = 1.32\ \mu A/cm^2$; d) $I_E = 3\ \mu A/cm^2$. For each panel 51 values for τ and 21 for g_{Sy} were considered.

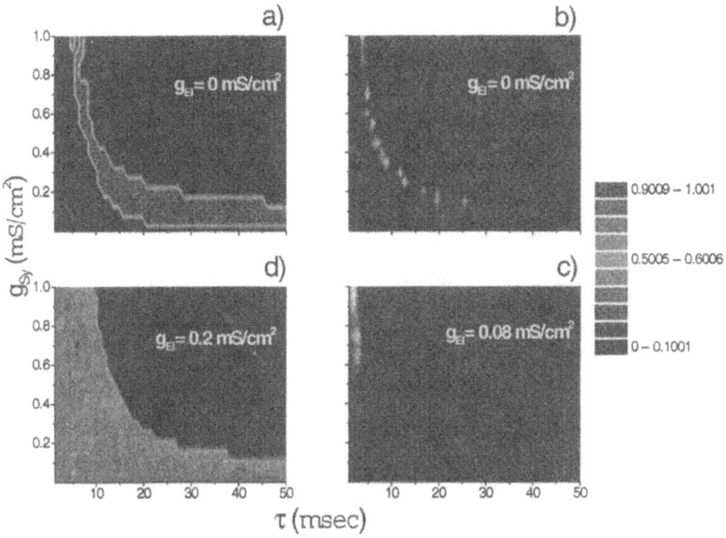

Figure 2. Coherence measure C of spike trains for two FS cells interacting by means of chemical and electrical synapses. Except panel a), for which it is $I_E(1) = I_E(2)$, the heterogeneity is introduced by assuming $I_E(1) = 1.32\ \mu A/cm^2$, $I_E(2) = 1.1\ \mu A/cm^2$. The intensities of electrical coupling are reported in the corresponding panels.

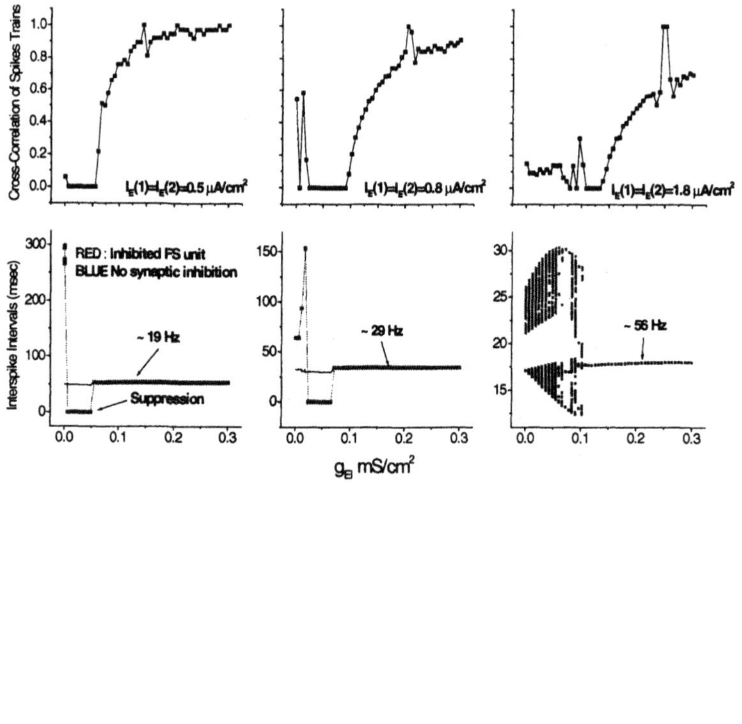

Figure 3. The coherence measure C of spike trains for two FS cells interacting by means of electrical synapses in the presence of unidirectional chemical coupling is reported in the top panels. The corresponding interspike intervals of both FS interneurons are plotted in the bottom ones.

THE FRACTAL BORDERLAND BETWEEN THE DETERMINISTIC ORDER AND THE UNPREDICTABLE CHAOS

Giuseppe Damiani
IDVGA-CNR,
Via Fratelli Cervi 93, 20090 SEGRATE (MI) - ITALY
damiani@idvga.mi.cnr.it

Abstract Many complex systems exhibit fractal structures near the intermediate stages of phase transitions. Sometimes, these critical situations define the boundary between the deterministic order and the unpredictable chaos. According to the "Binary Theory of Everything", this universal behaviour is due to the interaction between the catabolic-entropic decay of electromagnetic energy and the anabolic-negentropic increase of matter configuration.

The goal of most of the modern scientists is to break the universe down into the simplest parts that will obey reproducible rules. Euclidean geometry and analytical calculus, typically in the form of linear differential equations, are the standard mathematical tools used for the description of simple phenomena with repeatable behaviour. But the real world is full of irregular and unforeseeable non-linear events. Is the indeterminable behaviour of quantum entities and chaotic systems produced by our ignorance of the "hidden variables" of a deterministic universe? Computer simulations have shown that even if a system is controlled by a set of deterministic rules, its behaviour may be unpredictable. As discovered by Feigenbaum, the breakdown of an ordered system into a chaotic one is often determined by a fractal pattern of rate doubling of the possible system states. Many complex systems exhibit fractal structures and self-similar properties. This is particularly true for systems poised at a 'critical point' where two or more phases become indistinguishable [15]. If fractal structures and power laws are so common [12], perhaps some universal processes are at work [2].

According to the "Binary Theory of Everything", very few and simple basic mechanisms and principles might have generated our universe starting from the fundamental state of the vacuum, called superspace [4, 6]. Experimental evidence indicates the vacuum is not empty but is filled with randomly fluctuating force fields having the zero-point isotropic spectrum. This spectrum is shape-invariant, while the magnitude of its intensity depends on Planck's constant. According to quantum field theory, the entire Universe might be a vacuum fluctuation [16]. As noted by Weyl [17, 18], Eddington [9], Dirac [8], Caldirola, Pavsic, and Recami [3] an elegant geometric description of the physical properties of our Universe may be obtained using two space-time metrics, one at the microscopic atomic level and the other at the macroscopic cosmological level. If these two scale factors have not a special state, then the superspace might be filled with a self-similar fractal-chaotic distribution of shape and scale invariant fluctuating vacuum sets. Different nonconnected Universes may be quantized vacuum sets with different scales emerging in the superspace like "bubbles" in boiling water.

A quantum mechanism of vacuum decay might have generated our Universe: a macroscopic fluctuation may be stabilised by resonant tunnelling with a complementary microscopic fluctuation (Fig. 1).

A phase transition towards an isotropic state between the macroscopic and the microscopic levels produces two opposite trends: an expansive process of binary dissociation and a complementary contractive process of binary association (Fig. 2).

The binary dissociation produces an expanding Peanian network of microscopic fluctuations similar to the Tartaglia-Pascal triangle. This process is coupled to the binary association of the macroscopic vacuum sets which breaks up as a Cantor dust. The feedback relation between the opposite autocatalytic binary processes can be described by a simplified version of the prey-predator equation discovered by Lotka-Volterra. The coupling between the complementary binary processes leads to their progressive damping. When the radius of the expansion of the microscopic entities reaches that of the initial macroscopic set a situation with inverted scales is produced, and the reversal processes of microscopic association and macroscopic dissociation take place. At the end of these processes the starting condition of the superspace is restored and the macroscopic and microscopic fluctuations may disappear.

The process of binary dissociation leads to an isotropic spread of the microscopic entities in the maximum available volume while the superspace reaction limits the surface of the expanding structures. The four-dimensional hypersphere is the minimal isotropic boundary region containing the maximum amount of Euclidean superspace. A cosmological

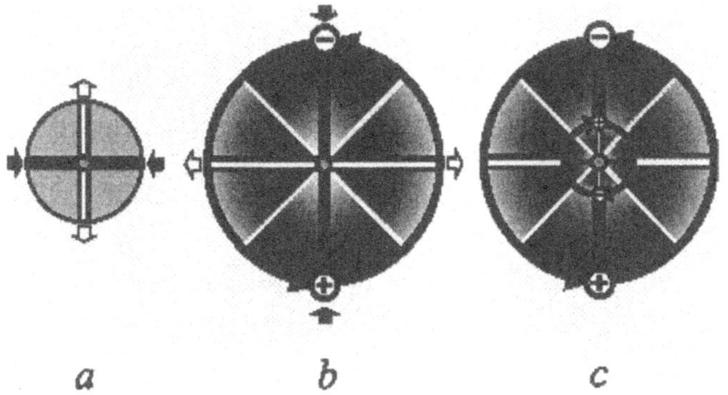

Figure 1. Schematic view of the vacuum decay. The geometric properties of the Euclidean superspace and three simple laws are needed to explain this process: the principle of conservation and of action and reaction of the Newtonian dynamics, and of principle of parsimony similar to the Maupertuis principle of minimal action. According to these principles the simplest vacuum fluctuation is an expanding monodimensional string (white bar) which produces a complementary contracting string (black bar) in the perpendicular dimension (a). The circle indicates the horizon of events of the "bubble-Universe". The "stresses" of the strings produced by the reaction of the disturbed superspace may induce a positive and a negative "charge" at the extremities of the strings indicated respectively by the circles with the plus and minus signs (b). This situation may induce the emergence of an internal microscopic fluctuation at a smaller scale (c) and the opposite rotation of the two fluctuations. The two opposite poles of the microscopic fluctuation move away towards the repulsive external regions, while the two poles of the macroscopic fluctuations move towards the attractive internal regions. At each rotation the microscopic fluctuations replicate themselves (binary dissociation) and the macroscopic fluctuations reduce into halves (binary association). The dynamics of the rotating fluctuations is the same of a photon model developed by Mac Gregor [13].

model of a constant curved four-dimensional space-time developed by De Sitter is related to a group discovered by Fantappié [10]. This group describes the possible rotational self-movements of a four-dimensional space-time and is the base for the development of the final relativity of Arcidiacono [1]. The vacuum decay produces four different intercon-

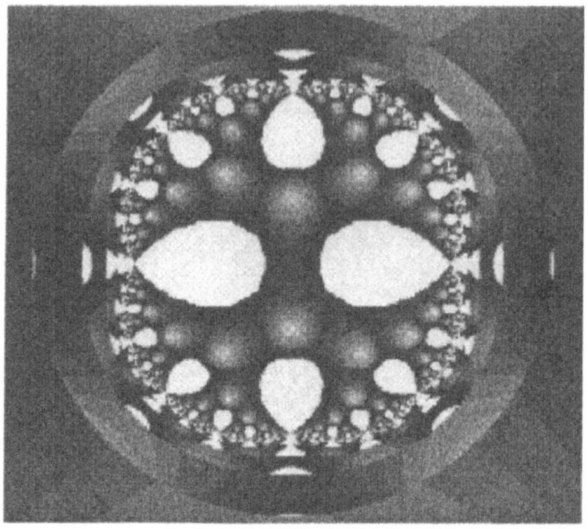

Figure 2. Computer graphic representation of an Ising model of the phase transitions in a magnetic system. This picture is a Julia set and provides an useful model of the processes described by the Binary Theory. An internal binary dissociation process of expansion is counterbalanced by an external binary association process of contraction. The border between these two coupled processes has a fractal shape. Many mathematical non-linear systems, and in particular the computer graphic representation of trigonometric and exponential functions in the complex plane, exhibit analogous fractal behaviour at the critical border of transitions between two phases. These analogies are the results of the topological properties of the Euclidean space, which allows only few types of patterns produced by iterative processes.

nected expansion and contraction fields which are related to the gravitational (FG), electric (FQ), strong (FC), and weak (FH) forces.

According to the suggested geometrodynamical origin of physical entities [7] and to a particle model proposed by Harari [11] and Shupe [14], the quantized microscopic vortices corresponding to the gravitational, electric, strong and weak charges are the fundamental components of the space-time and of the microscopic particles (Fig. 4).

All the fundamental mathematical relations and numerical coincidences of classic and quantum physics and the Harari-Shupe rules for the construction of the observed particles can be deduced from the geometric structure of the proposed model, the length units of the microscopic and macroscopic fluctuations, and the present status of the binary processes.

As proposed by Weyl [17, 18], the Compton radius of an electron $r_e \approx 3.86 \cdot 10^{-13}$ [m] is the fundamental spatial unit of the microscopic entities. This unit is related by the light velocity c to the unit of the

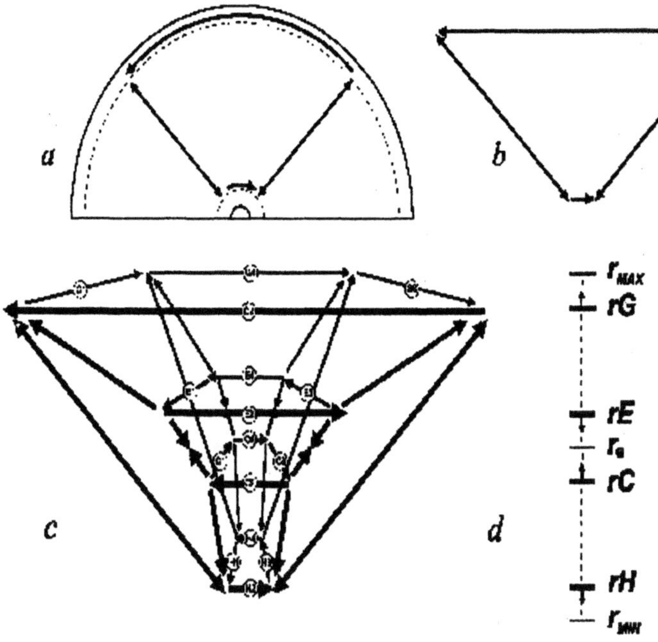

Figure 3. Graphic representation of the relations between the rotational self-movements, the charges, the force fields, and the horizons of events. The rotations at the macroscopic and microscopic level (a) are transformed respectively in big and small horizontal lines (b) while the oblique lines represent the contraction and expansion processes between the two levels. A deformed hypercube (c) shows the symmetries of our Universe. Only one of the two possible orientations of the rotations is shown. The four biggest lines correspond to the space-time of everyday life while all the other dimensions are of the Kaluza-Klein type. The G, E, C and H entities, correspond to the gravitational, electric, colour, and hypercolour charges and force fields. On the side of the hypercube (d) the different levels and their characteristic distances are shown: $r_{MAX} = r_e \cdot z$ is the radius of the maximum expansion of our Universe, $rG = r_e \cdot n$ is the radius of the visible part of the Universe corresponding to about 137 times its present Schwarzschild radius, $rE = r_e \cdot z/n$ is the radius of the hydrogen atom, r_e is the Compton radius of the electron, $rC = r_e \cdot n/z$ is the Compton radius of the nucleon, $rH = r_e/n$ is approximately 137 times the Schwarzschild radius of the nucleon, and $r_{MIN} = r_e/z$ is the smallest radius of the microscopic particles.

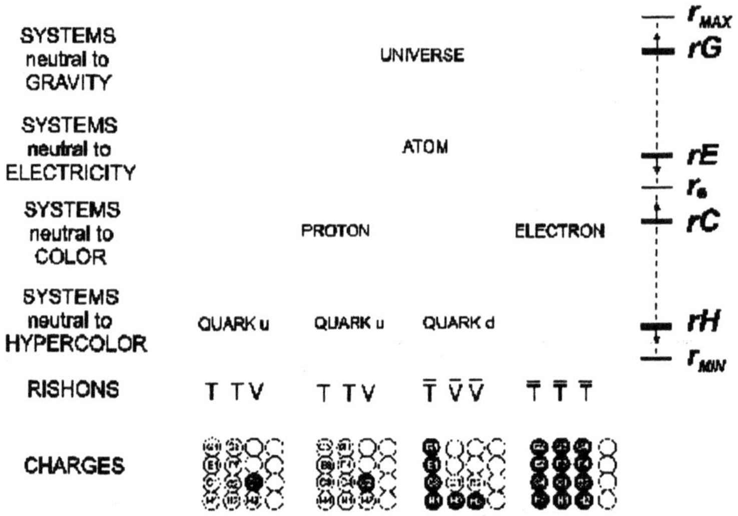

Figure 4. Model for the composition of the most common microscopic particles based on the relations described in figure 3. The G, E, C and H entities, correspond to the gravitational, electric, colour, and hypercolour charges. The values of rG, rE, rC, and rH are described in the legend of figure 3.

temporal quantum $t_e = 2 \cdot r_e/3 \cdot \pi \cdot c \approx 2.73 \cdot 10^{-22}$ [sec]. Starting from these quantized values of space and time and going backwards along the steps for the construction of the international measure system, we can calculate the values of the quantized unit of mass $m_e = t_e^2/2 \cdot r_e \approx 9.11 \cdot 10^{-31}$ [kg] (m_e is the electron mass), angular momentum $J_e = t_e \cdot r_e = h/2 \cdot \pi \approx 1.05 \cdot 10^{-34}$ [kg m sec^{-1}] (h is the Plank constant), energy $E_e = m_e \cdot c^2 \cdot 3 \cdot \pi/2 = r_e \approx 3.86 \cdot 10^{-13}$ [kg m^2 sec^{-2}], and thermodynamical entropy $S_e = E_e/T \approx k \cdot \log_e 2$ [kg m^2 sec^{-2} K^{-1}] (T is the temperature, k is the Boltzmann constant). The values of the forces and charges increase or decrease according to the number n of cyclic iterations of the binary processes. The strengths of the forces are correlated, in a complementary way, to the radii of the horizons of events which are the borders of expanding or contracting binary processes

$$FG \cdot rG = FE \cdot rE = FH \cdot rH = FC \cdot rC = r_e \tag{1}$$

The values of rG, rE, rH, and rC are described in the legend of figure 3. Starting from experimental data, it is possible to calculate the approximate values of $n \approx 1 \cdot 10^{39}$ and $z \approx 1 \cdot 10^{41}$ where $n/z \approx 1/137$ is the fine structure constant and the ratio between the present age of our Universe and its total length of life.

The proposed model suggests that the entire Universe is a complex dissipative system, in non-equilibrium state. The evolution of a macroscopic dissipative system is directed by the catabolic-entropic degradation of the electromagnetic energy and by the anabolic-negentropic increase of information content of matter. The behaviour of a macroscopic system near the conditions of thermodynamic equilibrium is generally deterministic. On the contrary a microscopic particle is a small universe oscillating between alternating collapsed and diffused phases with probabilistic and unpredictable trajectories, according to the Heisenberg's principle. Even if it is not possible "to see" inside a single microscopic quantum, its evolution may be the results of a deterministic mechanism, as in the Bhom theory. However there is some unavoidable chaotic behaviour in the superspace fluctuations which influence the collapsed particles during each regeneration cycle. The main question concerns the role of these fluctuations in the creation instant and in every quantum phenomena: are they produced by a "blind chance" or by a "God's will"?

The physical laws are not incompatible with the existence of a partially free choice of God or of our conscience. God and our soul can play dice: a single result can be chosen even if the sum of the results must be in accordance with the probabilistic laws. Therefore, it may be that in some parts of our brain the "soul" is partially free to influence quantum events without violating any physical laws. These "conscious" microphysical events can be amplified to macroscopic level by the Lorenz's "butterfly effect". This scenario is supported by our knowledge of biological systems. Most of the chemical reactions and physiological processes of living organisms work near their critical and unstable equilibrium points, which define the borderline between "deterministic" order and "unpredictable" chaos [19, 5]. The fractal-chaotic state is characterized by a long-range correlation between the different elements of the system and by very high sensitivity to small perturbations. A small pulse at a critical time or place can direct a reproducible transition of the system towards a defined state. The amplification of small events using the long-range correlation of fractal-chaos may be a very useful tool for an economical and efficient control of complex systems. With this mechanism the conscious soul can influence the activity of the brain. Even if the existence of metaphysical entities is unnecessary and useless for the probabilistic previsions of natural phenomena, it cannot be ruled out and is important in the search for the missing understanding of consciousness.

References

[1] ARCIDIACONO G., *Relatività e cosmologia*, Libreria Veschi, Roma 1979.

[2] BAK P., *How nature works: the science of self-organized criticality*, Springer, New York 1996.

[3] CALDIROLA P., PAVSIC M., RECAMI E., *Explaining the large numbers by a hierarchy of "Universes": a Unified Theory of strong and gravitational interactions*, Nuovo Cimento **48B** (1978), 205–271.

[4] DAMIANI G., *Il gioco della vita, la teoria binaria dell'Universo fisico*, Editrice Italiana Audiovisivi, Roma 1984.

[5] DAMIANI G., DELLA FRANCA P., *Morphé and evolution*, Biology Forum **90** (1997), 227–266.

[6] DAMIANI G., *Evolution of life in a fractal Universe* in "Fractals in biology and medicine" Vol. 2, Losa G.A., Merlini D., Nonnenmacher T.F. and Weibel E.R. eds., Birkhauser-Verlag, Basel 1998, 169–187.

[7] DE SABBATA V., SIVARAM C., *Torsion, string torsion, and topological origin of charge and mass*, Found. Phys. Lett. **8** (1995), 375.

[8] DIRAC P.A.M., *Cosmological models and the Large Number hypothesis*, Proc. Roy. Soc. A **333** (1973), 403–418.

[9] EDDINGTON A.S., *Fundamental Theory*, Cambridge Univ. Press, Cambridge 1946.

[10] FANTAPPIÉ L., *Opere scelte*, Ed. Unione Matematica Italiana, Bologna 1973.

[11] HARARI H., *A schematic model of quarks and leptons*, Phys. Lett. B **86** (1979), 83–86.

[12] MANDELBROT B., *The Fractal Geometry of Nature*, Freeman, New York 1982.

[13] MAC GREGOR M.H., *Model basis states for photons and "empty waves"*, Found. Phys. Lett. **8** (1995), 135–160.

[14] SHUPE M.A., *A composite model of leptons and quarks*, Phys. Lett. B **86** (1979), 87–92.

[15] STANLEY H.E., *Power laws and Universality*, Nature **378** (1995), 554.

[16] TRYON E.P., *Is the Universe a vacuum fluctuation?*, Nature **264** (1973), 395–397.

[17] WEYL H., *Eine neue Erweiterung der Relativitatstheorie*, Ann. der Phys. **59** (1919), 101–133.

[18] WEYL H., *Philosophy of Mathematics and Natural Sciences*, Princeton University Press, Princeton 1949.

[19] YATES F.E., *Fractal applications in biology: scaling time in biochemical networks*, Methods in Enzymology **210** (1992), 636–675.

EMERGENT PROPERTIES AND COMPLEXITY FOR BIOLOGICAL THEORIES*

Paolo Freguglia
Università degli Studi dell'Aquila - ITALY
fregugli@univaq.it

In memory of Martino Rizzotti

Abstract In this contribution we put into evidence some meanings of the concept of *emergent property* with particular reference to the construction of biological theories. To this aim, we propose a biological synthetic theory which concerns evolutionary and genetic aspects. In this work we consider our previous contributions about this subject (see References).

1. The theoretical approach to Biology

The study of Biology, particularly today, can be realized also from a theoretical point of view. Obviously this approach must not leave behind the experimental analysis and results. The theoretical examination of biological themes often, but not always, needs to use the mathematical language and hence the mathematical techniques.

If we want to offer a general view of the mathematical applications to Biology, we must first consider the fact that these have been until now restricted and a little importance in comparison to the modern developments of biological research, because the construction of biomathematical models often bears a strong analogy with the methodology that we use in physics. On the other hand, this point of view, for instance, has allowed us to solve and study some interesting dynamical problems which we find in biology, like the dynamics of the growth of a population. Hence we can actually consider the following kinds of models and theoretical attitudes:

*This work flows from discussions and reflections with Vieri Benci and Nicola Guglielmi.

- Models of growth and propagation of populations, that is logistic (differential or finite difference) equations, Lotka-Volterra models, epidemic models, etc.
- Stochastic models
- Game dynamic models
- Computer models of biological structures
- Physiological and biophysical models
- Axiomatic approach to genetic and evolution theories

However, biological theories (mathematical or non-mathematical in general) are not adequate to explain all the properties which regard the living, for example some aspects of life are not completely explainable or are not shown by any theory. As a matter of fact, we say that biological theories are *complex* (see, for instance, [2]). In the other words, we can say that a system or a part of reality (the whole) is complex if it has very numerous parts or members and if it presents very numerous interactions between these parts which are not easily reducible to well defined, circumstantial and not too numerous interactions (see [7] and [11]). We can presume that a complex part of reality cannot be described only by one model or theory or have only one explanation. That is, we cannot give a mono-casual and deterministic explanation of it (see [7]). We are now going to examine some kind of properties which are very interesting and useful in understanding the scientific theories based also on the experimental data. We will try afterward to give an epistemological justification to the notion of emergent property which can be connected to the notion of complexity.

2. Emergent properties: the properties of synthesis

There are some properties which theoretically flow or, as we say, emerge from a metatheoretic analysis of some given theories. This is a case where we can individualize a kind of emergent properties. For instance, the properties which we can find when we try to unify different theories or models. When, bearing these in mind, we can put our considerations on a higher epistemological level.

Now we provide some easy examples from mechanics and from biology.

Example 1 (Example from Mechanics). We consider (see [1]) the following mechanical system (see Fig. 1), where F and P are two forces,

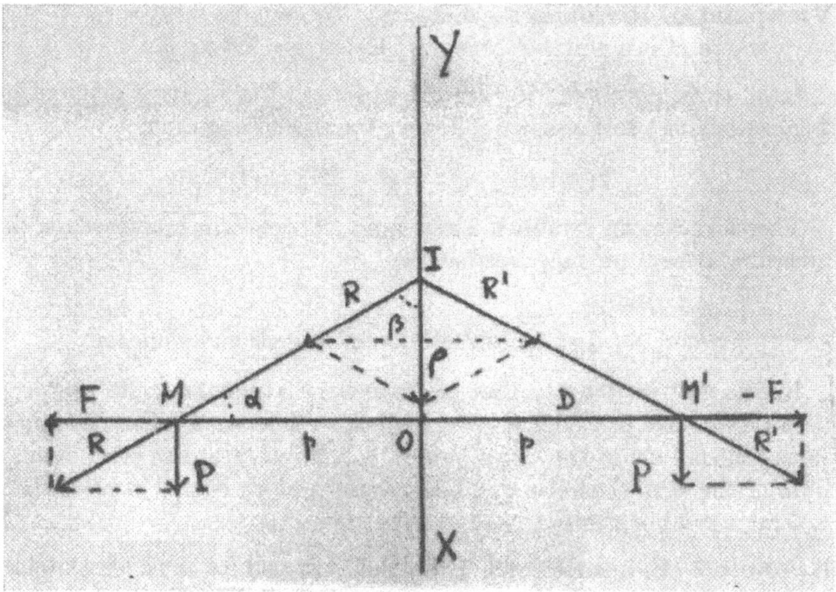

Figure 1. Scheme of a mechanical system proposed by J. Andrade.

R and ρ are resultant forces and $|R| = |R'|$, we must determine the expression of ρ.

When $\beta < \frac{\pi}{2}$ we have $\rho = 2R\cos\beta$ and we know for $\cos x$ the following property:

$$\cos(x+y) + \cos(x-y) = 2\cos x \cos y$$

Examine the case when $\beta \to 0$ and hence $\alpha \to \frac{\pi}{2}$. In studying this case we can choose three geometrical points of view.

Viewpoint of Lobatchevsky's geometry: $0 < \Pi(p) < \frac{\pi}{2}$, where $\alpha = \Pi(p)$ is the angle of parallelism. We can suppose:

$$\Psi(p) = \frac{1}{\sin\Pi(p)} = \cosh\frac{p}{k}$$

where k is a suitable constant and we have $\Psi(p) > 1$. Hence $\rho = 2R\cosh\frac{p}{k}$.

Viewpoint of Euclid's geometry: $\Pi(p) = \frac{\pi}{2}$ and $\Psi(p) = \cos\beta = \sin\alpha = \sin\frac{\pi}{2} = 1$.

We have $\rho = 2R\cos\beta$.

Viewpoint of Riemann's geometry: We suppose $\Psi(p) = \cos\frac{p}{H} < 1$, where H is a suitable constant. Hence $\rho = 2R\cos\frac{p}{H}$.

In all three points of view we can observe what we have denoted by $\Psi(p)$ satisfying the Poisson's following functional equation:

$$\Psi(x+y) + \Psi(x-y) = 2\Psi(x)\Psi(y)$$

Therefore we can establish a new point of view which transcends the preceding ones if we suppose that:

$$\begin{cases} \rho = 2R\Psi(p) \\ \text{where } \Psi(p) \text{ satisfies the Poisson's functional equation} \end{cases}$$

In this way we can say that we have done a *metatheoretic synthesis* because we have unified a property which we find with different expressions respectively in the three models which interpret the same reality. Indeed, the metatheoretic synthesis enables us to obtain a new, more general, model or theory.

Example 2 (Example from Biology). Let us now consider a biological example which concerns the evolution theory (see [6]). We can have the following points of view.

Biospherocentric viewpoint: The fundamental idea of this approach is the Biosphere considered as a whole constituted by many parts which evolve together. This is also an attempt to apply the system theory (by Von Bertalanffy [11]) to evolution.

Organismocentric viewpoint: This theory has the landscape metaphor as conceptual basis: evolution develops in a landscape of valleys and hills which are determined by auto-organization. The harmony of the organization of the organisms (and inside an organism as well) is not explained by the action of natural selection.

Genocentric viewpoint: The fundamental entity of evolution is the gene (as a sequence of DNA). Genes evolve freely according to their intrinsic laws and to natural selection.

Because a biological system is a complex system, a single description of it is not possible. Every evolutionary phenomenon or event α (see Fig. 2) can be explained and justified by simultaneously utilizing those parts which, between them consist of the three preceding points of view. In this way, for every α we establish a synthesis which in regard to the considered points of view, is a *synthesis of co-operation among theories* which concerns only the single event α without the construction of a general theory.

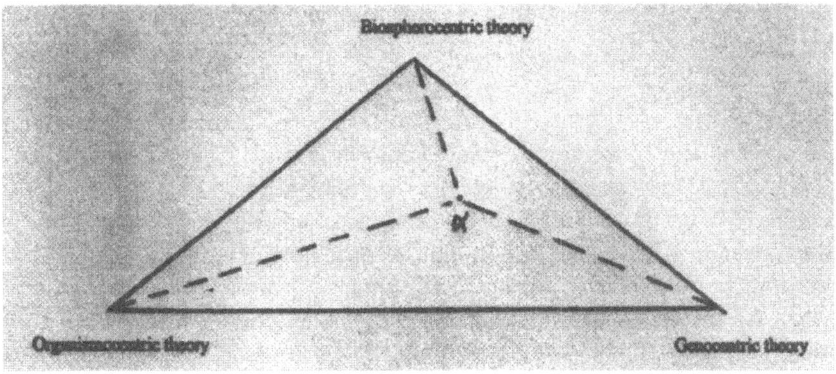

Figure 2. Metaphorical scheme of synthesis of co-operation among theories: triangle scheme.

3. A biological theory

We will now consider an example of biological theory, from which we will see, in the next paragraph, how the axioms can be considered emergent properties. We express the theory also and as an example by the logic of first order predicates (with some constant monadic and dyadic predicates and by appropriate functionals). See [5] and for another logical approach see [3]. In fact we can use other types of logic (second order, modal, etc.). So we can describe formally and more rigorously the evolutionary theory, which however in reality is independent of its logical descriptions. The first axioms flow from synthetic considerations about the events which concern the evolutionary ideas. The formal language $\mathcal{L}$, which we are going to use, must contain the appropriate biological terms.

Let $(c_1, c_2, \ldots, c_n)[t]$ be the set of measurable (on $\mathbb{Q}$) parameters, the number of which determines the dimension of a biological determined context, where an n-tuple represents a temporary state in the life of an individual. The set $(c_1, c_2, \ldots, c_n)[t]$ is relative and not absolute, which means that it is functional, consistent and suitable to a phenomenological analysis. We can propose the following axioms [5]:

Axiom 1. *Let x be a biological individual, x can verify or not, in any instant t of the period of its life, the parameters $(c_1, c_2, \ldots, c_n)[t]$.*

Using the first-order mathematical logic of predicates, symbolically we have:

$$\forall x \, \forall t \, (U(x) \wedge T_x(t) \longrightarrow \forall i \; (J(i) \longrightarrow (\pi_i f(x,t)) = 1 \longrightarrow \mu_i(x,t) > 0))$$

where U is the constant predicate which we read "x is a biological individual". The concept of *biological individual* (or *biological object*) is a general concept. It denotes, for instance, whether it is a mammal or a zygote. T_x is the constant monadic predicate which we read "t is a temporary state of the period of life of x", J is the constant monadic predicate which we read "i is an index", f is a suitable functional, π_i is the functional "choice of the i-th element of a n-tuple" and μ_i is another appropriate functional. The $f(x,t)$ associates to an individual x, at a determined temporary state in its life, a n-tuple $\varepsilon_1, \ldots, \varepsilon_n$, formed by 0 and 1, which corresponds to verifying (1) or not (0) the respective parameters $c_1, c_2, \ldots, c_n$. If the i-th value of the n-tuple $\varepsilon_1, \ldots, \varepsilon_n$ is 1 (that is $\pi_i f(x,t) = 1$) then we can associate to this value the $\mu_i(x,t) > 0$, that is a rational positive number which is the measure of the i-th parameter.

Hence x can verify none or some or all of the parameters. A *biological individual* is a complex thing which cannot be determined only by $(c_1, c_2, \ldots, c_n)$. With this axiom we must admit also the *empty individual* as regards the set of parameters $(c_1, c_2, \ldots, c_n)$. Now we give two important definitions.

Definition 1. *A biological individual x is said* to belong to a type or variety *if sooner or later it verifies at least one of the parameters.*

The verified parameters determine a type or variety. In fact the Definition 1 defines a constant monadic predicate:

$$V_j(x) =_{def} U(x) \wedge \exists t \; (T_x(t) \longrightarrow \tau(f(x,t))$$

where τ is a constant dyadic predicate which we read as follows: "in this temporary state t of life of x, some parameters are positively verified and measurable". Now we can define when an individual x belongs to a species S_j.

Definition 2. *A biological individual x is said* to belong to a species *if it belongs at least to one of the types which determine a species.*

That is we define the predicate

$$S_i(x) =_{def} V_1(x) \vee \cdots \vee V_r(x) \, .$$

We must consider the following further axioms.

Axiom 2 (Law of existence of species). *As regards the set* $(c_1, c_2, \ldots, c_n)$, *for any biological individual* x *we can always establish at any time whether* x *belongs to a species or not.*

Symbolically:

$$\forall x\ (U(x) \longrightarrow \vee_{i \in I} S_i(x))$$

Because we express this theory by means of the first-order logic of predicates, we can quantify only the subject-variables and it is not possible to quantify the predicates. Thus, if we want to say "A species exists for any biological individual x", we cannot write $\forall x\ \exists S\ (U(x) \longrightarrow S(x))$. In this case, our theory would be expressed by the second-order logic of predicates.

It is possible to determine an *average individual* among all individuals which have the values of their parameters into an established domain which determines (according to Ax. 2) a species (or a type). Indeed these domains characterize the types V_s. The average individual, which we denote by $\tilde{x}$, can or cannot really exist. The set of individuals very near to $\tilde{x}$, that is, which have the values of all parameters very near to those of $\tilde{x}$, will be denoted by the functional $I(\tilde{x})_{V_s}$. The individuals closer to $\tilde{x}$ are more numerous.

Axiom 3 (Law of the regular generation). *If an individual* x, *which belongs to a particular type* V_s, *breeds another individual* y, *then in general the two individuals are very similar, that is they have very similar characters (that is parameters). The values of all parameters of* x *and of* y *are near to* $\tilde{x}$ *of* V_s.

Symbolically:

$$\forall x\ \forall y\ (U(x) \wedge U(y) \wedge G(x,y) \longrightarrow (A(x,y) \wedge x \in I(\tilde{x})_{V_s} \wedge y \in I(\tilde{x})_{V_s})$$

where G is a constant dyadic predicate which we read "x breeds y" and $G(x,y) \neq G(y,x)$ and $G(x,y) \longrightarrow\ \urcorner\ G(y,x)$. A is another constant dyadic predicate which we read "x and y are very similar". From Axiom 3, the theorem follows immediately:

Proposition 1. *If* x *breeds* y *and* z *breeds* y *then* x *and* z *belong to the same type.*

Symbolically:

$$\forall x\ \forall y\ \forall z\ (U(x) \wedge U(y) \wedge U(z) \wedge G(x,y) \wedge G(z,y)$$
$$\longrightarrow (x \in I(\tilde{x})_{V_s} \wedge z \in I(\tilde{x})_{V_s}))$$

Now we introduce the important notion of *generation distance* $\gamma(x, y)$ between two individuals x and y such that $G(x, y)$ so:

$$\gamma(x,y) =_{def} \mathcal{F}(\|c_{xi} - c_{yi}\|)$$

where, as regards the logical language, γ is a functional. In particular $\mathcal{F}(\|c_{xi} - c_{yi}\|)$ can be $(\sum_{i=1}^{n} \|c_{xi} - c_{yi}\|)^{\frac{1}{2}}$. We have then the following

Axiom 4. *The* generation distance *between two individuals x and y depends on the biological situation of x.*

Symbolically:

$$\forall x\, \forall y\, (U(x) \wedge U(y) \wedge G(x,y) \longrightarrow D(\gamma(x,y), x))$$

where D is a constant dyadic predicate which we read "...depends on ...".

The evolution theory which we have proposed in [5] and [4] has also two other axioms (**Ax. 5:** *Law of immanence of the standard type or of the exceptional generation* and **Ax. 6:** *Fundamental law of speciation*).

In [5] and [4] we have already shown the

Theorem 1. *The theory determined by the preceding Axioms 1, 2, 3, 4, 5, 6 has a geometrical interpretation in $\mathbb{R}^n$, where n is the number of the parameters $(c_1, c_2, \ldots, c_n)$.*

But Axiom 4 and Axiom 6 enable the construction of a dynamical model by means of a finite difference equation. As a matter of fact, if we denote by $\mathbf{x}_t$ the vector which determines the parent individual x in $\mathbb{R}^n$ and by $\mathbf{x}_{t+1}$ the vector which determines the son y, we can give the following system which interprets the Axiom 4 by means of the Theorem 1. Axiom 6 implies the existence of a *threshold* like the number of successive generations from a given individual of a species (either the evolution goes toward the speciation or stays into the species), there is the system:

$$\begin{cases} \|\mathbf{x}_{t+1} - \mathbf{x}_t\| = k\frac{\|\mathbf{x}_t\|^h}{\|\mathbf{x}_t\|^2+m} & \text{with } h \in \mathbb{Q};\ m,\ k \in \mathbb{R}_+ \\ angle(\mathbf{x}_t, \mathbf{x}_{t+1}) = \text{rand}\left(\left[hypercone\left(-\frac{\pi}{s}, \frac{\pi}{s}\right)\right]\right) & \text{with } s \in \mathbb{N} \end{cases} \quad (1)$$

where h, m and k are suitable coefficients and $angle(\mathbf{x}_t, \mathbf{x}_{t+1})$ is, into an assigned hypercone, the random deviation of the vector (from $\mathbf{x}_t$ to $\mathbf{x}_{t+1}$) which represents $G(x, y)$ in $\mathbb{R}^n$. The dynamics of (1) asymptotically diverges, but we can establish the number (*threshold*) of steps. Now we consider the 2D case, that is when we have only two independent parameters (c_1, c_2). This case is not particularly close because c_1 and c_2

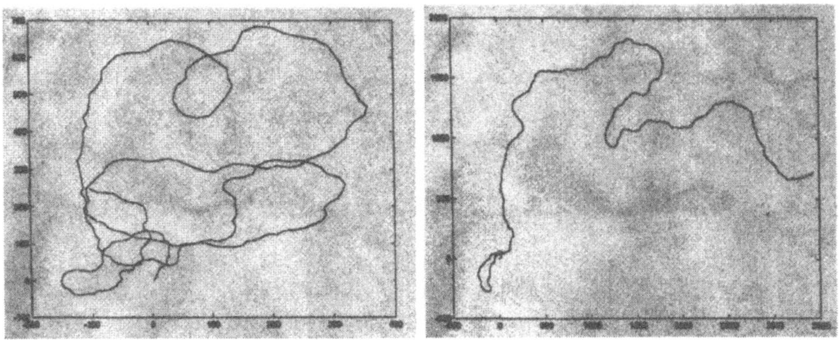

Figure 3. Numerical simulation (when $m = 1$, $h = \frac{5}{2}$, $k = 1$ and $s = 3$) according to model (1). The two graphs in $\mathbb{R}^2$ represent two among infinite possibilities of individual generation sequences which depend too much on the random part of (1) and on the choice of s. One of these infinite possible sequences, but we cannot know it, could coincide with one which is real into an hypothetical species. This model must be considered as a beginning of a working hypothesis.

will be two parameters definitely independent between them, and each of them can synthetically express other dependent parameters. According to Theorem 1 in this case we consider $\mathbb{R}^2$. Hence we have the following results of numerical simulation when $m = 1$, $h = \frac{5}{2}$, $k = 1$ and $s = 3$. By virtue of the random position of $angle(\mathbf{x}_t, \mathbf{x}_{t+1})$ we have "a priori" infinite possibilities of evolution dynamics like the two represented in figure 3. This for the chance of the directions of breeding. Hence, the model (1) is valid if we think an evolution with a lot of possibilities for an individual which breeds, even if the generation distance is determined for every individual. In an upcoming next work we will consider a case where, for the evolution dynamics, also $\tilde{x}$ changes its situation, and we will establish the dependence of the generation distance also from an appropriate expression of the density $\rho(\mathbf{x}_t)$ of individuals round $\mathbf{x}_t$.

4. Emergent properties as axioms

The previous theory, which is inspired by G.V. Schiaparelli's mathematical translation of Darwin's original ideas (1898) (see [10] and [4]), considers the generation, but is it possible, in an exact and formal exposition, to express, by some constant predicates, this concept connected to genetic concepts? The answer is positive (see [8, 9, 13]) and so we can individualize emergent properties. Hence we can construct a unique theory which regards either evolutionary concepts or genetic concepts.

So the language $\mathcal{L}$ of this theory includes also terms like *allele* [constant monadic predicate E], *chromosome* [constant monadic predicate C], *zygote* [constant monadic predicate Z], *parental zygote* [constant monadic predicate P], *child zygote* [constant monadic predicate F], and by $C_z(y)$ we denote that y is a chromosome into a zygote and by $C_P(x)$ we denote that x is a chromosome into a parental zygote, analogously $C_F(x)$. Besides, we need the following constant dyadic predicates:

$\approx$ expresses, according to Woodger's genetic ideas [12], that two individuals cannot be equal if they are not the same individuals, while two individuals can be identical even if they are not the same individual;

$\tilde{\in}$, that is $x\tilde{\in}y =_{def} \exists z\,(z \approx x \wedge z \in y)$. This predicate says that x and y are correlated when an individual which is identical to x belongs to y;

$H_z(x,y)$ which we read "the chromosomes x and y are homologous into the zygote z";

$H(x,y)$ simply means "x and y are homologous".

By means these and other terms of $\mathcal{L}$ we can express some properties about the *generations* which cannot be deduced by the preceding axioms. Therefore these properties can be considered "emergent" in regard to the preceding biological theory, which is characterized by the Axioms 1,...,6. It seems obvious, natural, to consider these emergent properties as further new axioms because these emergent properties are independent from the Axioms 1,...,6. Here are some emergent properties which become axioms [8]:

Axiom 7. *For every zygote, every chromosome has a unique homologous chromosome.*

Symbolically:

$$\forall z\,\forall y\,\exists x\,[(U(z) \wedge U(y) \wedge U(x) \wedge Z(z) \wedge C_z(y) \wedge C(x) \\ \longrightarrow H_z(y,x)) \wedge (\forall w(U(w) \wedge C(w) \longrightarrow H_z(y,w))) \longrightarrow x = y]$$

Axiom 8. *The two parental zygotes take part each by means of a chromosome for every pair of homologous chromosomes of a child zygote.*

Symbolically:

$$\forall x\,\forall y\,\forall z\,\forall w\,\forall w'\,[(U(x) \wedge U(y) \wedge U(z) \wedge U(w) \wedge U(w') \wedge P(x) \wedge P(y) \wedge \\ \wedge F(z) \wedge C(w) \wedge C(w') \wedge G(x,z) \wedge G(y,z) \wedge H_z(w,w')) \\ \longrightarrow ((w\tilde{\in}x \wedge w'\tilde{\in}y) \vee (w\tilde{\in}y \wedge w'\tilde{\in}x))]$$

Axiom 9. *For every chromosome of the parental zygote, a homologous chromosome of the child zygote exists.*

Symbolically:

$$\forall x \forall y \forall z \forall w \, [(U(x) \wedge U(y) \wedge U(z) \wedge U(w) \wedge$$
$$\wedge P(x) \wedge P(y) \wedge F(z) \wedge C_P(w) \wedge G(x,z) \wedge G(y,z))$$
$$\longrightarrow \exists w' \, (U(w') \wedge C_F(w') \wedge H(w,w'))]$$

But if we consider the other laws or properties of the classic genetics, it is necessary to enrich the language $\mathcal{L}$ to express them. Then we will have other emergent properties and therefore we will assign other axioms.

Of course, we can find other biological properties which are results of the scientific research and of the experimental discovery and they are not explainable and demonstrable by this system of axioms, even if they can be expressed by the language $\mathcal{L}$ of the same system. Thence we will have more new emergent properties.

But where does this process of iteration of emergent properties or this consistent expansion of a theory go? How far does it reach? Does it stop or not?

5. Conclusion

We do not really know whether this process of iteration emergent properties stops or not and when it stops. The metatheorical limiting results (Gödel and Chaitin) can support our opinion. We think that it is not possible to construct an absolute theory which can explain all biological phenomena. However, we can propose different theories to explain, formally or not, the same set of biological phenomena. Therefore a possible definition of "emergent property" and of "complex part of reality" (for the Biology) can be the following:

> *Given a theory T, an* emergent property *for T is a property which can be expressed by means of the language $\mathcal{L}$ of T or by means of a conveniently enriched $\mathcal{L}$, but it cannot be shown by the general principles (axioms) and the logical rules of T.*
>
> *Also we define a part of reality as* complex *if every theory T which describes this part of reality always presents emergent properties.*

Indeed we think that in order to study the biological phenomena we cannot be determinist but surely a little reductionist: the complexity of the biological reality is a datum which is difficult to deny, but to construct some form of biological theory it is necessary to start from a conceptual reduction and adaptation which must be as much accepted as possible.

References

[1] ANDRADE J., *Leçons de Mécanique Physique*, Soc. d'édition scientifiques, Paris 1898.

[2] DEL RE G., MARIANI E. EDS., *Caos, ordine e complessità*, I.P.E., Napoli 1994.

[3] FORTI M., FREGUGLIA P., GALLENI L., *An axiomatic approach to some biological themes*, in: "The Application of Mathematics to the Sciences of Nature. Critical Moments and Aspects", P. Cerrai, P. Freguglia and C. Pellegrini eds., Kluwer Academic/Plenum Publishers, New York 2002.

[4] FREGUGLIA P., *Considerazioni sul modello di Giovanni V. Schiaparelli per una interpretazione geometrica delle concezioni darwiniane*, in: "Modelli matematici nelle scienze biologiche", P. Freguglia ed., Quattro Venti ed., Urbino 1998.

[5] FREGUGLIA P., *For an Axiomatic Theory of the Evolutionary Darwinian Ideas: A Proposal*, in: "The Application of Mathematics to the Sciences of Nature. Critical Moments and Aspects", P. Cerrai, P. Freguglia and C. Pellegrini eds., Kluwer Academic/Plenum Publishers, New York 2002;
Per una teoria assiomatica delle concezioni darwiniane II, in: "Atti e Memorie della Accademia Petrarca di Lettere, Arti e Scienze", n.s., **LXI**, Arezzo 1999.

[6] GALLENI L., *Aspetti teorici della biologia evoluzionista*, in: "Modelli matematici nelle scienze biologiche", P. Freguglia ed., Quattro Venti ed., Urbino 1998.

[7] ISRAEL G., *La mathématisation du réel*, Éditions du Seuil, Paris 1996.

[8] RIZZOTTI M., ZANARDO A., *Assiomatizzazioni della Genetica. Aspetti non estensionali in alcune leggi elementari della genetica classica*, Memorie Lincee, Scienze Fisiche e Naturali s. IX **II** (4) (1993).

[9] RIZZOTTI M., ZANARDO A., *Axiomatization of genetics. 1. Biological meaning*, J. Theor. Biol. **118** (1986).

[10] SCHIAPARELLI G.V., *Studio comparativo tra le forme organiche naturali e le forme geometriche pure*, Hoepli ed., Milano 1898.

[11] VON BERTALANFFY L., *General System Theory*, George Braziller, New York 1968.

[12] WOODGER J.H., *The Axiomatic Method in Biology*, Cambridge University Press, Cambridge 1937.

[13] ZANARDO A., RIZZOTTI M., *Axiomatization of genetics. 2. Formal development*, J. Theor. Biol. **118** (1986).

IGNORING COMPLEX INTERACTIONS IN NATURAL ECOSYSTEMS: THE CASE OF TERMINATOR TECHNOLOGY

Manuela Giovannetti
Dipartimento di Chimica e Biotecnologie Agrarie,
Università di Pisa,
and
Center for Soil Microbiology, C. N. R.,
Via del Borghetto 80, 56124 PISA - ITALY
mgiova@agr.unipi.it

Abstract Potential risks and benefits of genetically modified organisms (GMO) released into the environment have been largely debated within the political and scientific communities. Available studies on the ecological impact of transgenic crop plants have shown that risks and benefits are difficult to assess in field experiments where environmental complexity hinder absolute certainty and clearcut, yes/no responses. Nevertheless, some authors reported "previously unsuspected hazards", stressing the need of further experimental work, both in the field and in the laboratory, to gather as many data as possible to evaluate environmental risks of GM crops. Since existing scientific literature reports contrasting results, the general view is to avoid generalisation and to make judgements on a case-by-case basis. Here I shall discuss available data concerning "unexpected events", with the aim of pointing out the risks associated with the cultivation of "Terminator" plants. The unpredictability of genetic events due to transformation, such as horizontal gene transfer among different organisms and harmful effects of some gene products on non-target organisms, raises serious concerns about the cultivation of Terminator seeds in agriculture. Terminator seed technology is a paradigmatic case which may help us to point out potential environmental problems and envisage possible disasters, since the nature of the genes involved poses such a threat that fundamental genetic and biological questions should be answered before any environmental release of the modified seeds. In fact, plants modified through this technology contain the genes which will cause the death of second generation seeds, thus protecting patented genes introduced in a crop against further utilisation. The awareness that scientists and technologists do not know in advance all the possible interactions between transgenes and differ-

ent components of complex ecosystems should foster long-term studies aimed at evaluating the environmental impact of particularly dangerous genes, such as Terminator, which interfere so heavily with fundamental life processes.

1. Introduction

Potential risks and benefits of genetically modified organisms (GMO) released into the environment have been largely debated within the political and scientific communities, though, experimental evidence published in peer-reviewed journals is still scanty. Available studies on the ecological impact of transgenic crop plants have shown that risks and benefits are difficult to assess in field experiments where environmental complexity hinder absolute certainty and clearcut, yes/no responses. Nevertheless, some authors reported "previously unsuspected hazards", stressing the need of further experimental work, both in the field and in the laboratory, to gather as many data as possible to evaluate environmental risks of GM crops. Accordingly, the scientific journal "Nature" published a briefing on the subject, advertised on the cover page by the relevant question "GM crops – how safe is 'safe'?" [1]. Since existing scientific literature reports contrasting results, variable in space and time and depending on the plant, the general view is to avoid generalisation and to make judgments on a casebycase basis [18].

Here I shall discuss available data concerning the environmental impact of transgenic pollen diffusion, the impact of toxins produced by transgenic crops on non-target organisms, horizontal gene transfer from GM plants to microorganisms and among microorganisms. The reported experimental evidences could represent the basis for *ex-ante* evaluations of the risks associated with the cultivation of transgenic plants, transformed with potentially dangerous genes, such as "Terminator". In fact, given the nature of the genes involved, which interfere with fertility and life, any risk of irreversible genetic events should be avoided and, in the absence of reliable scientific data, the environmental release of transgenic "Terminator" organisms should not be permitted.

2. Environmental risks

2.1 Impact on non-target organisms

Most studies have been carried out on plants modified to produce insecticidal proteins, such as *Bacillus thuringensis (Bt)* toxins, which affect Lepidoptera (butterflies and moths) and Coleoptera (beetles). A much quoted example is represented by the laboratory based work on pollen from maize engineered with *Bacillus thuringensis (Bt)* insecticide genes. Such pollen has been found to be harmful not only to the target organism, the European corn borer, but also to the non-target monarch butterfly (*Danaus plexippus*) [14]. It has been reported that the *Bt* toxin is expressed in pollen in most hybrids and that maize pollen may be dispersed by wind over 60 meters and deposited from *Bt* maize fields on milkweed leaves growing nearby. Since monarch larvae feed exclusively on milkweed, the research findings of Losey et al. show that even GM plants previously claimed to have negligible environmental impact may represent a potential hazard to non-target organisms and a threat to the conservation of biodiversity.

Another example refers to an unexpected event described by Saxena, Flores and Stotzky [15]. The authors showed that *Bt* insecticidal toxin is released by *Bt* corn roots into the rhizosphere soil and that it retains insecticidal activity when bound to soil particles. This finding indicates that further investigations are necessary on phenomena occurring in the soil environment, where the amount of *Bt* toxin may reach high levels due not only to the incorporation of crop residues after harvest but also to its release in root exudates during corn growth. The high levels of insecticidal toxins, together with their persistence in the soil environment, may increase the probability of negative effects on non-target organisms, since *Bt* toxin is able to actively bind to soil particles, where it is protected against microbial degradation. Moreover, it has been found that the persistence of active *Bt* toxin in the soil is of at least 234 days and increases in clay and acid soils.

2.2 Pollen dispersal

Another important unexpected event concerns potential pollen dispersal and the spread of transgenes through hybridization with closely related wild species or subspecies. Such gene flow mightrepresent a risk when herbicide-tolerant crops are released into the environment. About 20 different crop species have been genetically modified to confer tolerance to herbicides, such as glyphosate, and range from maize and soybean to oilseed rape, sugar beet, cotton, lettuce, tomato. Large-scale cultivation of these crops, with the inevitable and unpredictable diffusion of their pollen, could lead to genetic pollution of natural gene pools, and creation of "superweeds" [11]. Published studies on oilseed rape (*Brassica napus*) genetically modified for glufosinate tolerance demonstrated not only intraspecific gene flow and hybrid formation but also interspecific gene flow with *Brassica rapa*, *Brassica campestris* (=*rapa*) and wild mustard [18].

Since many important food crops, including wheat, rice, maize, barley and soybean, are capable of natural hybridization with wild relatives [11], the fundamental problem about 'gene escape' is that if the genes involved pose a threat, the distance of pollen dispersal is not a relevant issue, since there is no way to stop pollen movement and cross-pollination.

2.3 Horizontal gene transfer

The fundamental environmental risk of transgenic crops relates to the fact that almost nothing is known of the fate of transgenes after their field release. In particular, we do not know the impact of GM crops and GM crop residues on organisms which feed on them, such as insects, animals and microbes. What we certainly know is that in nature, as well as in the laboratory, genes can be transferred from one organism to another [12, 13, 2] and that they have been transferred from plants to bacteria also during evolution [17, 4]. Recent studies have shown that engineered genes can be transferred from GM plants to soil bacteria [8]. The authors challenged the technology used for the production of transgenic plants, consisting in the co-introduction, together with the gene of interest, of a gene conferring antibiotic resistance, which allows the detection of transformed cells incorporating the transgenes. Since the antibiotic resistance genes are bacterial genes, the large quantities of these genes released in the environment with transgenic crops raise concerns on the possibility that they are taken up by indigenous soil bacteria. Although industry and FDA claimed that the novel genes transferred to plants could not be incorporated back by bacteria [US FDA, 1995], investigations carried out by soil microbiologists showed horizontal gene transfer between GM plants and a bacterial species belonging to the genus *Acinetobacter* [8]. In their transformation experiments, performed under optimized laboratory conditions, 2 μg of transgenic sugar beet DNA yielded bacterial transformants with a transformation frequency of 5.4×10^{-9}. Interestingly, also plant homogenate of transgenic sugar beet leaves showed transformation capacity, though at lower frequencies, 1.5×10^{-10}. This kind of transformation event has still to be demonstrated to occur under natural conditions, yet laboratory results suggest that the occurrence of gene transfer from plant chromosomes to competent native soil bacteria carrying homologous sequences is possible. These data raised concerns over the dissemination of transgenes into the environment by means of indigenous terrestrial microorganisms, in particular when considering the wide use of antibiotic resistance genes as marker genes in GM plants and bacteria. The risk of horizontal transfer of such genes from transgenic plants to soil bacteria and then from the latter to an indefinitely large number of other bacteria had been previously considered highly improbable, but these recent findings suggest that the issue must be reconsidered. When bacteria are concerned, the meaning of "improbable event" has to be assessed taking into account the fact that bacteria, given enough food and space, reproduce very quickly and, for instance, may reach the number of one billion of billions in about 30 hours. On the other hand, considering that about one billion of bacteria live in one gram of fertile soil, we can easily calculate that one billion of billions of bacteria may be found in about 3000 square meters of soil [10]. These calculations suggest that in the microbial world, whatever event can occur, it will. Thus, the fundamental problem about 'gene escape' is that if the genes involved pose a threat, the low probability of an event is no relevant issue.

2.4 Other unpredicted outcomes

The introduction of new genes into an organism have in some cases led to "previously unpredicted" outcomes, related to unplanned increase or decrease in the content of secondary metabolism compounds or alterations in crop chemistry (see [6]). Besides these experimental findings, the claim that single-gene changes lead to well-characterised plant responses has to be reconsidered, since when using transformation techniques, based on random insertion of new constructs into genomes, we are unable to predict positional effects and rearrangements of genes, which raise the risk that silent genes are switched on (see [16]). Accordingly, the risks of exotic genes isolated from different organisms, such as viruses and bacteria, or non-food plant species, and introduced into food species, often with viral promoters inserted to make genes overexpress continuously, should be adequately evaluated.

The wide utilisation of the conventional probabilistic approach in risk assessments and the "safe until proven otherwise" claim supported the common view that absence of data on harm equals absence of harm and safety of products or processes: on the contrary, it often means just that experiments have still to be carried out. By contrast, the "precautionary approach" takes into account the probability that a certain technology will represent a risk and weighs benefits against potential costs, considering the irreversibility of putative harmful events [7]. The awareness of our incomplete knowledge and the possibility of unpredictable events may help us to avoid the occurrence of large-scale health and environmental disasters.

3. Terminator technology

The unpredictability of genetic events due to transformation, such as the location of the transgenes on chromosomes, gene silencing or activation by the introduced genes, horizontal gene transfer among different organisms, and the occurrence of previously unexpected events such as harmful effects of some gene products on non-target organisms, raise serious concerns about the cultivation of Terminator seeds in agriculture. Terminator seed technology is a paradigmatic case which may help us to point out potential environmental problems and envisage possible disasters, since the nature of the genes involved poses such a threat that fundamental genetic and biological questions should be answered before any environmental release of modified seeds [9].

In March 1998 the seed company Delta and Pine Land, together with the United State Department of Agriculture, obtained the US Patent Number 5.723765: Control of Plant Gene Expression, later dubbed "Terminator Technology" by Rural Advancement Foundation International (RAFI). The plants modified through this technology contain the genes which will cause the death of second generation seeds. Thus, any new patented gene introduced in a crop can be protected against further utilisation by also introducing a group of genes which will kill the seeds in the second generation: in this way farmers could not save and replant seeds, and should buy new seeds every year.

We know that in modern agriculture some crops such as maize are not grown from saved seeds, but are planted as hybrid seeds which are bought every year from big seed companies. On the contrary, other important crops, such as rice, wheat, soybeans and cotton, are not always grown from hybrid seeds, so that farmers, especially poor farmers in developing countries, may save and replant the seeds of their crops. This will not be possible any more if Terminator technology is utilised for engineering crop plants to kill their second generation seeds. I will describe how Terminator works, taking into account the elegant explanation published by Martha Crouch [3]. She made the example of cotton seeds engineered for herbicide tolerance. Such seeds will develop normally until maturity, when a peculiar toxin will be produced, only in the seeds, which will kill second generation seeds.

Terminator genetic construct is composed of:

1. A promoter activated late in seed development, called LEA (Late Embryogenesis Abundant), fused to the coding sequence for a toxin. In this way the gene will produce the toxic protein only in mature seeds, but it will not kill any other part of the plant. The proposed toxin is a

Ribosome Inhibitor Protein (RIP), originated from *Saponaria officinalis*, which inhibit protein synthesis.

This will allow Terminator plants to grow until maturity and to produce seeds, but since the seeds will not be viable, seed companies would not be able to sell their products. Thus, the toxin coding sequence should be activated only after the production of viable seeds to be sold to farmers, and exactly during the development of second generation seeds. The genetic engineering allowing such event is represented by:

2. A DNA piece inserted in between the LEA promoter and the toxin coding sequence, blocking the production of the toxin. The DNA blocking piece carries at both ends two DNA sequences recognised by a recombinase enzyme, which is capable of cutting and removing the DNA at the outside of the DNA piece, so that the cut ends of the DNA fuse and the LEA promoter is next to the toxin coding sequence. This ingenious genetic construct will obtain the same result as before, i.e. the production of the toxin at the end of seed development, and the problem of large production of seeds will remain unsolved. To multiply plants and obtain large seed quantities, another clever piece of genetic engineering has been devised:

3. The recombinase coding sequence is put next to a constitutive promoter, which is repressed, but which can be de-repressed by a chemical compound. Such chemical may be added to the seeds before selling them to the farmers, just before sowing. The antibiotic tetracycline is proposed as the compound controlling the repressible promoter system of recombinase. Thus, without tetracycline, the recombinase gene will be repressed and the toxin will not be produced.

When the seeds are treated with tetracycline, the cascade of genetic activation of the different constructs placed in the plant genome may be described as follows:

Tetracycline will interact with the repressor protein which will not interfere with the recombinase constitutive promoter. Recombinase will be produced, which will cut out the DNA piece blocking the toxin gene. Toxin will be produced at seed maturity.

This elegant patent will allow seed companies to treat seeds just before selling them, activating the production of recombinase in growing plants, the production of the toxin killing second generation seeds, and the protection of any patented gene introduced in the crop.

4. Risks associated with environmental release of Terminator seeds

When considering what already discussed about the fact that the location of the transgenes on chromosomes is an unpredictable event, that many of the introduced genes may be unpredictably silenced or activated, and that they may be transferred among different organisms, the environmental risks associated with the cultivation of Terminator plants in agriculture should be taken seriously.

A major problem posed by Terminator technology concerns the large quantities of the antibiotic tetracycline necessary to start the genetic cascade leading to toxin production in second generation seeds. The large-scale use of antibiotics in agriculture, and even in medicine, has been claimed to be one of the causes of the increase and spread of antibiotic resistance genes in bacteria [5]. Also the World Health Organisation [19] stressed the risks associated with the production of transgenic organisms which often carry antibiotic resistance marker genes in their genome. The huge amount of tetracycline reaching the soil through treated seeds would further increase the occurrence and spread of antibiotic resistance in soil microorganisms, and could have a negative ecological impact on all soil organisms.

Another serious risk is represented by 'gene escape' through transgenic Terminator pollen carrying the activated toxin producing gene. This escape cannot be avoided, since, as already discussed, there is no way to stop pollen movement and cross-pollination. The spread of Terminator genes through hybridisation with closely related wild species might represent a risk, in particular when major food crops, such as wheat, rice, maize, barley and soybean, which are able to hybridise with wild relatives, are concerned [18]. Terminator pollen would provoke the death of second generation seeds, but this would become evident late, after the farmer has planted the saved seeds. This means that Terminator genes would not spread any further, but it means also that an unpredictable number of saved seeds would be dead, representing a threat to the crops cultivated nearby Terminator crop plants.

Our inability to stop gene escape through pollen has been recognised also by biotech companies, since they claimed that Terminator technology could prevent the spread of transgenes in the environment. This would not be the case. In fact, the efficiency of tetracyclin treatments on every single seed out of millions of kilograms cannot be guaranteed, and recombinase may remain inactive in some seeds. Such seeds would carry all the Terminator genetic construct, the plants would grow and produce second generation seeds which will not die, but will develop into viable plants producing pollen, and then seeds, carrying both Terminator genes and the patented transgenes to be protected. In this way the genetic construct devised to kill second generation seeds would enter the complex chain of interactions characterising real life: how could we prevent birds and bees from spreading Terminator pollen and seeds, and soil bacteria from undergoing transformation and further horizontal gene transfer? As discussed above, a growing number of diverse "unexpected events" should be seriously evaluated, before claiming the environmental safety of genetic constructs dangerously interfering with life.

In addition, the phenomenon of gene silencing should not be underestimated, since introduced genes have in some cases unexpectedly failed to work. Moreover, the different components of the complex Terminator genetic construct could reshuffle during plant sexual recombination, the toxin would become active and kill seeds in unpredictable ways. Accordingly, in my view, the major threat is posed by a general application of reductionistic epistemology, which is at the base of genetic engineering.

The awareness that scientists and technologists do not know in advance all the possible interactions between transgenes and different components of complex ecosystems should foster long-term studies aimed at evaluating the environmental impact of particularly dangerous genes, such as Terminator, which interfere so heavily with fundamental life processes.

In conclusion, I would like to mention another fundamental environmental risk associated with Terminator, concerning more than one billion poor people whose main food source is based on replanting second generation seeds. The introduction of death genes in crops such as rice or wheat would have a great impact on the fate of millions of people: considering them non-target organisms, the negative impact of Terminator raises to unacceptable levels.

References

[1] Butler D., Reichhardt T., *Long-term effects of GM crops serves up food for thought*, Nature **398** (1999), 651.

[2] Courvalin P., *Gene transfer from bacteria to mammalian cells*, C. R. Acad. Sci. Ser. III Sci. Vie **318** (1995), 1207.

[3] Crouch M.L., *How the Terminator terminates: an explanation for non-scientists of a remarkable patent for killing second generation seeds of crop plants*, The Edmonds Institute, Edmonds, Washington 1998.

[4] Doolittle R.F., Feng D.F., Anderson K.L., Alberro M.R., *A naturally occurring horizontal gene transfer from a eukatyote to a prokaryote*, J. Mol. Evol. **31** (1990), 383.

[5] Ferber D., *Superbugs on the hoof?*, Science **288** (2000), 792.

[6] FIRN R.D., JONES C.G., *Secondary metabolism and the risks of GMOs*, Nature **400** (1999), 14.

[7] FOSTER K.R., VECCHIA P., REPACHOLI M.H., *Science and the precautionary principle*, Science **288** (2000), 979.

[8] GEBHARD F., SMALLA K., *Transformation of Acinetobacter sp. strain BD413 by transgenic sugar beet DNA*, Appl. Environ. Micriobiol. **64** (1998), 1550.

[9] GIOVANNETTI M., *Piante transgeniche, ecosistemi e geni della morte*, Il Ponte **40** (1999), 104.

[10] GIOVANNETTI M., *Environmental impact of transgenic crops: potential risks associated with "Terminator" technology*, in: "5th International Conference on Biotechnology, Science and Modern Agriculture: a New Industry at the Dawn of the Century", Ravello, Italy, June 15–18th 2002, 47.

[11] GRAY A.J., RAYBOULD A.F., *Reducing transgene escape routes*, Nature **392** (1998), 653.

[12] HEINEMANN J.A., SPRAGUE G.F., *Bacterial conjugative plasmids mobilize DNA transfer between bacteria and yeasts*, Nature **340** (1989), 205.

[13] HOOYKAAS P.J.J., *Transformation of plant cells via Agrobacterium*, Plant Mol. Biol. **13** (1989), 327.

[14] LOSEY J.E., RAYOR L.S., CARTER M.E., *Transgenic pollen harms monarch larvae*, Nature **399** (1999), 214.

[15] SAXENA D., FLORES S., STOTZKY G., *Insecticidal toxin in root exudates from Bt corn*, Nature **402** (1999), 480.

[16] TWEEDIE S., BIRD A., *Mutant weed breaks silence*, Nature **405** (2000), 137.

[17] WAKABAYASHI S., MATSUBARA H., WEBSTER D.A., *Primary sequence of a dimeric bacterial haemoglobin from Vitreoscilla*, Nature **322** (1986), 481–483.

[18] WOLFENBARGER L.L., PHIFER P.R., *The ecological risks and benefits of genetically engineered plants*, Science **290** (2000), 2088.

[19] WHO, *Health aspects of marker genes in genetically modified plants. Report of WHO Workshop*, Ginevra 1993, WHO/FNO/FOS 93, 6.

A COMPRESSION ALGORITHM AS A COMPLEXITY MEASURE ON DNA SEQUENCES

Giulia Menconi
Centro Interdisciplinare per lo Studio dei Sistemi Complessi,
Università di Pisa,
Via Bonanno 26/b, PISA - ITALY
menconi@mail.dm.unipi.it

Abstract A new compression method has been used to prove the existence of long range correlated repetitive sequences in some complete genomes within the three domains of life. We defined the computable complexity of a sequence. The consequent complexity analysis both allowed to distinguish the functional regions of the genome and to identify the lowest complex regions which match with noncoding regions.

1. Information of finite sequences

We look at genomes as symbolic sequences whose alphabet is the alphabet of nucleotides $\{A, C, G, T\}$.

The *Information content* of a sequence is the amount of *information units* which are necessary to describe the sequence. It is a quantity which depends only on the intrinsic features of the sequence under examination and not on the statistical properties of the source which generated the sequence.

As an example of this point of view, consider a constant sequence $s =' AAAAAAAAAA'$: its Information content can be represented by the number of bits of the following sentence:

repeat ten times the symbol 'A'.

This is the intuitive meaning of the ideal information content: the Algorithmic Information Content $AIC(s)$ of the sequence s, first described by Kolmogorov and Chaitin ([5]). For a generic sequence s, the quantity $AIC(s)$ is the length in bits of the shortest program which outputs s. The AIC is infinite if there are no programs which output s. We require that our computing machine is universal. Roughly speaking, a computing machine is called *universal* if it can simulate any other machine (for a precise definition see any book of recursion). In particular if U and U' are universal then $I_{AIC}(s, U) \leq I_{AIC}(s, U') + const$, where the constant depends only on U and U'. This implies that, if U is universal, the AIC of s with respect to U depends only on s up to a fixed constant.

The AIC is the optimal encoding of a sequence, but unfortunately its computation for a sequence cannot be performed by any computer. The AIC is called a not computable number.

A way to overcome the computability problem is to use a different measure for the Information: we will adopt the compression algorithms to measure the Information content of the genomes.

Let us consider a lossless compression algorithm Z, which is a function $Z : \Sigma \to \{0,1\}^*$ where Σ is the space of the sequence of any length and whose symbols are taken in the set of nucleotides and $\{0,1\}^*$ is the space of the binary sequences of any length. The compression algorithm Z encodes the original sequence s into an encoded sequence $Z(s)$. We will indicate the length of the sequence s with $|s|$.

Definition 3. *Given a compression algorithm Z, the Computable Information Content of a sequence $s \in \Sigma$ with respect to the algorithm Z is the length of the encoded sequence: $CIC_Z(s) = |Z(s)|$.*

It turns out that this definition of Information content is a computable number, because its the result of a performed computation.

It is useful to compare the Information content of a sequence with its original length, so that we can define as follows the compression ratio of the sequence, which we will call Computable Complexity of the sequence.

Definition 4. *Given a compression algorithm Z, the Computable Complexity of a sequence $s \in \Sigma$ with respect to the algorithm Z is the compression ratio $K_Z(s) = \frac{|CIC_Z(s)|}{|s|}$.*

The computable complexity of a sequence is then the mean number of bits which are necessary to generate any symbol of the sequence.

We define in the same way the *algorithmic complexity* of a finite sequence s as $K_{AIC}(s) = \frac{|AIC(s)|}{|s|}$. The two notions are related by the following result, whose details and hypotheses are described in [3].

Theorem 2. *If the symbolic source Σ is ergodic and σ is an infinite sequence drawn from Σ then*

$$Prob\left(\lim_{n\to\infty} K_Z(\sigma^n) = \lim_{n\to\infty} K_{AIC}(\sigma^n) = \hat{H}(\sigma)\right) = 1$$

where $\hat{H}(\sigma)$ is the Shannon entropy of the sequence σ and σ^n is the sequence containing the first n digits of σ.

1.1 The algorithm CASToRe

In this paper, we will analyse the Computable Complexity of genomes using a compression algorithm called *CASToRe*, which has been presented in [2] and [3]. In the following, we will always refer to the algorithm *CASToRe*, so that we will omit to indicate Z in the CIC and K formulas.

We now give a short description of the algorithm. CASToRe is a compression algorithm which is a modification of the LZ78 algorithm ([7]). Its theoretical advantages with respect to LZ78 are shown in [4] and [2]: it is a sensitive measure of the Information content of sequences. That's why is called **CASToRe**: **C**ompression **A**lgorithm, **S**ensitive **To** **Re**gularity.

The algorithm CASToRe is based on an adaptive dictionary.

At the beginning of encoding procedure, the dictionary is empty. In order to explain the principle of encoding, let's consider a point within the encoding process, when the dictionary already contains some words.

We start analyzing the input stream, looking for the longest word W, which is already contained in the dictionary, matching the stream. Then, we look for the longest word Y (already contained in the dictionary and not necessarily different form the word W) where the joint word WY matches the stream.

The output file contains an ordered sequence of pairs (i_W, i_Y) such that i_W and i_Y are the dictionary index numbers corresponding to the words W and Y, respectively. The pair (i_W, i_Y) is referred to the new endoded word WY and has its own index number.

The following example shows how the algorithm CASToRe encodes the input stream

$$s =' AGCAGAGCCAG' .$$

The first column is the dictionary index number of the codeword which is showed in the same line, second column. For an easier reading, we add a third column which shows each encoded word in the original stream s, but it is not contained in the output file:

$$\begin{array}{lll} 1 & (0, A) & [A] \\ 2 & (0, G) & [G] \\ 3 & (0, C) & [C] \\ 4 & (1,2) & [AG] \\ 5 & (4,3) & [AGC] \\ 6 & (3,4) & [CAG] . \end{array}$$

1.1.1 The window segmentation.

A modified version of the algorithm CASToRe can be useful in studying correlated and repetitive sequences within the genomes. We performed a window segmentation option which partitions the stream into segments of a fixed size and proceeds to encode them separately, as the following scheme shows:

1 choose a window range L;

2 construct the set $\{W_i\}$ of all the frames of size L within the sequence s under examination;

3 calculate the $CIC(W_i)$ for each frame W_i;

4 calculate the **complexity at range L**, $K(L) = \langle K(W_i)\rangle$, which is defined as the mean computable complexity over all the frames W_i.

At this point, there are two possible results which can be extracted from the window segmentation performed on the sequence s:

(i) we can study the mean complexity $K(L)$ as a function of the range L and obtain a sort of *spectrum* of the sequence;

(ii) we can study how the complexity $K(W)$ of the windows W varies along the sequence, at a fixed window range L, this way identifying the regions at lowest complexity.

2. Complexity of complete genomes

We have analysed the complexity of 13 complete genomes within the three domains of life: Archaea, Prokaryotes and Eukaryotes, together with chromosomes II and IV of *Arabidopsis thaliana.*

The table shows the results: we have to remark that the maximal complexity is achieved when the occurrences of the symbols are independent from each other and its maximal value equals $\log_2 \#\mathcal{A}$, where $\mathcal{A}$ is the alphabet and $\#\mathcal{A}$ is the number of symbols in the alphabet. For quaternary sequences, like the genomes, this mean maximal complexity is 2 bits per symbol.

In the case of the genomes, we can recognise that the values of the complexity K are significantly different from 2, so that the genomes are not equivalent to random sequences. It is also possible to clearly recognise that some genomes have very low complexity (smaller than 1.84 bits per symbol), which means that their internal structure presents mid and long range correlations. Another reason to justify the low complexity is the presence of repetitive sequences in the genome: the algorithm CASToRe is forced by its own nature to find out previous occurrences of words, so that the Information content decreases. In section 3, we will investigate where the low complexity regions are located.

Finally, we underline that the algorithm CASToRe has not specialistic features for the analysis of genomes, so this result is quite impressive and highlights the differences with previous application of compression methods on genome analysis ([1], [6], [8], [9]), which were focused not on complete genomes.

Genome	Complexity K
Methanococcus Jannaschii	1.79411
Archeoglobus fulgidus	1.90939
Methanobacterium autrophicum	1.90707
Pyrococcus	1.90068
Aquifex aeolicus	1.88282
Escherichia coli	1.89303
Bacillus subtilis	1.87038
Haemophylus influenzae	1.86589
Mycoplasma genitalium	1.84845
Rickettsia prowazekii	1.82346
Thermotoga maritima	1.89342
Arabidopsis thaliana (chr. II and IV)	1.89176
Caenorhabditis elegans	1.77681
Saccharomyces cerevisiae	1.88927

3. Long range correlated repetitions

In order to identify the amplitude of the repetitive sequences within a genome and their role in the Information content behaviour, we have studied the value of the complexity K as a function of the window range L, performing the window segmentation (see section 1.1.1).

The *spectrum* shows the extent of the correlated repetitions within the genome. As an example, consider figure 1 on the left. The graph shows the function $K(L)$ versus the window range L for the genome of the yeast *Saccharomyces cerevisiae.* As the window range increases, the mean complexity decreases. It follows that the extent of correlated repetitive sequences increases, so that we can conclude that the yeast genome presents correlations at a range of at least 2×10^6 base pairs.

This analysis has been repeated for all the 21 genomes and this way it has been possible to detect how long the correlations can be in those genomes.

Figure 1 on the right shows that complexity behaves in clearly different ways in the three different functional regions of the genome of *Saccharomyces cerevisiae*: coding (exons), intergenic and intronic regions. The coding regions are the less correlated and the most random (the complexity decrease is early interrupted), while both noncoding regions present longer correlated repetitions, which are more evident in the intronic regions.

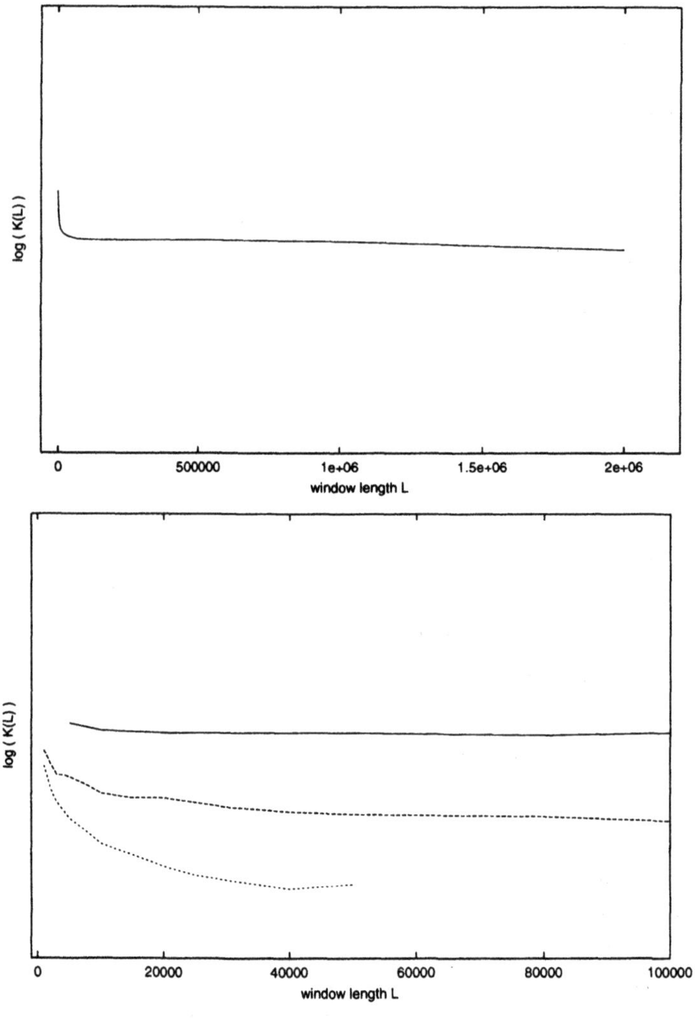

Figure 1. Top: the complexity K(L) vs. the window range L for the complete genome of *Saccharomyces cerevisiae*. Bottom: the complexity K(L) vs. the window range L for the different functional regions of the genome of *Saccharomyces cerevisiae*. The solid line is the complexity of the coding regions, the dashed line is the complexity of intergenic regions, the dotted line is the complexity of intronic regions.

Finally, we remark that, even in the cases where the complexity K of the complete genome is not significantly small, the analysis of the window complexity along the genome, at a fixed window range, reveals interesting characteristics of the genome and allows to localize the regions which present lower complexity.

Figure 2 on the left shows the behaviour of the complexity $K(W)$ as a function of the windows W with fixed window range $L = 10000$ base pairs, for the complete genome of *Methanobacterium autotrophicum*. The complexity of the whole genome is 1.90707 bits per symbol, but the window analysis highlights two regions where the complexity intensely decreases.

The enhancement pictured in figure 2 on the right is referred to the first low complex region (from 960000 bp to 1040000 bp). The dotted lines at level 1.6 and at level 1.2 respectively represent the noncoding regions and the coding regions. It is quite impressive that the low complex region matches a noncoding region with the greatest extent. This result can be found also in the other genomes.

On conclusion, this technique can be useful in identifying different functional intervening regions and can be easily applied to complete genomes with no need of specialistic tools.

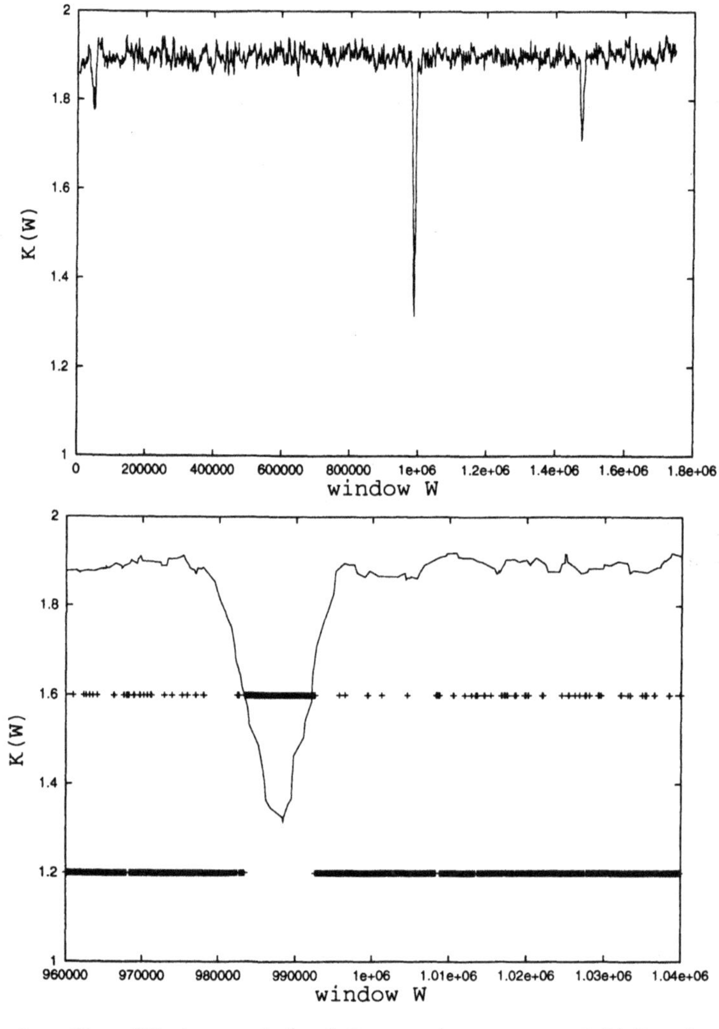

Figure 2. Top: Window analysis of the complete genome of *Methanobacterium autotrophicum*. The window range is fixed at $L = 10000$ bp. Bottom: the enhancement of the first region where the complexity rapidly decreases, compared with its functional meaning (see text).

References

[1] ALLISON L., STERN L., EDGOOSE T., DIX T.I., *Sequence complexity for biological sequence analysis*, Comput. Chem. **24** (2000), 43–55.

[2] ARGENTI F., BENCI V., CERRAI P., CORDELLI A., GALATOLO S., MENCONI G., *Information and dynamical systems: a concrete measurement on sporadic dynamics*, Chaos, Solitons and Fractals **13** (2002), 461–469.

[3] BENCI V., BONANNO C., GALATOLO S., MENCONI G., PONCHIO F., *Information, complexity and entropy: a new approach to theory and measurement methods*, http://www.mathpreprints.com (2001).

[4] BONANNO C., MENCONI G., *Computational information for the logistic map at the chaos threshold*, arXiv E-print no. nlin.CD/0102034 (2001).

[5] CHAITIN G.J., *Information, randomness and incompleteness. Papers on algorithmic information theory*, World Scientific, Singapore 1987.

[6] GUSEV V.D., NEMYTIKOVA L.A., CHUZHANOVA N.A., *On the complexity measures of genetic sequences*, Bioinformatics **15** (1999), 994–999.

[7] LEMPEL A., ZIV J., *Compression of individual sequences via variable-rate coding*, IEEE Transactions on Information Theory IT **24** (1978), 530–536.

[8] MILOSAVLJEVIC A., JURKA J., *Discovering simple DNA sequences by the algorithmic significance method*, CABIOS **9** (1993), 407–411.

[9] RIVALS E., DELGRANGE O., DELAHAYE J.-P., DAUCHET M., DELORME M.-O., HENAUT A., OLLIVIER E., *Detection of significant patterns by copression algorithms: the case of approximate tandem repeats in DNA sequences*, CABIOS **13** (1997), 131–136.

REDUCTIONISM AND HISTORY: THE BIOLOGY BETWEEN SCYLLA AND CHARYBDIS

Renzo Morchio
Dipartimento di Fisica,
Università di Genova - ITALY

Abstract This paper analyses and carefully discusses the two terms: *Reductionism* and *History*, related to biology. In fact they are often used but never defined in biology. The results of this analysis is that the two quoted terms have no meaning when used in biology. This analysis asserts the autonomy of biology as a science and its possibility to formulate its own laws.

Often in science there are terms that are used whose meanings are tacitly accepted, without ever having been codified or defined. In the following paper I would like to analyse the current use of two terms: *Reductionism* and *History* (related, of course, to biology).

1. The Reductionism

Biologists generally interpret this term as "reductionism to physics" and often dispute this point of view. Basically, the dissent derives from the conviction that biology is an autonomous science, and therefore not reducible to any other discipline.

I was already involved in reductionism [5], but it is a topic worth returning to. Essentially, its ambiguity is due to the fact that it is used with different meanings in different contexts. I discovered while speaking to biologists who use this term, that can be used to represent at least three ideas, **factual reductionism**, **methodological reductionism** and **reductionism in a strict sense**. Although we are particularly interested in reductionism in a strict sense, I will briefly discuss all three interpretations.

Factual reductionism is when a biologist with a definite biological problem considers that physics has already solved a formally similar although substantially different problem. He can use the solution and the derived equations, as they are or with necessary adjustments, and in so doing, solve the biological problem. In this case I heard biologists speaking about the "reduction" of the biological problem to physics. I feel that this is an inappropriate expression. Obviously a biologist interested in hemodynamics is forced to look at what physicists interested in hydrodynamics have done. Similarly, in order for a neurobiologist to explain passive asson conduction, he must use the differential equation called the submarine cable equation. In such cases, we are not dealing with a type of reductionism, but rather with the obvious use of already formulated general laws of matter, or with the assumption of different but formally similar problems that have already been solved in order to help solve the actual problem.

If hemodynamics had preceded hydrodynamics, I do not think that physicists would have stated that physics had been "reduced" to medicine. Of course I never heard of Darwin "reducing" biology to sociology or economy just because he knew Malthus and was inspired by him. Finally, I mention [6] that in Padua, at a seminar of mine, prof. Azzone, a well-known biologist, told me that the "reduction" of biology to physics could probably be accomplished more easily in physiology than in other branch of biology. I agree that physiology has a lot to do with physical and chemical processes and that, as a consequence, many physiological problems could find help in certain fields of physics, but I disagree completely with the use of the term "reduction" in this context.

In general, *Methodological Reductionism* means that, in front of a complicated biological problem, we try to use all the options the scientific method offers to us, including the laws and mathematical relations that mathematics and physics put at our disposal. When the laws of physics that we refer to are the general laws of matter, the quoted methodological statement is an assertion of the validity of scientific method. However, when we refer to particular physical laws, which relate to some physical topic with some formal similarity to the biological problem we are studying, it should only be because we are searching for some help in solving our problem. In this case, we would have a form of factual reductionism. In other words, what I have called methodological reductionism is nothing but an assertion of the validity of the scientific method. A tautological sentence for a discipline asserting itself to be a science.

The third meaning of the term "reductionism" (*Reductionism in a strict sense*) concerns us closely, since it is true reductionism. In fact, it aims to "reduce" biology, all branches of biology, to the physical sciences. It therefore deserves a careful definition. In regards to this, we refer to the definition given by the american physicist and epistemologist E. Nagel [8]. Actually, no important new facts have been added since then. Nagel characterises reductionism by the two following properties: a) the typical terms of the reduced science must be translated using terms of the reducing science; b) the laws and principles of the reduced science have to be translated into sentences by only using terms that have also been translated into the science in which the reduction was done, as indicated in a). Now the problem seems to be: can all biological terms, laws and principles be translated into physical terms without remainders? The answer is no, as can be shown theoretically and empirically. Theoretically, we can ask ourselves the question: do biological terms say something more than a possibly long sentence written in physical terms? If the answer is yes, then the translation will not be possible, otherwise it will be possible. In other words, atoms form molecules through reactions that follow physical and chemical laws. From these reactions, complex molecular structures are formed which produce processes exclusively interpretable in terms of chemical and physical laws. In this case, it is possible to say that, in a classical way, the whole is exactly the sum of its parts. Therefore, the biological terms cannot contain more than the physical and chemical concepts do and, as a consequence, the translation is possible. Otherwise such a translation cannot be successful. But if biological terms cannot be translated into physical terms, then neither can biological laws be translated. Given the fact that too many reasons prevent us from asserting that a biological system is only the sum of its parts, the only logical conclusion seems to be that a reductionist view is untenable.

Empirically, by actually trying to do the translation described above, we come to the same conclusion, that the translation is not possible. Ayala [3] tried without success and I too failed in this attempt. The only general conclusion seems to be that reductionism is not defensible.

Someone could object by stating that we have only dealt with reductionism to physics, and in theory a reduction to some other discipline could be practicable. Objection seems untenable since "reducing" a discipline studying natural objects, made by matter, to another discipline studying different natural objects is different than "reducing" to a discipline that studies laws and properties of the same matter in its more general form. In the first case, we are only dealing with the search for a suggestion or help. All this is legitimate, but it is not a "reduction".

2. Physical laws and autonomy of biology

The conclusions just obtained could give rise to some misunderstandings. I must therefore specify that the fact that biology is not reducible to physics does not mean that physical laws, valid everywhere, are not valid in biology. What we have said only indicates that each chemical reaction, each enzimatic reaction, each process implying physical or chemical mechanisms respect physical and chemical laws, but the cell shows events that are not reducible only to commonly used physical and chemical laws. In fact, laws of general validity have different weight and meanings in different systems whose matter has reached different levels of organization. I refer to *internal rotation* as an example that clarifies what I mean. It is present in all molecules, and the relative laws are valid for small and large molecules. However, in a small molecule, if an atom, as a consequence of thermal agitations, rotates round the simple bond which binds it to an other atom, the results are poor. Instead, in a protein, rotation round a simple bond drags the remaining part of the chain. If several of these rotations occur round several bonds the phenomenon of protein **conformation** takes place. I must remind you that conformation determines the function of the protein and other important phenomena such as **denaturation**, which arises, for example, due to an increase in temperature.

We could give several additional examples. But it would always be clear that the degree of organization of the matter that forms the considered system can give rise to seemingly emergent properties. These properties can also be described by using specific laws, typical of the discipline which deals with the considered system, but coherent with and eventually derived from physical and chemical laws. Therefore, they are always homogeneous to all of science in general.

The refusal of reductionism permits an additional consideration. This refusal, in fact, implies that biology is an autonomous science; able to formulate its own autonomous laws. We could demonstrate, in fact, that the ability to formulate specific laws is a condition for the autonomy of a discipline and for its possibility in having both theoretical and experimental fields together [7].

Yet another problem concerning the structure and the status of biological theories arises here, but given the limitation of room, I will save this discussion for another time and limit myself to noting that biological laws are unlikely to be deterministic, in the sense of classical mechanics. I would expect biological laws to be statistical in nature. Of course many examples from statistical mechanics remind us that only probabilistic microscopic situations could provide quasi-deterministic laws at a macroscopic level. Finally, I remind you that we could be dealing with complex dynamics, deterministic chaos, and so on.

As I have already said, physiology turns to physics and chemistry often, because physical and chemical processes are more important. This cannot be considered reductionism since physiology can provide biological laws as well. However, considering that it is more likely that typical biological laws flourish in other chapters of biology, I would like to discuss them further in the following pages. To be particularly concrete I will refer to a specific case: the theory of evolution.

Now we really encounter a large problem, which seems to offer a different perspective of the problem of reductionism: the reductionism to the history.

3. Biology and history

Once again, it is important to mention two opinions that I have previously quoted elsewhere. They are very significant opinions because such different people: Agazzi [1], a philosopher, and Ageno ([2]), a physicist, expressed them. Their conclusions are the same even if the reasons, official or not, are very different. They both agree to regard biology as a *historical science*, but Agazzi draws his conclusion from general philosophic considerations, while it wouldn't amaze me if Ageno (also because I knew him in person) had reached his conclusion by considering the rigour of physic and from persuasion that if a problem is correctly stated, then with patience and knowledge of the theoretical and experimental means given by physics, a solution to every problem can always be found. But when a problem seems to be truly insoluble, the best you can do is leave it to humanistic sciences, which are more approximative and less rigorous. In any case, both authors believe that there is no way to give a reason for the fact that in a given geological era the horse appeared, a real horse and not a different living system. But then again, Agazzi says that the appearance of the horse is a contingent event and, as such, cannot be justified; it can only be ascertained. But if we can only ascertain and describe the appearance of the living systems, one after the other, we can only do their history.

First of all, I would like to note that such an interpretation implies two consequences: 1) the refusal (or impossibility) to explain a number of problems; 2) to return to descriptive methods (or to their revaluation) in biology. Such consequences are exactly contradictory with respect to what biological researchers involved in different biological problems are doing daily. In any case, they continue to attempt to find rational explanation to the problems they face.

Secondly, what is it meant when we declare that "biology is a historical science"? Is it maybe that biology studies evolution in time of given systems (biological systems)? If this is the answer, then it is not convincing. In fact, all sciences study the time evolution of systems in their field. Mechanics studies how varies in time the spatial position of a moving body. Thermodynamics is interested in time variation of the state of a system. Geology too, and so on. Therefore, in order to be consistent, we must say that all disciplines are historical sciences. Meaning, "if everything is history, then nothing is". So the initial sentence of this paragraph would be useless. I would like to add that the current meaning of the term "history" contains the idea that we cannot provide general laws of history. All this while biologists and geneticists work hard to give an explanation as to the appearance of given biological forms using biological laws.

The fact remains that it is impossible to formulate deterministic motivations for the appearance of the horse in a given period. But: why should we do that? Why do we accept the crisis of deterministic laws in every field, except in the field of biology? It is too easy to close a discussion with the apparent solution that at that moment the horse appeared, and this fact was a historical event. It is not possible to exclude that, by intensifying our studies, we will arrive, goodness knows when, to assert, for example, that in that period a given genetic system had to be in a state of instability and it was very likely therefore that it followed a different structure of that genetic inheritance. This is not, of course, a deterministic assertion, but (see what was said above about the structure of biological laws) it is possible that biological laws could be statistical. Are Mendel's laws perhaps deterministic? What about quantum mechanics? Still nobody has said that it is a historical science.

Actually, I suspect that we do not like to admit that today our knowledges do not permit us to explain the appearance of the horse in some exact way. So, by recurring to history we accept the idea that biology has cognitive holes.

The use of the term "history" in biology prompts some further comment. First of all, we must remember that in this context we do not give the term "history" a defined meaning that is different from the one we give it in the ambit of historical science. One of the most authoritative italian dictionaries, that of Devoto and Oli [4], defines history in the following manner: "The inquiry or critical research relative to an ordered reconstruction of human events connected in conformity with a unitary line of development (transcending the pure chronological succession typical of the chronicle, for example)". In biology the subjects are not the "human events" anymore, as historical sciences always underline with great attention, but the biological systems. The distinction between chronicle and history becomes vague, to the limits of nullity. Forgetting other aspects, I note that those who use the term "history" tied to biology neither define the exact new meaning of the term, nor are anxious to demonstrate the legitimacy of the consequent generalisation.

So, what is meant by the affirmation that biology is a historical science? Often it is done in reference to geology, which is also said to be a historical science. We neglect however that geologists in some ways have tried to clarify what they mean. I refer to the principle of "actualism" formulated by Joseph Hutton (1726–1797) and then specified by the great Charles Lyell (1797–1875). The principle affirms that past events can only be explained by using the forces that are still operating today, and therefore they can be studied and known because they are still present. Actually a geologist only studies the present and draws, from traces of events, the idea that these events have to be assigned to the past. Virtually, geology deals with the attribution of a temporal coordinate to the considered event, i.e. a number. Generally speaking, biology does not do anything very differently. Today, biologist finds traces of past events (fossils, for example) and so these fossils become data found in the present time but indicators of past events. He interprets them and, within the limits of uncertainty of the dating methods used, gives them a temporal coordinate, i.e. a number. Often we can indicate only a "before" and "after". It is not necessary to discuss Einstein and the theory of relativity. It is however a fact that in this situation the contemporaneity is very uncertain. Particularly, in some cases, stabilising temporal correspondence (contemporaneity) between two layers of two given series is very difficult. We also have to consider the fact that, in all these problems, the historian lacks a very important tool, that of the testimony of contemporaries. Even after all of this, biologists continue to try to build up sequences of biological forms. What I do not understand is the validity of the current behaviour in confirming such sequence of forms using dating, but then, since there is no other choice, confirm the datings through the coherence of the sequence of forms. In conclusion, it seems to me that biologists do not make *history* and so cannot say that they do. They only try to order events of the past by dating them, in order to provide an interpretation.

There are many other problems associated with this argument however; I limit myself to noting only one. Biologists find traces of past events (fossils, for example). These traces in themselves are true; they are what they are. Only their interpretation could be erroneous. On the contrary for a historian not only could the interpretation of a trace be erroneous, but the very trace itself could also be incorrect. You can think back to the accounts given by some ancient historians.

To avoid misunderstanding, I would like to emphasise that I did not say that biologists should not study the past. Biologists must study the time evolution of living beings; they have to search for the rules determining such an evolution and the relations between organisms and their changing environments. But in "history" there are no rules, and so it makes no sense to search for them. There is a great difference between doing something at random and doing something following a strange idea you have in your mind. In evolution we could find randomised events, but this is a wellknown situation in science. Of course, the laws of evolution can be probabilistic and not deterministic. The fact remains that to renounce searching for something ("history") is very different from searching for something without success.

To conclude, the biologist, as every scientist, describes time evolution of the systems he is interested in and tries to formulate their laws. The last goal would be impossible if he were making "history", where, as we have already noted, it seems impossible to formulate laws. But the goal of the biologist, convinced of the autonomy of biology, is only to search for the autonomous laws of his discipline.

4. Conclusions

The idea that turning to history could be a good antidote for biology against reductionism is a false. History could be at most a different form of reductionism. The alternative to reductionism (whatever type of reductionism) is only autonomy of biology and the consequent search for autonomous laws of the quoted discipline. Such laws will or will not be found, but we have the duty to search for them, whatever their structure could be.

The fact remains that we must not fear the situation that in biology there are so many things we do not know. But the two reductionisms are and remain for biology a sort of Scylla and Charybdis.

References

[1] AGAZZI E., *L'universo come problema scientifico e filosofico*, Conferenza Internazionale "La cosmologia nella cultura del '900", Torino, Nov. 1990.

[2] AGENO M., *Dal mondo fisico al mondo della vita*, in: AA. VV., "SCIENTIA. L'immagine e il mondo", Atti 80° compleanno della Rivista, Milano 1989, 199.

[3] AYALA F.J., *Biology as an Autonomous Science*, American Scientist **56** (3) (1968), 207.

[4] DEVOTO G., OLI G.C., *Il dizionario della lingua italiana*, Le Monnier, Firenze 1995.

[5] MORCHIO R., *Reductionism in Biology*, in: "Reductionism in Science", E. Agazzi ed., Kluwer, Dordrecht 1991, 149–160.

[6] MORCHIO R., *Biologia teorica: motivazioni e prospettive*, Seminar and discussions, Univ. of Padua 1997.

[7] MORCHIO R., *Conditions for a Theoretical Biology: is Biology truly an Autonomous Science?*, Rivista di biologia/Biology forum, in press.

[8] NAGEL E., *The Structure of Science*, Harcourt, Brace and World, New York 1961.

A CHARACTERIZATION FOR A SET OF TRINUCLEOTIDES TO BE A CIRCULAR CODE

Giuseppe Pirillo
IAMI CNR,
Viale Morgagni 67/A, 50134 FIRENZE - ITALY
and
Université de Marne-la-Vallée,
5, boulevard Descartes, Champs sur Marne
77454 MARNE-LA-VALLÉE Cedex 2 - FRANCE
pirillo@hermes.math.unifi.it

To Piero Mangani for his 65th Birthday

Abstract The Arquès-Michel autocomplementary circular codes are 216 sets of 20 trinucleotides. They have a lot of very interesting properties. Studying them, we found the following result: a set X of trinucleotides is a circular code if and only if it has no 5-necklaces.

Alphabets and words in biological sequences. Let β be an *alphabet.* The set of *non empty words* (resp. *words*) on β is denoted by β^+ (resp. β^*). The *letters* (or *nucleotides* or *bases*) of the *genetic alphabet* β_4 are A, C, G, T. The set of 16 words of length 2 (or *binucleotides* or *dominoes*) over β_4 is β_4^2. The set of 64 words of length 3 (or *trinucleotides*) over β_4 is the *universal genetic code* β_4^3.

Circular code. ([4]) *A language* X *in* β_4^+ *is a* circular code *if, for each* $n, m \geq 1$ *and for each* $x_1, \ldots, x_n, x'_1, \ldots, x'_m$ *in* X, $p \in \beta_4^*$ *and* $s \in \beta_4^+$, *the conditions*

$$sx_2 \cdots x_n p = x'_1 \cdots x'_m$$

and

$$x_1 = ps$$

imply $n = m$, $p = 1$ *and, for* $i = 1, \ldots, n$,

$$x_i = x'_i.$$

Conjugate words. *Two words u and v are conjugate if there exist words s and t such that $u = st$ and $v = ts$.*

Proposition 1. *A circular code cannot contain two distinct conjugate words.*

The Arquès-Michel maximal circular code consists of the following 20 trinucleotides (see [1, 2]):

$$AAC, AAT, ACC, ATC, ATT, CAG, CTC, CTG, GAA, GAC,$$
$$GAG, GAT, GCC, GGC, GGT, GTA, GTC, GTT, TAC, TTC.$$

They occur preferentially in frame 0 in a large bank of gene sequences. Arquès and Michel discovered other 215 codes similar to X_0 (see [3]) finding first all the (maximal) circular codes on β_4 and after selecting the autocomplementary ones. A possible application of the Arquès-Michel code is considered in [5] and an algebraic approach for a study of the 216 codes *à la* Arquès-Michel can be found in [6].

Necklaces.

Definition 1. *Let $l_1, l_2, l_3, l_4, \ldots, l_n, l_{n+1}$ be letters in β_4, and let $d_1, d_2, d_3, \ldots, d_n$ be dominoes in β_4^2. We say that the ordered sequence*

$$l_1, d_1, l_2, d_2, l_3, d_3, \ldots, l_n, d_n, l_{n+1}$$

is a $n+1$-necklace *for a subset $X \subset \beta_4^3$ if*

$$l_1d_1, l_2d_2, l_3d_3, \ldots, l_nd_n, d_1l_2, d_2l_3, d_3l_4, \ldots, d_nl_{n+1} \in X\,.$$

The following Proposition does not require any hypothesis of autocomplementarity.

Proposition 2. *Let X be a subset of β_4^3. The following conditions are equivalent:*

a) X is circular code;

b) X has no 5-necklace.

Proof. a) $\to$ b). Let X be a circular code. By contradiction, suppose that

$$l_1, d_1, l_2, d_2, l_3, d_3, l_4, d_4, l_5$$

is a 5-necklace for X. Since β_4 is a 4-letter alphabet, for some $i, j \in \{1, 2, 3, 4, 5\}$, $i < j$, we have

$$l_i = l_j\,.$$

Put $x_i = l_i d_i, \ldots, x_{j-1} = l_{j-1}d_{j-1}$ and $x'_i = d_i l_{i+1}, \ldots, x'_{j-1} = d_{j-1}l_j = d_{j-1}l_i$.

If $j = i+1$ then x_i and x'_i are both in X and are conjugate. This contradicts the above Proposition.

So assume that $j > i+1$. We have

$$\begin{aligned} d_i x_{i+1} \cdots x_{j-1} l_i &= d_i(l_{i+1}d_{i+1}) \cdots (l_{j-1}d_{j-1})l_i \\ &= (d_i l_{i+1})(d_{i+1}l_{i+2}) \cdots (d_{j-1}l_i) \\ &= x'_i \cdots x'_{j-i}\,. \end{aligned}$$

Since $x_i = l_i d_i$ and l_i, being a letter, is non empty, X is not a circular code. This contradicts the definition of circular code.

b) $\to$ a). Let X be without 5-necklaces. By contradiction, assume that X is not a circular code. All elements of X have length 3, therefore there exist $n \geq 1$, $x_1, \ldots, x_n, x'_1, \ldots, x'_n$ in X, $p_1, s_1 \in \beta_4^+$, such that

$$s_1 x_2 \cdots x_n p_1 = x'_1 \cdots x'_n$$

and

$$x_1 = p_1 s_1\,.$$

Moreover, there exist $p_2, \ldots, p_n \in \beta_4 \cup \beta_4^2$ and $s_1, \ldots, s_n \in \beta_4 \cup \beta_4^2$, such that

$$\begin{aligned} &|p_1| = |p_2| = \cdots |p_n| \\ &|s_1| = |s_2| = \cdots |s_n| \\ &|p_1| + |s_1| = |p_2| + |s_2| = \cdots = |p_n| + |s_n| = 3 \\ &p_1 s_1 = x_1\,,\ p_2 s_2 = x_2\,,\ \ldots,\ p_{n-1}s_{n-1} = x_{n-1}\,,\ p_n s_n = x_n \end{aligned}$$

and

$$s_1 p_2 = x'_1\,,\ s_2 p_3 = x'_2\,,\ \ldots,\ s_{n-1}p_n = x'_{n-1}\,,\ s_n p_1 = x'_n\,.$$

Since we can exchange the x_i's with the x'_i's, without loss of generality, we can suppose

$$|p_1| = |p_2| = \cdots |p_n| = 1$$

and consequently

$$|s_1| = |s_2| = \cdots |s_n| = 2\,.$$

If $n \geq 5$, then $p_1, s_1, p_2, s_2, p_3, s_3, p_4, s_4, p_5$ is a 5-necklace.
If $n = 4$, then $p_1, s_1, p_2, s_2, p_3, s_3, p_4, s_4, p_1$ is a 5-necklace.
If $n = 3$, then $p_1, s_1, p_2, s_2, p_3, s_3, p_1, s_1, p_2$ is a 5-necklace.
If $n = 2$, then $p_1, s_1, p_2, s_2, p_1, s_1, p_2, s_2, p_1$ is a 5-necklace.
If $n = 1$, then $p_1, s_1, p_1, s_1, p_1, s_1, p_1, s_1, p_1$ is a 5-necklace.
In any case we have a 5-necklace, a contradiction. □

References

[1] ARQUÈS D.G., MICHEL C.J., *A possible code in the genetic code*, in: "STACS 95 - 12th Annual Symposium on Theoretical Aspects of Computer Science, Munich, Germany, March 2–4, 1995 Proceedings", E.W. Mayr and C. Puech eds., Lecture Notes in Computer Science, Vol. 900, Springer Verlag 1995, 640–651.

[2] ARQUÈS D.G., MICHEL C.J., *A complementary circular code in the protein coding genes*, Journal of Theoretical Biology **182** (1996), 45–58.

[3] ARQUÈS D.G., MICHEL C.J., personal communication.

[4] BERSTEL J., PERRIN D., *Theory of Codes*, Academic Press, London 1985.

[5] PIRILLO G., *Maximal circular codes and applications to theoretical biology*, in: "Mathematical and computational biology (Aizu-Wakamatsu City, 1997)", Lectures Math. Life Sci. **26**, Amer. Math. Soc., Providence, RI 1999, 187–190.

[6] PIRILLO G., *Remarks on the Arquès-Michel code*, Biology Forum **94** (2001), 327–330.

DETERMINISTIC AND RANDOM COMPONENTS OF OVER TIME EVOLUTION OF MILK PRODUCTION ANALYSED BY TIME SERIES METHODOLOGY

Giuseppe Pulina
Dipartimento di Scienze Zootecniche,
Università di Sassari,
Via De Nicola 9, 07100 SASSARI - ITALY
gpulina@ssmain.uniss.it

Corrado Di Mauro
Dipartimento di Scienze Zootecniche,
Università di Sassari,
Via De Nicola 9, 07100 SASSARI - ITALY

Niccolò Macciotta
Dipartimento di Scienze Zootecniche,
Università di Sassari,
Via De Nicola 9, 07100 SASSARI - ITALY

Aldo Cappio-Borlino
Dipartimento di Scienze Zootecniche,
Università di Sassari,
Via De Nicola 9, 07100 SASSARI - ITALY

Abstract Daily milk yield of dairy species shows a peculiar evolution over time from parturition to dry-off. In order to make correct comparisons among animals, lactation of standardised length are considered: in the dairy cow, for example, the standard lactation length has been fixed at 305 days. In official dairy recording schemes milk yields are measured on a large number of animals usually once a month: therefore, the maximum number of measurements per cow is 10, although lactations that end before 305 days or that show several missing data are quite frequent. A main problem in analysing the evolution of milk production over time is the separation of the regular and continuous component of the phenomenon from the random component, usually related to environmental short-term perturbations.

Actually, the deterministic component refers to genetic and physiological mechanisms that underlie milk secretion process, whereas the knowledge of the magnitude of the random component allow to define the range of an accurate forecasting for future yields or missing data, which represent a relevant technical problem for both the genetic improvement of the trait and the farm management.

Classical methodologies of time series analysis, particularly the box and Jenkins ARMA modelling, are able to identify the structure of the deterministic component and to estimate the relative magnitude of the random component. A major diagnostic tool is the autocorrelation function which is very useful to describe the nature of a process through time. Inference based on this function is often called an analysis in the time domain. At the same time, the analysis in the frequency domain, based on the spectral density function which describes how the variation in the time series may be accounted for by cyclic components at different frequencies, can be developed. Results of these analytic tools can be useful guidelines for the development of forecasting models able to reconstruct the original series in its pure deterministic component and in stochastic quota that follows Autoregressive (AR) or Moving Average (MA) probability models, whereas no inferences can be obviously made on the completely random (white noise) component.

1. Introduction

The over time evolution of milk production represents an interesting topic in animal breeding. In lactation studies, data are usually Test Day yields recorded at regular intervals of time from parturition. They can be regarded as genetically different variables and, consequently, analysed with multivariate methodologies. More common is, however, the univariate approach, according to which milk yields taken at different times along each lactation are considered as repeated measures on the same experimental unit [10]. Test day measures therefore represent the base data for the calculation of total milk yields per lactation which are the key variables for the estimation of the breeding values of animals candidate to selection [6] and for management decision at herd level.

The typical evolution over time of test day records is quite irregular and different among animals.

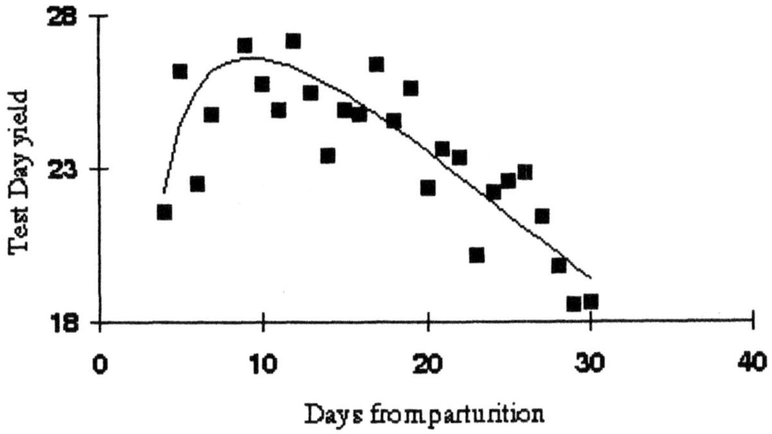

Figure 1. Typical over time evolution of daily milk yield (■) and data predicted by fitting a mathematical function of the lactation curve (—).

The classic method used to analyse lactation data disentangles the deterministic component of the phenomenon, called the "lactation curve", by fitting the experimental data to a time function, continuous and differentiable in the whole interval of time that corresponds to the lactation length (Fig. 1). Several linear or linearisable and non linear mathematical models of lactation curve have been suggested [9]. However, most of mathematical functions fit well average lactation curves of homogeneous groups of animals but the fitting is scarce for individual lactation curves. Furthermore, few data are usually available per animal (for example 5–6 for sheep and goats, 7–8 for cows) and there is a tendency to further reduce the number of official records which represents the main source of expense in genetic programs. Finally, a serious drawback of lactation curve modelling by continuous functions of time is that residuals are to be treated only as a noise component.

As an alternative, we proposed the modelling of test day measures by times series analysis methodologies [7]. The succession of test day yields within a lactation can be regarded as a time series, consisting of a set of values of a continuous casual variable ordered according to a discrete index variable represented by intervals of time measured in a suitable scale. The fundamental property of time series models is the ability to recognize the main structural components of time evolution of milk yield. Particularly, they take account of the shape of the average lactation curve of homogeneous groups of animals, in analogy with the above mentioned mathematical functions, but also they can fit the pattern of variance-covariance of residuals. Actually the relationship between TD pairs within lactation can be classified as stochastic but with a probability structure which is partly conditioned by a knowledge of past values [3]. Its structure varies according to the temporal distance among test day measures within lactation (lag) and can be regarded as expression of the individual variation around the mean lactation curve due to environmental short-term effects such as random variations from dietary quality and intake, weather, minor injuries, estrus [2] but also genetic variations. Models belonging to the large family of time series models, such as Autoregressive Moving Average (ARMA) models, are able to take account of the temporal dependence of the (co)variance structures by means of autoregressive parameters (AR), and the capacity to adjust predictions by considering the residuals of previous estimates with moving average (MA) parameters.

2. Time series methodologies

The Box-Jenkins methodology [1] to analyse and model time series, consists essentially of three steps: i) model identification; ii) parameters estimation; iii) model validation.

In the first step, two main analytical tools are commonly used to detect the general structure of the time series analysed: the spectral analysis and the autocorrelation function, that analyse the original series in the frequency and in the time domain respectively [8]. Spectral analysis is essentially a modification of Fourier analysis so as to make it suitable for stochastic rather than deterministic functions of time. Results of spectral analysis are synthesised by the periodogram, i.e. the plot of the intensity of the elementary waves in which the original series can be decomposed. The autocorrelation function at lag k of the original series x_t is the linear correlation coefficient between x_t and x_{t-k}, calculated for $k = 0, 1, 2$. These two methods supply useful indication for the general structure of a suitable ARMA model, i.e. the eventual occurrence of a periodic deterministic component (usually named seasonality) and the order of the AR and MA processes. Moreover, the autocorrelation function can be used to asses the goodness of fit of the ARMA model by check the White Noise nature (WN) of the residuals.

The most reliable method to estimate ARMA parameters is the Maximum Likelihood (ML) methodology, that is asymptotically correct for time series. General statistical softwares of good level and specific packages for time series analysis usually possess routines for ML estimation of ARIMA parameters according to the Box-Jenkins procedure. Estimators are usually sufficient, efficient and consistent for Gaussian distributions and are asymptotically normal and efficient also for several non Gaussian distribution families.

The validation of the goodness of fit can be developed by evaluating the statistical significance of parameters, analysing the autocorrelation function of residuals, checking the WN connotation of residuals by means of appropriate tests and by calculating the quantity:

$$R^2 = 1 - \frac{Var(\epsilon_t)}{Var(x_t)} \quad (1)$$

where $Var(\epsilon_t)$ is the residual variance and $Var(x_t)$ is the original variance of the data, by analogy to the determination coefficient of the multiple regression analysis.

Finally, the ARMA model can be used to make predictions of future values of the series general in order to validate its general reliability.

3. Building time series by milk test day yields

Fitting of time series models requires a sufficient large data set. Actually, time series models have been used to model TD data of complete individual lactations in order to detect animals affected by diseases [4]. This constraint can be overcome by constructing an artificial succession of values in correspondence with the succession of natural numbers that can be analysed with time series methods.

In this study, data consisted of 1000 milk test day records of 100 third parity Holstein cows with 10 TD measures each. Records were arranged in a single vector and ordered according to the cow $(1, 2, \ldots, 99, 100)$ and, within each cow, to their distance from parturition. Therefore TD records resulted organised in a succession $(1, 2, \ldots, 1000)$ (Fig. 2) where the index variable had the following meaning: the index value 4, for example, corresponds to the fourth TD of the first cow of the data set, the index value 23 corresponds to the third TD of the third cow $((10 \times 2) + 3 = 23)$ of the succession. This series could be analysed by ARMA models because it addresses the criterion of the correspondence between the index variable and the succession of TD records [5]. Each data set was then split in an estimation data set, consisting of the first 50 lactation, used for the parameter estimation and in a validation data set, consisting of the next 50 lactations, which was used for making predictions by using previous parameter estimates.

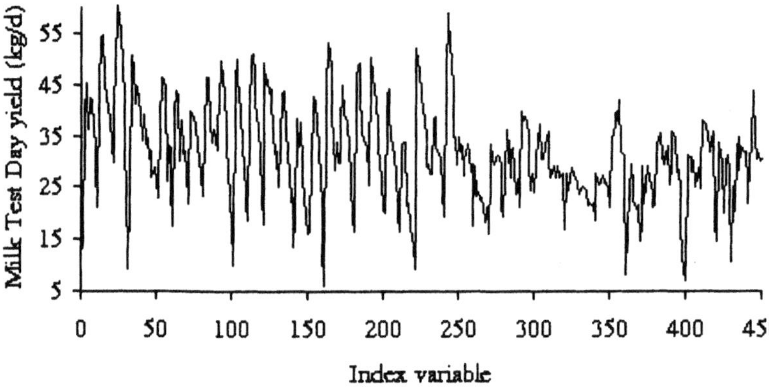

Figure 2. Pattern of raw data of the succession of milk test day yields of third calving Holstein cows.

4. Model identification and parameter estimation

Results of Fourier transform, reported in the periodogram of figure 3, clearly show the occurrence of well defined peaks at equally spaced intervals of the angular frequencies $\omega_i = n * 2\pi/T$, where T is the period of 10 lags of the index variable. This pattern underlines the existence of a periodic deterministic component that occurs in all lactations and that can be identified with the amount of variability explained by the average lactation curve of the succession.

Furthermore, the periodogram shows several less defined peaks, located at low angular frequencies, that evidence a residual deterministic linkage among TDYs within each lactation.

The Autocorrelation function (Fig. 4) further confirm results of spectral analysis as far as the identification of a suitable ARMA structure is concerned. The relevance of the periodic component of the variability (average lactation curve) is evidenced by the values significantly different from zero at lags multiple of ten whereas the linkages within lactation are highlighted by significant values at lags 1 and 2.

On the basis of the aforementioned considerations, the most suitable model for the successions under study seems to be an ARMA $(1,1) \times (1,1)_{10}$ structure.

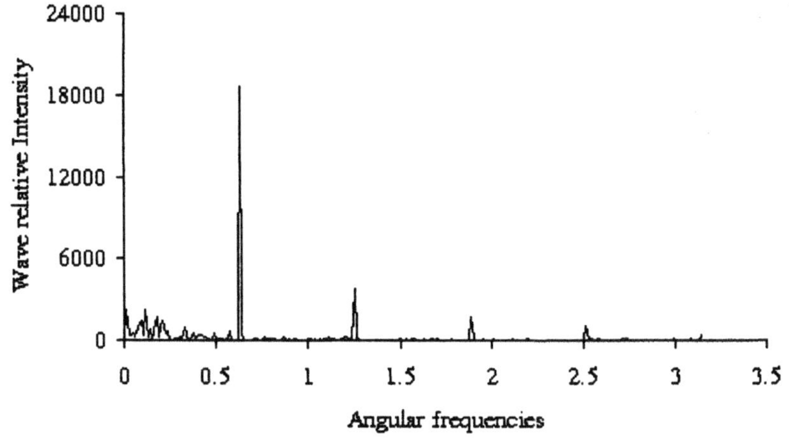

Figure 3. Periodogram of the result of Fourier transformation of the succession shown in figure 2.

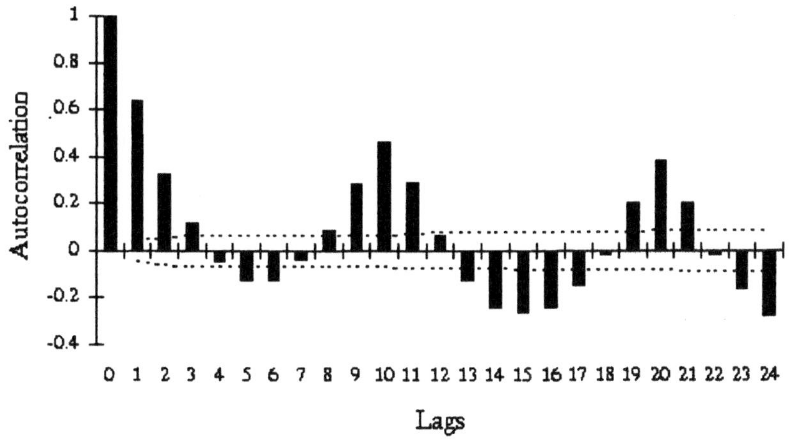

Figure 4. Autocorrelation function of the original data. Dotted lines are plotted at zero plus and minus twice the standard errors for each coefficient.

Parameter	Estimate	Standard error	T
Mean	34.031	6.652	5.12
AR non seasonal	0.768	0.040	18.90
AR seasonal	0.998	0.001	918.5
MA non seasonal	0.158	0.062	2.51
MA seasonal	0.946	0.023	41.29
Residual variance	35.122		
Original variance	114.575		

Table 1. ARMA parameter estimates.

The goodness of fit is highlighted by the statistical significance of all parameters (Table 1) that exceed the critical value of the T statistic ($p \leq 0.001$, except the nonseasonal moving average, and by the amount of original variance explained, about 70%. Moreover, residuals showed a white noise nature, as evidenced by the absence of autocorrelation.

5. Model validation

An indication of the forecasting power and of the limitations of ARMA modelling can be inferred from figures 5 and 6 which report two selected cases of prediction of individual lactation pattern: the continuous line represent the actual lactation curve (all 10 actual TD record) whereas the dotted line is the predicted curve obtained by using 6 actual (the first 6) and 4 predicted records. Whereas in the first case the ARMA model is able to reproduce very efficiently the lactation pattern, in the second case predictions are rather lower in comparison with actual data.

The main reason for such a difference can be found in the lactation pattern that has the typical shape in the first animal whereas is clearly atypical and, therefore, largely unpredictable in the second case.

However, the model is highly reliable when used to predict the total milk yield for lactation. In table 2 are reported the statistics of predictions obtained by comparing total milk yields calculated by using all 10 actual test day records (actual yields) or 6 actual and 4 estimated yields (estimated yields).

The correlation between actual and predicted yields is remarkably high and it is of the same order of those obtained by using more complicated prediction methods based on the projection of incomplete lactations by regression coefficients. The rank correlation is higher than Pearson correlation, showing that the use of predicted instead of actual lactation yields does not result in a significant change of the relative merit of the animals. Moreover, the standard deviations of actual and

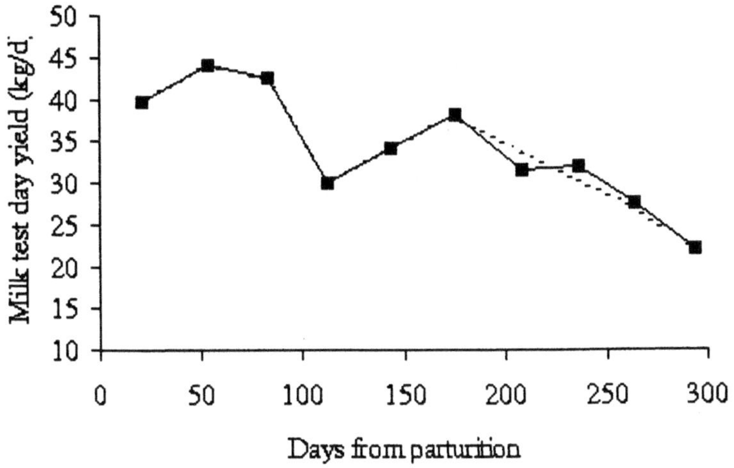

Figure 5. Actual (■ = 10 actual Test Day records) and predicted (— = 6 actual and 4 predicted Test Day records) patterns of an individual lactation curve.

Parameter	Value
Mean actual yields (AY) kg	8785
Mean estimated yields (EY) kg	8887
Correlation (AY, EY)	0.96
Rank correlation (AY, EY)	0.94
Standard deviation AY kg	1808
Standard deviation EY, kg	1578
Average (AY-EY) kg	−102
$\frac{\sigma(AY-EY)}{AY} * 100$	5.74

Table 2. Statistics of predictions for total lactation yields

estimated yields are of the same magnitude. This is an interesting result because usual methodologies to predict lactation yields by using multiplicative coefficients tend to markedly compress the variance of predicted yields with a subsequent bias in genetic evaluations. The average bias is negligible and it underlines a slight tendency of the model to overestimate lactation yields. Finally, the last index, obtained as 100 times the ratio of the standard deviation of residuals, was below eight which is considered a maximum threshold for a good prediction [11].

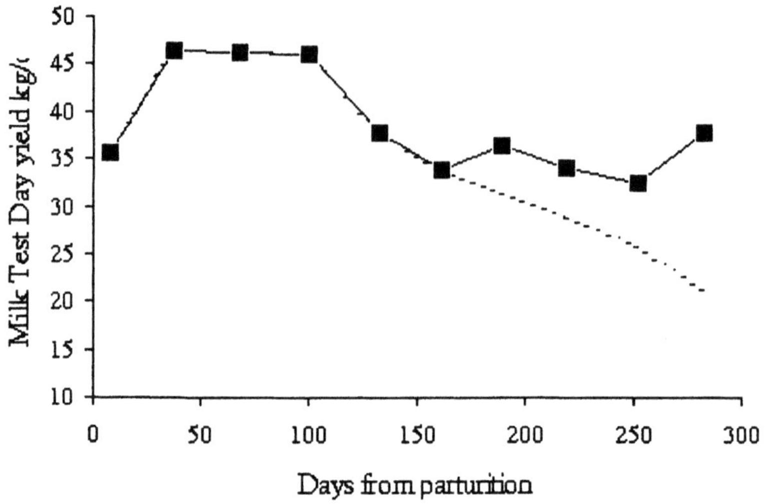

Figure 6. Actual (■ = 10 actual Test Day records) and predicted (— = 6 actual and 4 predicted Test Day records) patterns of an individual lactation curve.

6. Conclusions

Although ARIMA models are the simplest ones in the field of time series analysis, results obtained in this work demonstrate that they are able to identify the shape of the average lactation curve and to evidence the autoregressive nature of Test Days residuals within each lactation. Such a conclusion is in agreement with many instances occurring in the literature in which low order autoregressive processes have been proposed as possible structures for modelling milk yield evolution.

Furthermore, ARIMA models provide a simple and flexible tool to forecast lacking records values, thus obtaining sufficient data to estimate the whole production for a standardised lactation length.

Adaptive signal processing techniques will become more important in the future, when dairy animals may have much longer official test intervals but much more frequent records by the farm owner. There will also be more farms with automated milk recording systems that could provide data throughout the lactation, both on milk yield and on several physical and chemical properties associated with milk quality. To analyse and to decode such a complex net of information, more powerful and sophisticated tools of signal processing will be required, able to manage at the same time several time series of different variables and with different recording frequencies. Methods based on cross correlations among time series, artificial neural networks and wavelet decomposition seem particularly appealing for these purposes.

Acknowledgments

Work funded by the Ministero dell'Università, dell'Istruzione e della Ricerca (grant 40%) of Italy.

References

[1] Box G.E.P., Jenkins G.M., *Time series analysis: forecasting and control*, Holden Day, S. Francisco 1970.

[2] Carvalheira J.G.V., *Development and application of an autoregressive process for estimating genetic parameters of daily milk yield in a herd of tropical lucerna cattle and in US Holstein herds*, Ph.D. Diss., Univ. Cornell, Ithaca 1996.

[3] Chatfield C., *The analysis of time series. An introduction*, Chapman & Hall, London 1999.

[4] Deluyker H.A., Shumway R.H., Wecker W.E., Azari A.S., Weaver L.D., *Modelling daily milk yield in holstein cows using time series analysis*, J. Dairy Sci. **73** (1990), 539–548.

[5] Hamilton J.D., *Time series analysis*, Princeton University Press, Princeton NY 1994.

[6] Olori V.E., Brotherstone S., Hill W.G., McGuirk B.J., *Fit of standard models of the lactation curve to weekly records of milk production of cows in a single herd*, Livest. Prod. Sci. **58** (1999), 55–63.

[7] Macciotta N.P.P., Cappio-Borlino A., Pulina G., *Time series Autoregressive integrated moving average Modelling of Test Day Milk Yields of Dairy Ewes*, J. Dairy Sci. **83** (2000), 1094–1103.

[8] Piccolo D., *Introduzione all'analisi delle serie storiche (Introduction to time series analysis)*, La Nuova Italia Scientifica Eds., Roma 1990.

[9] Rook A., France J., Danhoa M.S., *On the mathematical description of lactation curves*, J. Agric. Sci. Camb. **121** (1993), 97–102.

[10] Van der Werf J., Schaeffer L.R., *Random regression in animal breeding*, Course notes, CGIL Guelph, June 25 – June 28 1997.

[11] Wilmink J.B.M., *Comparison of different methods of predicting 305-day milk yield using means calculated from within-herd lactation curves*, Liv. Prod. Sci. **17** (1987), 1–17.

TOWARD CREATING LIFE IN A TEST-TUBE

Martino Rizzotti
Dipartimento di Biologia,
Università di Padova,
Via Bassi 58/B, 35131 PADOVA - ITALY
martino.rizzotti@unipd.it

Abstract Attempts to dominate minimal life from the conceptual as well as the experimental viewpoints would take advantage of better understanding of the properties of life in general. Reproduction is usually considered as its most basic property, but detailed examination leads us to the conclusion that this property derives from two more basic ones, namely, overproduction and modularity. From the experimental viewpoint, the requirements of a minimal cell are being currently investigated both from subtracting components from the simplest known organism and from adding components to a fully synthetic cell-like object. However, even the concept of a minimal cell is poorly defined.

1. Is reproduction the basic property of life?

Identification of the basic properties of living things is often taken as a philosophical exercise, but is now going to assume operational value partly because of planetary explorations, partly because modern biotechnology is on the way to build artificial cells. To this end, reliable definitions are essential, as it is almost nonsense to build something that is only vaguely described. Moreover, the problem is not to draw a blueprint of a machine where virtually all interactions are described in detail, but to identify the minimal set of properties that an object must display in order to be accepted as a living object in spite of the impossibility to describe in detail the immense number of interactions between the involved molecules.

Reproduction is usually supposed to be the most basic property of living things because it is thought to ensure their temporal continuity. This idea is questioned by some authors, but without any true deepening. For instance, Fleischaker [6] claims that cell reproduction is "already subsumed in cell metabolism", whereas Varela [46] points out the "logical and ontological priority" of identity, that can be summarized with the fact that living things have a structural and functional boundary. Mahner and Bunge [27] quibble on the fact that "many biosystems are simply not capable of self-reproduction", but cannot deny that all biosystems appear thanks to reproduction.

In order to analyze the concept of reproduction, we have better only take into account the true living portion of living things, not the dead portions such as the bark in trees. Even single cells have parts that are metabolically inactive, i.e., that are not involved in rapid chemical reactions. We call living matter only the portions of the cells that are involved in these rapid reactions. An operational definition of life can but regard, in the first place, precisely the metabolism, even if we miss any evidence of phenomena the detection of which would require relatively long periods (usually impossible to be investigated) such as reproduction and, to a much greater extent, evolution [6, 25].

Sometimes the metabolism undergoes interruptions, in nature, due to either normal resistance stages, e.g., spores and seeds, or contingent situations, e.g., viable bacteria trapped in salt deposits [47, 30] or ice formations [39]. The same occurs in artificial situations such as deep-freezing of gametes and embryos or freeze-drying of yeast cells and animal cysts. These phenomena do not reduce the defining importance of metabolism because living things cannot only consist of dormant stages, whereas, if they lack them, they do not lose their living character. This means that a definition of life is allowed to ignore the possible suspension of metabolism.

2. Continuity of life is ensured by overproduction

We might assume that living matter could ensure its temporal continuity without reproduction, i.e. by simply maintaining itself unmodified, like many inorganic objects, for example in the form of suspended life. However, in these conditions, the maintenance of living matter is not ensured because it cannot resist many causes of degradation that also affect inorganic objects, such as high temperatures, low-wavelength radiations or fast particles. Moreover, degradation (albeit extremely slow) of dormant cells is also promoted by entropy increase, because they are not at the thermodynamic equilibrium [42].

In fact, living matter undergoes continuous *degradation* (Fig. 1) that renders some living matter of a species no longer living. When the rapid reactions of metabolism are irreversibly interrupted, the matter becomes available for other species or for the inorganic world. The degradation of living matter consists of two aspects: one is *metabolic*, which implies the elimination of substances to the exterior of its boundary or their inactive deposition inside the organism; this process puts an end to that part of the metabolism that is also called catabolism. Of course, this aspect of degradation is absent in suspended life. The other aspect of degradation is *non-metabolic* and consists of the effects of physical and chemical factors, such as destructuring mechanical action (as in many kinds of macroscopic predation), thermal motion, high-energy radiations, fast particles, and aggressive chemical actions (like in many kinds of microscopic predation), which modify the structures and mechanisms of living matter leading to reduction and, possibly, permanent interruption of its metabolism (also called death). Non-metabolic degradation can also affect any form of suspended life. The non-metabolic aspect of degradation can be both natural and artificial.

Living matter can maintain itself only if degradation is compensated by a process with the opposite effect, namely *assimilation* (Fig. 1). Assimilation usually refers to the metabolic production of the constituents of living matter also called anabolism while, more in general, it also includes the uptake of mass and energy from the environment, the construction of constituents that are similar (hence "assimilation") to those already present [42], the correct emplacement of the constituents, and any process which restores living matter in its capability to carry on further assimilation. It must be pointed out that, while degradation partially depends on factors that are not due to intrinsic processes of living matter, assimilation completely relies on its internal activities, i.e., it is only metabolic. This means that there is no symmetry between the constructive and destructive sides of living matter. If anabolism limits itself to compensating catabolism, living matter diminishes because of non-metabolic degradation, while if assimilation also compensates non-metabolic degradation, the living matter of a species is maintained constant. It is well known that living things undergo cyclic changes. These changes also involve precise molecules. For instance, the proteins present in the living matter of a species change stage by stage during the life cycle. This means that assimilation is not qualitatively exact and could thus be called "quasi-assimilation".

The term "overproduction" (of living matter) may be adopted to mean the effect of assimilation when it exceeds metabolic degradation. Hence, a species must overproduce its components in order to maintain itself: its temporal continuity is ensured by some rate of overproduction. This overproduction of living matter is obscured by the gross constancy of living matter itself, but if non-metabolic degradation were to stop, it would

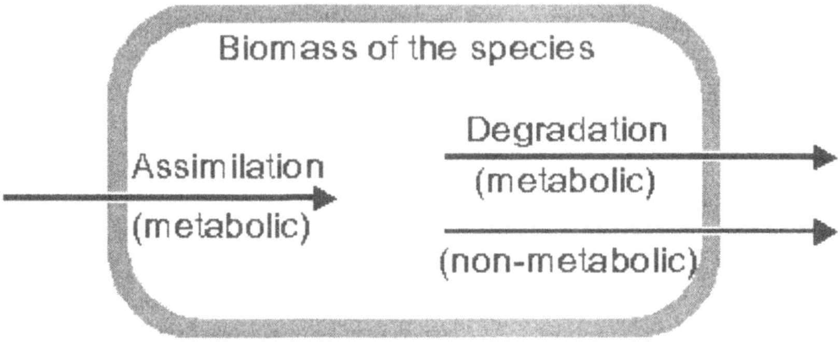

Figure 1. Global budget of a species.

really result in a steady increase of the living matter of the species. The living matter of a species increases or decreases depending on life cycles, environmental cycles, and erratic phenomena. But we may assume that the living matter of most species is relatively constant over long time periods so that, on average, assimilation equals degradation. Thus, temporal continuity and the near-constancy of living matter of each species are ensured by a rate of overproduction that is exactly sufficient to compensate for non-metabolic degradation.

In any case, since failing to compensate non-metabolic degradation over a sufficient period of time leads to extinction, extant lineages are those that fulfilled two conditions: (1) on average, they compensated for non-metabolic degradation, and (2) the instantaneous amount of the living matter never dropped to zero.

3. Control of diffusion imposes modularity

We might assume that living matter could last in time as lumps of objects of randomly varying sizes that are undifferentiated above the molecular (including macromolecular) level, except for the presence of a boundary. In sharp contrast with this theoretical possibility, many structural and functional units may be described above the molecular level, the absolutely constant ones being the cells. Cells are usually presented as a discovery of the microscopic investigation that led to the enunciation of the "cell theory" [49]. However, at the present stage of biological knowledge, it is perhaps legitimate to expect that we are able to go beyond this history-oriented conception of the cell theory, and begin to base it on a proper explanation. In fact, the cellular organization is likely to be unavoidable, owing to the particular molecular constitution of living things and their environment. In particular, an open system that is chemically active can only sustain itself if it controls the diffusion of its constituent molecules [17, 48], otherwise their concentrations fall to sub-effective values by dilution into the environment. In order to control the diffusion of a large number of different molecules, a compartmentalizing boundary is the obligatory solution. As the cell content is primarily aqueous, this boundary must be essentially hydrophobic in nature, although its selective permeability toward hydrophilic molecules can be ensured by means of regulated or even energy-driven hydrophilic pores.

But such an open system remains relatively constant if it is controlled by a homeostatic reference. In a very complex molecular system, the most economic way of providing the reference is by means of a very stable molecule with a large number of possible states: this may be achieved with a long polymer where the different states are distributed along its length. DNA genomes may be regarded (together with many other meanings) as quite stable homeostatic references.

The indirect molecular interactions between the homeostatic reference, on one side, and the boundary structure, on the other side, (i.e., the genome and the plasma membrane) limit the upper cell size mainly due to diffusion problems. It is well known that in the largest cells, such as ostrich eggs, only a microscopic part around the genome has the usual water concentration that allows the diffusion of hydrophilic molecules. The largest sizes are reached in organisms where the diffusion over long distances is controlled by a sophisticated organization, like in plants, where a whole tree can virtually have its living matter bounded by a single plasma membrane. In very stable environments, where the controls can slow down, large cells are possible, such as those of the marine protists Xenophiophora [29]. The lowest cell size is determined by the minimal set of cellular constituents [17] and the smallest active cells are those with the smallest genomes [9]. Smaller cell sizes may be found either in bacterial endospores or in starving bacteria, but in these examples the extremely low water content does not allow the usual diffusion, and thus the metabolism, to take place. Perhaps, there are metabolically active bacteria that are smaller than *Mycoplasma genitalium* and endowed with a smaller number of protein-coding genes [18]. A few of them, provisionally dubbed nanobacteria, have been cultivated in mammalian cell culture medium and classified among Proteobacteria [15], but their genomes are still unknown.

Although a cell is, in essence, a membrane surrounding a genome and a complex set of molecules interacting (directly or indirectly) with these two features, there are apparent exceptions. Genomes are absent in some cells, such as the red blood cells of mammals and the sieve-tube cells of flowering plants. However, these kinds of cells are terminal stages that lose their genomes while still maintaining a role in the organism. As regards the plasma membrane, it does not always line every individual genome and its surrounding components. This necessarily happens in the initial stages, during cell fission, because the welding of the plasma membrane around each genome cannot be instantaneous. In some cases, this situation is more or less lasting, be it the result either of incomplete closure of the plasma membrane or fusion of previously separated cells, but this does not prevent the whole community of genomes from being lined by a common plasma membrane. It is well known that these examples are not considered as exceptions to the cell theory, and diffusion aspects can be recovered as in normal cases to justify cell size and the space of competence of each genome. At present, no life form is known to exist without cells, nor is it expected that any living thing lacking the cellular organization will ever be found.

In any case, as all living matter seems to be necessarily organized into discrete units or modules, it may be said to have *modular organization*. This does not mean that modules are equal to each other: even cells that are usually considered to be indistinguishable (such as the individual bacteria of a colony or the individual cells of a homogeneous tissue) differ slightly from one another. Thus, their quasi-modularity has nothing to do with the precise periodic arrangement of matter in some inorganic objects, such as crystals; even crystal defects do not prevent many elementary "cells" to be exactly equal. This precise modularity only occurs either in single molecules, even in large ones such as proteins and some RNA molecules, or in assemblies of them made up of a definite number of molecular components (sometimes called "closed crystals") such as ribosomes or the smallest viruses, but it does not occur in structures made up of a very large number of molecules lacking definite spatial relationships with one another in all their portions, such as membranes or vacuole contents.

4. Reproduction derives from overproduction and modularity

According to the discussion above, reproduction turns out not to be a logically primitive property, but to be the consequence of two more basic ones, namely, overproduction and modularity. Because of modularity, overproduction implies reproduction (of modules). In the modular organization of living matter, non-metabolic degradation partially regards single portions of modules that can recover their living state, but partially entire modules; overproduction replaces these by means of reproduction of pre-existing ones. If living matter would not have modular organization, no reproduction would take place, only growth. The reproduction of modules is obvious in unicellular species and only a little more complicated in multicellular ones. In unicellular species, reproduction basically consists in the duplication by fission of pre-existing modules. As these are similar to each other, but not exactly equal, duplication is actually quasi-duplication. Quasi-duplication is the basic property that any elementary object, natural or artificial, that is claimed to be living, must possess, at least in one environment, although the set of quasi-duplicating objects also includes non-living things [42].

Overproduction and modularity should be typical of any material system that we call life, even the minimal one, be it made up starting from a natural cell (*subtractive approach*, section 5) or a mixture of chemicals (*additive approach*, section 6). In the laboratory, the non-metabolic degradation is virtually ruled out so that overproduction leads to increase of organic matter. Non-metabolic degradation could be simulated by means of programmed extraction of excess production. In any case, overproduction depends on efficient assimilation and metabolic degradation, and these two properties cannot be efficient unlike modularity reduces diffusion problems. As a consequence, modularity is always preserved, and thus quasi-duplication. The maintenance of this process would be considered as sufficient evidence that we are still dealing with a living thing; actually, duplication is the usual evidence in microbiological laboratories that an organism is living, although metabolism can also go on in the absence of duplication.

5. The subtractive approach to minimal life

A way to realize the hypothetical minimal cell is to subtract some dispensable functions from a natural cell. This may be achieved by artificially shortening the genome of a natural species. The subtractive approach takes advantage of the knowledge of complete sequences of prokaryotic genomes. The shortest genome ever found in an "autonomous" cell is that of *Mycoplasma genitalium*: it is 580,070 bases long [9] and includes 480 protein-coding genes [14]. Random inactivation of many genes has been achieved by means of the insertion of a transposon into them, so that the genome is functionally reduced although it is physically elongated. It has been demonstrated that this species still undergoes duplication after its active genes are reduced to 350–265, depending on the experiment. The damaged cells can be defined as partially artificial [43], like all the genetically modified cells, but there is no doubt that they can still be considered as living things. There is a project to really shorten the genome of this species in order to better identify its "minimal" gene set [14, 3]. Up to now, no subtractive investigation has been carried out on nanobacteria.

Species endowed with very reduced genomes (if any) have been detected and deeply investigated also in their genome sequences: they are the plastids and the mitochondria. As the endosymbiotic origin of these organelles is no longer questioned [36], they may be considered as prokaryotic or eukaryotic unicellular organisms [40] that, at some moment in their evolution, began to take up proteins from their host cells. This means that their genomes have not all the genes to make up their proteomes. Mitochondria with reduced genomes, but still able to duplicate, have been obtained many times. Some anaerobic mitochondria that are called hydrogenosomes usually have no genomes at all and import all their proteins [2]. The organelles of some species even import from their host cells some RNA molecules, so that their genomes, in addition, have not all the portions to make up all their transcriptomes. Precisely because of the insufficiency of their genomes, all these living units are qualified as semi-autonomous, as organelles, because the other organelles have no autonomy from their hosting cytoplasm. However, as cells, they should be qualified as non-autonomous or dependent on protein (or also RNA) import (Table 1). In any case, they are considered useless for the subtractive approach precisely because of their import needs.

So far no subtractive investigation has been carried out on autotrophic cells, in spite of the fact that, on a metabolic basis, they are the most independent units of the biosphere: they depend on the other species only for the maintenance of the useful values of some environmental parameters such as the partial pressure of CO_2, but also geological phenomena could give rise to the same outcomes. The genome of the sole autotrophic species (a cyanobacterium) that is known in all its sequence is 3,573,470 bases long and has 3168 protein-coding genes [16]. An autotrophic species that only relies on the availability of CO_2, water, some minerals, and light (or other energy sources), is also likely to bear dispensable portions of its genome such as non-coding stretches, redundant genes, or genes that are only useful in particular situations but are useless in suitable laboratory conditions. In any case, the limit of genomic reduction is expected to be much higher than that of a heterotrophic species because the metabolites are all synthesized by the cell. This limit could only be overcome if the cell is modified to take up some metabolites from the environment, i.e., if it is turned from autotrophic into heterotrophic. This possible metabolic transition renders it evident that simplification of the cell is to some extent an illusion because it is compensated by complexification of the environment. For autotrophic non-autonomous cells, i.e., the plastids, the environment is required to have not only small organic molecules but also specific macromolecules.

In conclusion, there is no defined minimal autonomous cell because its condition is related to its environment [3]. If we succeed in inducing an organism to take up organic molecules that are absent either from the natural environment or from the usual culture medium, we can further reduce the genome. Moreover, if we take off from the environment of a species a few interfering molecules, we can leave it with less selective and perhaps fewer membrane channels for organic molecules, thus further reducing its genome. To sum up, the lowest limit of genome reduction, if we want a heterotrophic cell to maintain its autonomy, is mainly dictated by the polynucleotide processing functions, i.e., replication, transcription, and translation, that are to some extent incompressible, and the core of metabolic transformations.

6. The additive approach to minimal life

Attempts to create an autonomous cell in a test-tube that completely detach from a living cell as regards the starting material have also begun. In this additive approach to minimal life, both overproduction and modularity are taken for granted, so that quasi-duplication also occurs. Moreover, growth of the duplicating system relies on a further factor, i.e., the absence of metabolic degradation, in addition to non-metabolic one. This depends on the fact that virtually no catabolism takes place because energy is entirely provided with the building blocks of the constituent molecules (such as the membrane lipids or the polynucleotides) and there is no need for chemical energy in any other process than their synthesis. The only sort of catabolism is due to the hydrolysis of the activating or protecting chemical groups during the one-step synthesis of the constituent molecules. If also metabolic degradation is nil, any production turns out to be overproduction.

Autotrophic projects for completely artificial cells would essentially be projects of autotrophic liposomes. We could call autotrophic all the organisms that only rely on a carbon source with a single carbon atom (Table 1). Usually, only those dependent on CO_2 are considered as autotrophic, because this molecule is released from inorganic sources such as volcanoes, in addition to organisms. But also the other molecules with a single carbon atom more reduced than CO_2, i.e., CO, HCOOH, H_2CO, CH_3OH, and CH_4, are sometimes known to exist in inorganic formations. For standard autotrophic cells relying on CO_2 as a carbon source, the explicit project of the environment is unimportant because all scholars could agree on a single minimal one, i.e., the one required for most unicellular cyanobacteria. In any case, the great diversity of natural autotrophic bacteria makes clear that the solutions can be many and many, even if we restrict our survey to oxygen-producing photoautotrophic ones, i.e., to cyanobacteria.

Heterotrophic projects are favored because they can be much simpler, as regards the endowment of enzymes. In any case, an autonomous cell bases its production on assimilation, and thus on constructive metabolism [42]; in other words, it must make its own macromolecules. Heterotrophic projects are based on the three current models of prebiotic aggregates, i.e., liposomes, microspheres and ribozymes [41] or mixtures of them. The investigations on them have been distinguished between top-down and bottom-up [26], but these terms have been interpreted in various ways [7, 31, 3]: this is the reason why we propose here to speak of subtractive and additive approaches, when we are referring to the creation of a minimal cell. In any case, the goal to build an artificial cell is so important that we can go on regardless of any constraint of the putative prebiotic chemistry [41].

A project based on coacervates of proteinoids [8, 37] does not show duplication regular enough to be truly demonstrated. A non-duplicating ultimate living unit [4] or an unbounded one [19] do not satisfy, but also the idea that a replicating molecule [38, 21] can be considered as life [13] has been rejected (e.g., [44]). Projects based on replicating ribozymes resulted to be also of difficult realization in spite of so many efforts [12]. By contrast, liposomes gave promising results [24, 25], so that most models inspiring the current research are based on combinations of liposomes and polynucleotides. The experiments carried out on quasi-duplicating liposomes that are implemented with polynucleotides and their replicase taken from the living world [33] or a simple translation machinery to the more so taken from a living cell [34] can be regarded as advanced attempts to probe the technical tractability of combined reaction systems. In these systems, respectively, the one of the dividing liposome includes a polymerizing one without any reaction that connects the liposome system and the polymerizing one. These two reaction systems only interact because one is encapsulated within the other, and this strict contact can also affect the reactions. But blueprints of minimal life require that the constituent molecules are metabolically linked [10, 1] because only this situation fulfills the current concept of the cell as a chemically integrated system, not only as an assembly of chemically separated systems. The mere presence of a replicating genome within a duplicating liposome (e.g., [20]) does not make up a cell: an essential performance of the genome is that of promoting the construction of specific catalysts needed for assimilation. To sum up, the integration is expected to be both metabolic, in the sense of linked reactions, and interactive, in the sense that the regulation of the reaction rates should also depend on the membrane extension, and this on the mass of the RNA, and so on. This general regulation in natural cells is a consequence of evolution, and is impossible to be achieved since the beginning in the additive approach because of the huge amount of information on molecular interactions it would require.

However, an integrated artificial cell that fulfills the requirements above does not yet realize "coded life", such as natural cells are. This kind of life is only realized if a polynucleotide is translated into protein enzymes. But is the supposed requirement of "information processing" [5] also fulfilled in the absence of translation? Is the supposed requirement of "meaning" fulfilled if the only meaning is the catalytic efficiency of a couple of ribozymes? Apparently, most scholars consider the translation as no longer needed to define life (e.g., [12]).

In any case, for a heterotrophic system, a project of the environment is as important as the project of the system itself. Hence, a project of a simple cell only makes sense if the environment is explicitly declared within the project itself. If the environment is more complex, a simpler cell, with less genes, can be made, presumably. So which is the actual minimal cell? Perhaps, the only meaningful minimization is that of the pair *cell* and *environment*.

7. Is the concept of a minimal cell sound?

At present, "minimal life" seems to be only a bridge that connects the distant banks that separates the subtractive and the additive approaches. That bridged distance can be roughly quantified with the lowest gene number empirically reached in *Mycoplasma genitalium*, i.e., 265 [14], and the number of "ribogenes" in the blueprint of a synthetic cell-like system endowed with an RNA genome, i.e., two: one for a replicase and one for a lipid convertase [1]. That bridge also has a few intermediate pillars [32] built both before and after the genomic divide, with the 38 miniprotein-coding genes of the minicell of Loomis [22] and the 105 protein-coding genes of the latest computational model of the non-duplicating cell of Tomita [45]. Maybe, instead of just discussing what is life, even the minimal one, we have better distinguishing between different systems that are close to natural heterotrophic life.

The first distinction is between integrated and simply combined systems; natural cells are, of course, integrated chemical systems. Integrated systems, in turn, can be genetically non-autonomous and autonomous, where the latter are endowed with constructive assimilation of macromolecules. Moreover, the supply of chemical energy to carry out the constructive assimilaton can entirely come from the environment or also from hydrolytic metabolic degradation: the latter solution is that of cells [10, 11, 42], and is necessary to ensure metabolic processes even in the temporary absence of nutrients. But there are energy parasites that draw part of their energy-releasing molecules from their hosts. The choice to make energy parasites is mandatory in order to minimize both natural and synthetic cells, but we must be aware of the fact that they can only work in an artificial environment where the nutrient supply, including ATP or a similar high energy-hydrolysis molecule, is warranted. We might call them "cells" in a very broad sense because the environment in which they could thrive is, at the same time, extremely rich, in that it provides all the needs of the duplicating system, and extremely poor, as it lacks any threat that abound even in the most friendly natural environment.

But a final remark is necessary on the robustness of these objects. Their quasi-duplication refers to both statistical variability in the number of copies of some constituent molecules, such as particular lipids or enzymes, and rare qualitative variations in polynucleotides, i.e., mutations. The tolerance toward the number of molecules is also extended to the shape of the membrane or even small damages of it and is an important factor of robustness. Another and more typical factor is a relative tolerance to mutations, in the sense that quasi-duplication is not prevented by every mutation; it is well known that some lucky mutations can even improve it. As mutations are inherited due to replication (or, better, they are quasi-inherited due to quasi-replication), they provide the novelties of evolution, whereas the differential tolerance toward nucleotide changes is the premise of differential quasi-duplication (i.e., of selection) in a non-competitive environment. However, evolution in vitro of an autonomous quasi-duplicating system does not guarantee evolution toward robustness: evolution toward simplification, thermodynamic equilibrium, and eventual stop of any metabolic activity remains a possible outcome. Neither the subtractive nor the additive approaches proceed *via* selection, but *via* engineering, so that there is no adaptation and only poor regulative interactions. We may try to select in some way any improvement of quasi-duplication, but this cannot be confused with the "artificial selection" of Darwin and Wallace because it is not applied to organisms that are the result of an evolutionary process. In conclusion, we cannot take it for granted that such "minimal" systems would be capable of evolution.

Acknowledgments

The author is grateful to R.C. Guimarães (Federal University of Minas Gerais, Belo Horizonte) for useful comments on a preliminary version of the manuscript, to P.L. Luisi (Swiss Federal Institute of Technology, Zürich) and M. Negrotti (University of Urbino) for brief but fruitful discussions, to P. Cordella and N. Salmaso (University of Padova), R. Grubbs and A. Pappelis (University of Carbondale), D. Lancet and D. Segré (Weizmann Institute, Rehovot), and F. Tani (University of Milan) for information; the author is also indebted to R. Mazzaro for computer drawing.

References

[1] BARTEL D.P., UNRAU P.J., *Constructing an RNA world*, Trends Cell Biol. **9** (1999), M9–M13.

[2] BIAGINI G.A., BERNARD C., *Primitive anaerobic protozoa: a false concept?*, Microbiology **146** (2000), 1019–1020.

[3] CHO M.K., MAGNUS D., CAPLAN A.L., MCGEE D., THE ETHICS OF GENOMICS GROUP, *Ethical considerations in synthesizing a minimal genome*, Science **286** (1999), 2087–2090.

[4] COLOMBO L., *The biogen: the ultimate living unit*, in: "Defining life", M. Rizzotti ed., University of Padova, Padova 1996, 39–66.

[5] DAVIES P., *Physics and life*, in: "First steps in the origin of life in the universe", J. Chela-Flores, T. Owen and F. Raulin eds., Kluwer, Dordrecht 2001, 13–20.

[6] FLEISCHAKER G.R., *A few precautionary words concerning terminology*, in: "Self-production of supramolecular structures", G.R. Fleischaker, S. Colonna and P.L. Luisi eds., Kluwer, Dordrecht 1994, 33–41.

[7] FOX S.W., *The emergence of life*, BasicBooks, New York 1988.

[8] FOX S.W., *Synthesis of life in the lab? Defining a protoliving system*, Quart. Rev. Biol. **66** (1991), 181–185.

[9] FRASER C.M., 28 COAUTHORS, *The minimal gene complement of Mycoplasma genitalium*, Science **270** (1995), 397–403.

[10] GÁNTI T., *The essence of the living state*, in: "Defining life", M. Rizzotti ed., University of Padova, Padova 1996, 103–117.

[11] GÁNTI T., *Biogenesis itself*, J. theor. Biol. **187** (1997), 583–593.

[12] GESTELAND R.F., CECH T.R., ATKINS J.F., *The RNA world*, 2nd ed., Cold Spring Harbor Lab. Press, Cold Spring Harbor 1999.

[13] HURST L.D., DAWKINS R., *Life in a test tube*, Nature **357** (1992), 198–199.

[14] HUTCHISON C.A. III,PETERSON S.N., GILL S.R., CLINE R.T., WHITE O., FRASER C.M., SMITH H.O., VENTER J.C., *Global transposon mutagenesis and a minimal mycoplasma genome*, Science **286** (1999), 2165–2169.

[15] KAJANDER E.O., CIFTCIOGLU N., *Nanobacteria: an alternative mechanism for pathogenic intra- and extracellular calcification and stone formation*, Proc. Natl. Acad. Sci. USA **95** (1998), 8274–8279.

[16] KANEKO T., 23 COAUTHORS, *Sequence analysis of the genome of the unicellular cyanobacterium Synechocystis sp. strain PCC6803. II*, DNA Res. **3** (1996), 109–136.

[17] KOCH A.L., *What size should a bacterium be? A question of scale*, Annu. Rev. Microbiol. **50** (1996), 317–348.

[18] KOONIN E.V., MUSHEGIAN A.R., RUDD K.E., *Sequencing and analysis of bacterial genomes*, Current Biology **6** (1996), 404–416.

[19] KUNIN V., *A system of two polymerases – a model for the origin of life*, Origins Life Evol. Bios. **30** (2000), 459–466.

[20] LAHAV N., *The RNA-world and coevolution hypotheses and the origin of life*, Origins Life Evol. Bios. **23** (1993), 329–344.

[21] LEE D.H., GRANJA J.R., MARTINEZ J.A., SEVERIN K., REZA GHADIRI M., *A self-replicating peptide*, Nature **382** (1996), 525–528.

[22] LOOMIS W.F., *Four billion years: an essay on the evolution of genes and organisms*, Sinauer, Sunderland MA 1988.

[23] LUISI P.L., *Self-reproduction of micelles and vesicles*, Adv. Chem. Phys. **92** (1996), 425–438.

[24] LUISI P.L., *Self-reproduction of chemical structures and the question of the transition to life*, in: "Astronomical and biochemical origins and the search for life in the universe", C.B. Cosmovici, S. Bowyer and D. Werthimer eds., Ed. Compositori, Bologna 1997, 461–468.

[25] LUISI P.L., *About various definitions of life*, Origins Life Evol. Bios. **28** (1998), 613-622.

[26] LUISI P.L., OBERHOLZER T., *Origin of life on Earth: molecular biology in liposomes as an approach to the minimal cell*, in: "The bridge between the big bang and biology", F. Giovannelli ed., CNR special volume, Roma 2001, 345–355.

[27] MAHNER M., BUNGE M., *Foundations of biophilosophy*, Springer, Berlin 1997.

[28] MANILOFF J., *The minimal cell genome*, Proc. Natl. Acad. Sci. USA **93** (1996), 10004–10007.

[29] MARGULIS L., SCHWARTZ K.V., *Five kingdoms*, 3rd ed., Freeman, New York 1998.

[30] MCGENITY T.J., GEMMEL R.T., GRANT W.D., STAN-LOTTER H., *Origins of halophilic microorganisms in ancient salt deposits*, Environ. Microbiol. **2** (2000), 243–250.

[31] MOROWITZ H.J., *Beginnings of cellular life*, Yale Univ. Press, New Haven 1992.

[32] MUSHEGIAN A., *The minimal genome concept*, Curr. Op. Genet. Dev. **9** (1999), 709–714.

[33] OBERHOLZER T., WICK R., LUISI P.L., BIEBRICHER C.K., *Enzymatic RNA replication in self-reproducing vesicles: an approach to a minimal cell*, Biochem. Biophys. Res. Comm. **207** (1995), 250–257.

[34] OBERHOLZER T., NIERHAUS K.H., LUISI P.L., *Protein expression in liposomes*, Biochem. Biophys. Res. Comm. **261** (1999), 238–241.

[35] ORGEL L.E., *Unnatural selection in chemical systems*, Acc. Chem. Res. **28** (1995), 109–118.

[36] PALMER J.D., *A single birth of all plastids?*, Nature **405** (2000), 32–33.

[37] PAPPELIS A., BAHN P., GRUBBS R., BOZZOLA J., COHEN P., *From inanimate macromolecules to the animate protocells*, in: "First steps in the origin of life in the universe", J. Chela-Flores, T. Owen and F. Raulin eds., Kluwer, Dordrecht 2001, 65–68.

[38] REBEK J. JR, *Synthetic self-replicating molecules*, Sci. Am. **271** (1994), 34–40.

[39] RIVKINA E.M., GILICHINSKY D.A., *Metabolic activity of permafrost microorganisms*, in: "The bridge between the big bang and biology", F. Giovannelli ed., CNR special volume, Roma 2001, 370–376.

[40] RIZZOTTI M., *Early evolution: from the appearance of the first cell to the first modern organisms*, Birkhäuser, Basel 2000.

[41] RIZZOTTI M., *Precellular organic aggregates: the bridge between the non-living and the living*, in: "The bridge between the big bang and biology", F. Giovannelli ed., CNR special volume, Roma 2001, 321–333.

[42] RIZZOTTI M., *Living things are far from equilibrium: which equilibrium?*, in: "Fundamentals of life", G. Palyi, C. Zucchi and L. Caglioti eds., Elsevier, Paris, in press.

[43] RIZZOTTI M., *Are humans the sole producers of artificial?*, Yearbook on artificial **1**, in press.

[44] SZOSTAK J.W., BARTEL D.P., LUISI P.L., *Synthesizing life*, Nature **409** (2001), 387–390.

[45] TOMITA M., *Whole-cell simulations*, Trends Biotec. **19** (2001), 205–210.

[46] VARELA F.J., *On defining life*, in: "Self-production of supramolecular structures", G.R. Fleischaker, S. Colonna and P.L. Luisi eds., Kluwer, Dordrecht 1994, 23–31.

[47] VREELAND R.H., ROSENZWEIG W.D., POWERS D.W., *Isolation of a 250 million-year-old halotolerant bacterium from a primary salt crystal*, Nature **407** (2000), 897–900.

[48] WELCH G.R., *The enzymatic basis of information processing in the living cell*, BioSystems **38** (1996), 147–153.

[49] WOLPERT L., *The evolution of the cell theory*, Curr. Biol. **6** (1996), 225–228.

Table 1. Ranking of organisms according to their needs of carbon-containing molecules (examples in brackets).

One carbon-atom molecules (CO_2, CH_4)	Two or more carbon-atom simple mol. (amino acids, sugars)	Large complicated molecules (vitamins, sterols)	High energy -hydrolysis molecules (ATP)	Specific proteins	Specific RNA molecules (tRNAs)	INSTANCES of ORGANISMS, including non-autonomous ones, i.e., organelles
–						Cyanobacteria, methane-oxidizing bacteria, autotrophic protists, autotrophic plants
–	–					Some autotrophic bacteria
	–					Most heterotrophic bacteria, most fungi, heterotrophic plants (*Epifagus*)
	–	–				Mycoplasmas, animals
	–		–			Chlamydias, rickettsias
–	–			–		Autotrophic plastids
	–			–		Most heterotrophic plastids, most mitochondria
	–			–	–	Some mitochondria and hydrogenosomes, some heterotrophic plastids
				–		Some anaerobic mitochondria and most hydrogenosomes

TUBULIN AND THE COMPLEXITY OF LIFE

Giovanni Santangelo
Dipartimento di etologia, ecologia ed evoluzione,
Via Volta 6, 56126 PISA - ITALY
sant@deee.unipi.it

Giampaolo Magagnini
Dipartimento di etologia, ecologia ed evoluzione,
Via Volta 6, 56126 PISA - ITALY

Abstract The tubulins are fascinating examples of biological complexity. These "brick proteins" polymerising, build microtubules. Microtubules in turn arrange themselves to form complex cellular structures like cilia, flagella, mitotic and meiotic spindles and others. Due to the huge number of tubulin structures and the interactions that these structures establish each other and with the other cellular components, an astonishing degree of complexity has been attained. All these structures complement one another perfectly to form the highly efficient and coordinated engines at work in eukaryotic cells. Such structural complexity can not be understood by reductionist approaches such as the simple analysis of DNA sequences codifying for tubulins.

1. Introduction

Arecchi and Farina [1] use the term "complexity" to refer to "a property inherent to some subject, depending on the number of its parts and on the links (interactions) between them". By this definition all biological systems may be considered to be complex and can moreover be ranked according to their increasing complexity into several levels of organization ranging from biomolecules to single cell, to organisms and to ecosystems [2]. Recent years have seen the emergence of a considerable body of literature dealing only with genome complexity (the amount of information stored in an organism's genome) [3]. In this short note we wish to recall that complexity is an intrinsic characteristic of living organisms, stemming from their organization. In order to stress this concept, we examine one of the main sources of biological complexity, the tubulins. These "brick-proteins" have been of utmost importance in the evolution of eukaryotic life. All functions governed by these structures (cilia, flagella, cytoskeletons, mitotic and meiotic spindles and more) are fundamental to the life of all eukaryote cells in both unicellular (protists) and multicellular organisms (metazoans, plants, fungi).

Due to the development of anti-tubulin immunofluorescence techniques and new biochemical and biomolecular tools, in recent years our knowledge of the intracellular structures made up of tubulin has been greatly enhanced [4]. The tubulin "microtubules", forming the mitotic spindles, are also vitally important to the cell division processes. Recent studies on the control of cell division have successfully investigated many important facets of tubulin involvement: most notably, the mechanisms controlling microtubule assemblage during cell division have been analysed by three renowned researchers, Paul Nurse, Leland Hartwell, and Timothy Hunt, who were awarded the 2001 Nobel prize [4, 5, 6]. In this regard, it seems worthwhile stressing that better knowledge of cell division process is likely to be extremely important to our understanding of cell proliferation and thus to the development of new anti-tumor therapies [7].

2. Tubulin and microtubules

Tubulin is a highly conservative, globular protein, found only in eukaryotic cells [8] . All efforts to reveal tubulin in prokaryotes have, to date, been fruitless [9]. A "family of genes" codes for this protein, and several isoforms (similar molecules coded for by different alleles on the same locus) exist. Basically, there are two different subunits of tubulin (α and β), whose assemblage leads to the formation of tubulin dymers, which in turn polymerise to form the tubular structures known as "microtubules" (Fig. 1). Tubulins may differ one from the other due to post-synthesis cytoplasmic modifications (post-translational processes). Whatever their precise make up, all proteins of this sort have the peculiar characteristic of polymerising into tubular structures (22–25 nm wide) which, in turn have the capability of assembling themselves into several different kinds of "macrostructures", such as cilia, flagella, nerve cells and fibres, the hyphae of fungi, cell cytoskeleton components, centrioles, meiotic and mitotic spindles and others. Microtubules are made up of several longitudinal rows (protofilaments) of tubulin dymers (Fig. 1). Generally, 13 protofilaments go to make up a microtubule, though this number may vary from 10 to 17.

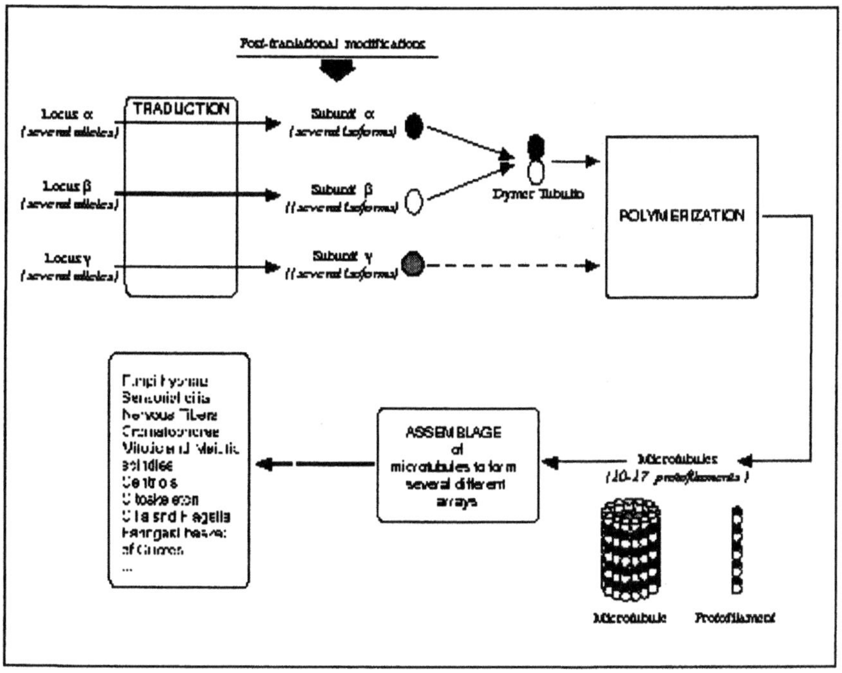

Figure 1. Scheme of tubulin and microtubule assembly.

Both the assembly and disassembly of microtubules are rapid processes. According to an early, classic model of microtubule dynamics, they are polymerised through the addition of new tubulin dymers to their distal (+) extremities and depolymerised by the removal of dymers from their proximal (−) ends. However, more recent data has shown that microtubules cannot shorten by loosing tubulin dymers from the proximal end, since this (−) end is frequently embedded in a "Microtubular Organization Centre" (MTOC). All polymerisation - depolymerisation processes must thus occur at the (+) end [10]. Whatever the mechanism of microtubule polymerisation and depolymerisation, they are able to elongate and shorten rapidly [10]. While "structural" microtubules (e.g., those of the cytoskeleton) are stable, "cytoplasmic" and "division spindle" microtubules are dynamically unstable.

Microtubules are sensitive to many specific drugs (e.g., Nocodazole, Colchicine, Taxol) which inhibit their assembly in different ways [13].

Such microtubule inhibitors are widely applied in cancer therapy to control cell proliferation.

3. Assemblages of the same bricks give rise to several different buildings

Many different cell structures are based on microtubules (see Section 1). In the following we shall describe some examples of the main microtubular arrangements.

A) Cytoplasmic and cortical microtubular networks. Together with microfilaments of varying composition, it is the cytoskeleton microtubules, variously arranged in the different kind of eukaryotic cells, that form the scaffolding of cells. Over the course of evolution this scaffolding has allowed eukaryotic cells to increase their size, while maintaining their shape.

Via the efficient transport system described below (Section 4, **A**) the extensive internal microtubular meshwork provides eukaryotic cells with an efficient communication and nutrient-distribution system spread throughout the cytoplasm, thereby allowing the nuclear DNA to control all regions of the cell.

Other microfilaments (e.g., actin and intermediate filaments) are also components of the internal meshwork and interact with microtubules, though such interactions are still poorly understood.

B) Mitotic and meiotic spindles. Mitosis is the division process by which eukaryotic cells typically multiply. According to Preston and co-workers, the mitotic spindle is a bipolar framework of microtubules assembled by the cell specifically to carry out the accurate segregation of identical sets of chromosomes between two daughter cells [10].

Chromosomes are attached to the spindle in a specialized region (kinetocore). If a chromosome lacks this region, it is lost during mitosis. Two microtubular structures, called "centrioles", preside over spindle assembly and are located at each pole of the spindle. The final goal of mitosis is to transport each complete, double set of replicated chromosomes to the cell poles so that they (via spindle elongation, spindle disappearance and cell division) will be shared equally between the newly formed daughter cells. This process proceeds slowly (1 μm/min), but with astonishing accuracy. The two sets of chromosomes and each single chromosome in a set move at exactly the same speed!

Though this description provides a general picture of the process, there are many variables. For example, the mitotic spindles may vary greatly in number of constituent microtubules (the number of microtubules attached to each chromosome kinetocores ranges between 1 and 100), and some types of cells lack true centrioles.

Meiosis, on the other hand, is the division process by which a eukaryotic cell halves the (diploid) chromosome number to produce haploid gametes. The meiotic spindle is similar to that described for mitosis in the foregoing.

Both mitosis and meiosis are fascinating processes whose mechanisms are known only in part. A major problem stems from the difficulty of collecting purified spindles in large enough amounts for biochemical study [10]. Even the *in vivo* observation of mitosis and, especially, meiosis is limited by the fact that spindles are single-copy organelles that are only present during a brief period of the cell cycle and, in the case of meiosis, only in some, peculiar cells (germinal cells) during a time-limited process (gametogenesis) [12].

C) Cilia and flagella. Microtubules can "self-organise" to form quite complex structures, such as cilia and flagella, which are able to move. These motile structures are present in protozoans (mostly in ciliates and flagellates), sperm and the ciliate or flagellate cells of multicellular eukaryotes (e.g., mucous membranes of the vertebrate breathing system, coanocytes of sponge flagellate chambers).

The main difference between cilia and flagella resides in their length and number (cilia being shorter and more numerous than flagella). Like tubulin, the structure of cilia and flagella is also highly conservative: in all living organisms they are always made up of 9 twin-paired microtubules surrounding 2 central microtubules. These microtubule pairs can slide one over the other through the action of mobile "arms" on their surfaces. These arm-like structures are made up of a mobile protein (dynein) that enables the cilium or flagellum to bend or wave [10]. The base of each cilium or flagellum bears what is known as a basal body, which is a cylinder composed of 9 triplets of short microtubules, similar to the centrioles of mitotic and meiotic spindles (see the following). These structures are needed in order for cilia and flagella to duplicate autonomously during cell division and are presumably involved in the control of microtubule production.

4. Different buildings perform different functions

Microtubular structures (e.g., cilia, flagella, cytoskeleton and cytoplasm networks, sensory cilia, nerve-cell microtubules, mitotic and meiotic mictotubular spindles) carry out different functions, all of which are indispensable for eukaryotic life (motility, feeding, maintenance of cell shape, cell traffic, sensorial and nervous activity, intracellular motility, cell division, sexual reproduction, etc.). In the following sections we shall examine some of these functions in more detail.

A) Intracellular motility. As microtubules are able to transport small structures on their surface, the cytoplasm microtubule network guarantees the cells a rapid, efficient intracellular transport network [13]. This intracellular "traffic control" is based on two different protein engines: kinesin and dynein.

Kinesin is a force-generating enzyme or "motor protein", which converts the energy of phosphate-ATP bonds into mechanical work [10], thereby ensuring that particles of different origins (ribosomes, vesicles, pigment granules, mitochondria, chloroplasts and others) move along the microtubules from the (−) to the (+) end (anterograde movement).

Another mobile protein (dynein) is associated to the microtubule surface, where it forms "arms" that move particles along the microtubule through ATP-fed strokes resembling a rowing action. Contrary to kinesin, dynein provides retrograde transport (from the + to − ends). Thus, while the former protein presides over transport from inside the cell to its outside, the latter ensures movement in the opposite direction. It is the action of these mobile proteins that enable microtubules to direct and control intracellular traffic throughout all regions of the cell [11]. Microtubules are responsible for the (axonal) transport of neurotransmitters from neurons to synapses along nerve fibres. Microtubules are also responsible for pigment granule motility in chromatophores (e.g., in melanocytes). By concentrating or dispersing pigment granules in these cells, microtubules provide for the colouring, mimic and warning behaviours of cephalopods, crustaceans, insects, fish, mammals and other metazoans.

B) Cell motility. Preston and co-workers judge cilia and flagella to be the most intensively studied locomotor organelles [10]. These structures project from the cell surface and in some protozoans cover the entire cell body surface. Both cilia and flagella have the same basic internal structure (described in Section 3 **C**) and can move in coordination one with the others according to changes in the cell membrane's electric potential. By bending and waving they provide for the cleansing of mucous membranes and the "swimming" motility of spermatozoa and protozoans (see protozoans in Section 5).

C) The role of microtubules in cell division and sexual reproduction. During eukaryotic cell division (both mitotic and meiotic) microtubules form microtubular spindles which impart movement to the chromosomes.

The centrioles, which like basal bodies are cylindrical structures made up of 9 microtubule triplets, organise the microtubules of the spindle, while they themselves remain at the poles. After duplication, the chromosomes move towards the "equatorial" cell zone (metaphase) and then move back (pole-wards) during anaphase and telophase. These movements, indispensable for chromosome distribution to the daughter cells, are due to shortening of the microtubules to which the chromosomes are linked and their sliding along the microtubules through the dynein-based mechanism described previously (Section 4 **A**). As meiosis is the division process leading to gamete formation, the origins of sexual reproduction is strongly linked to the emergence of tubulin [8].

Ultimately, the entire cell division process is controlled by specialised proteins (enzymes) that govern spindle microtubule synthesis and dynamics. Several enzymes are involved in controlling these processes, and thereby in regulating the stages of mitosis (as well as the second meiotic division), the main ones being Cycline and CDC2 kinase, two enzymes considered to be the main factors in a "universal mechanism controlling eukaryotic cell cycle" [4, 7].

5. Protozoans are equipped with a wide variety of complex microtubule arrays

Protozoans exhibit a wide variety of microtubular arrays: cilia, flagella, axopodia, cytostomes and pharyngeal baskets (the cellular mouth of ciliates), cytoskeletons, cytoplasmic networks and sensory cilia are all composed, at least partially, of microtubules. Ciliates in particular possess an incredible variety of microtubular arrays and networks [14].

Microtubules are found not only in the cilia (which cover the cell surface), but in their basal bodies, in the transverse cortical ribbons linked to them, in cell mouth structures (cytopharyngeal ribbons, post-oral fibres), as well as in at least 7 other intracellular arrays. Microtubules are involved in locomotion (membrane polarization and depolarization processes control ciliary movement in ciliates) [14], food capture (cilia or flagella) and a wide variety of life functions. For example, the Loxodidae species possess, apart from the usual tubulin-based structures of ciliates, a peculiar gravity sensor [15]. These subcellular organelles (a few μm wide) are composed of small sensory cilia (once again tubulin at work!) and tiny globular bodies (the statolites). Movement of the statolite due to gravity produces different pressures on microtubules connected to a sensory cilium, thus informing the ciliate of its spatial position (geotaxis)!

Other small sensory cilia, common on the dorsal surface of some ciliates (the "dorsal bristles" of Hypotrichida), function as specialised hydrodynamic-force sensors. This adaptation allows these ciliates, which live on marine and freshwater sand grains, to adhere more tightly to the grain surface when the water flow intensity increases and thereby avoid being washed off the sandy bottom.

The ciliary system of many ciliates is organised into complex arrays made up of thousands of cilia that surround their cytostome with several orders of ciliary rows or membranelles. Such ciliary arrays are efficient filters for collecting suspended bacteria or other microorganisms [16].

Cortical and adoral tubulin arrangements are strongly linked to the phylogeny of ciliated protozoans. Thus, their morphology, revealed by optic, SEM or TEM microscope, may have taxonomic significance [17, 18].

6. Conclusion

The appearance of tubulin is tightly linked to eukaryotic cell evolution, marking the passage from a world completely dominated by prokaryotes to one in which eukaryotic life flourished and diversified. Moreover, by enabling meiosis and thus sexual reproduction to occur, tubulin synthesis led to a dramatic increase in natural selection and evolution rates.

In this short note we have tried to present tubulin as an example of biological complexity, whereby the same basic components were successfully employed to construct progressively more complex arrays, which, in turn, generated a cascade of complex and highly diversified structures in different organisms. This process began (about two billion years ago) with the simplest tubulin dymers, progressing through the complex microtubular arrays of ciliated protozoans and finally leading to highly diversified natural communities dominated by eukaryotes.

We have attempted to follow tubulin's development during these stages of evolution, ignoring the numerous interactions between microtubules and other cellular structures (and a great number of enzymes) which have generated further complexity. Due to the huge number of tubulin structures and the interactions that these structures establish within small eukaryotic cells, an astonishing degree of complexity has been attained. All these structures complement one another perfectly to form the highly efficient and coordinated engines at work in eukaryotic cells.

Along with Nurse [19], we believe that such "structural" complexity cannot be completely understood by reductionist approaches, such as the simple analysis of the DNA sequences (e.g. those coding for tubulins), or by the mere, albeit exhaustive, description of all the molecules and reactions occurring within a living cell.

References

[1] ARECCHI T.F., FARTINA A., *Lexicon of complexity*, Istituto Nazionale di Ottica, Arcetri, Studio Editoriale Fiorentino, Firenze 1996.

[2] FORESTIERO S., *Complessità*, in: "Enciclopedia Italiana. Appendice 2000", Istituto Enciclopedia Italiana, Roma 2000, 405–410.

[3] ADAMI C., OFRIA C., COLLIER T.C., *Evolution of biological complexity*, PNAS 97 (2000), 4463–4468.

[4] NURSE P., MASUI Y., HARTWELL L., *Understanding cell cycle*, Nature Medicine **4** (1998), 1103–1106.

[5] HARTMANN J.L., GARVIK B, HARTWELL L., *Cell biology. Principles for the buffering of genetic variation*, Science **291** (2001), 1001–1004.

[6] GOOLDBERG N.M., GANNON J., HUNT T., MORGAN P.R., *The localization of p34 (cdc2) in normal, hyperplastic, and malignant epithelial and lymphoids tissues of the oral cavity*, J. Pathol. **78** (1996), 422–428.

[7] GOODGEN N.M., GANNON J., HUNT T., MORGAN P.R., *Cell cycle regulatory proteins. An overview with relevance to oral cancer*, Oral Oncology **33** (1997), 61–73.

[8] MARGULIS L., *Nuclei, Mitosis and undulopodia*, in: "Symbiosis in cell evolution", 2nd edition, L. Margulis ed., Freeman, S. Francisco 1993.

[9] BERMUDES D., HINKLE G., MARGULIS L., *Do procariotes contain tubulines?*, Microbiological Review **53** (1994), 387–400.

[10] PRESTON T.M., KINGS C.A., HYAMS J.S., *The cytoskeleton and cell motility*, Chapman and Hall, New York 1990.

[11] ALLEN N.S., *The microtubule as an intracellular engine*, Sci. Am. **256** (1987), 26–33.

[12] SANTANGELO G., BRUNO P., *An immunofluorescence technique for staining ciliated protozoans: highlighting cytoplasmic microtubular arrays and stages of micronuclear meiosis*, Micron **31** (2001), 207–210.

[13] SANTANGELO G., NAVARI S., BRUNO P., *A highly specific microtubule depolimerizing agent (Nocodazole) affecting meiosis activation in the protozoan Blepharisma japonicum*, Ital. J. Zool. **64** (1997), 227–234.

[14] GORTZ H.D., *Paramecium*, Springer-Verlag, Berlin 1996.

[15] FENCHEL T., FINLAY B., *The structure and function of Müller vesicles in loxodid ciliates*, J. Protozool. **33** (1986), 69–76.

[16] FENCHEL T., *Ecology of protozoa: the biology of free-living phagotrophic protists*, Sci. Tech. Pub., Springer-Verlag, Berlin 1987.

[17] EISLER K., FLEURY A., *Morphogenesis and evolution in ciliates. Protistological actualities*, in: "Proceedings II European Congress of Protistology. Clermont-Ferrand 1995", G. Brugerolle and J.P. Mignot eds., Imprimerie County, Clermont-Ferrand 1995.

[18] ADOUTTE A., FLEURY A., *Cytoskeleton of ciliates*, in: "Ciliates. Cells as organisms", K. Hausmann and P.C. Bradbury eds., Fisher, Stuttgard 1996.

[19] NURSE P., *Reductionism. The ends of understanding*, Nature **387** (1997), 657.

PHYLOGENIES AND THE NEW EVOLUTIONARY SYNTHESIS

Francesco Santini
Department of Zoology, University of Toronto,
TORONTO, M5S 3G5, CANADA
fsantini@zoo.utoronto.ca

Abstract The neo-Darwinian theory of evolution is based on the concept that the processes of genetic variation and natural selection, as seen in modern populations, are sufficient to explain the large-scale patterns of diversification of life on Earth. While a neo-Darwinist model treats polygenic traits statistically, as if they resulted from the additive effects of a large number of genes of equivalent phenotypic impact, it describes large-scale phenomena primarily in a historical manner, such that the forces responsible for the origin and persistence of basic body plans, the major changes in structures and ways of life, and the influence of abiotic factors on critical events in the history of life receive scant consideration. This model persists despite evidence from the fossil record and geological dating demonstrating that large-scale patterns and past rates of evolution are not compatible with those extrapolated back from modern populations. In addition an examination of molecular development shows that long-term evolution is not the result of selection of alternative alleles controlling specific traits, or the progressive accumulation of new mutations in an additive fashion, as models used in quantitative genetics suggest. Evolutionary developmental biology and historical biogeography are two disciplines that can allow us to move past this erroneous neo-Darwinian view of evolution. Their success depends on the availability of robust phylogenetic hypotheses and a rigorous application of the comparative method. It is now finally being recognized as the only way to fully explain the patterns and processes that have led to the present diversity of life is a more holistic approach based on robust phylogenies, sound biogeographical data, a good fossil record and molecular developmental information.

1. Introduction

Darwin's theory of evolution, and its subsequent development known as neo-Darwinism or the Modern Synthesis, was based on the concept that the processes of genetic variation and natural selection, studied in modern populations, can explain the large-scale patterns of diversification that have occurred throughout the history of life on earth [38]. In the attempt to understand large-scale phenomena little consideration has been given to the forces responsible for the origins and long-term perpetuation of basic body plans, major changes in structures and ways of life, or to the influence of abiological factors on critical events in the history of life [4]. Research in many disciplines during the last four decades has, however, demonstrated that the patterns, processes and forces of evolution are much more diverse than hypothesized by Darwin and the creators of the Modern Synthesis. In fact:

1 The fossil record and geological dating demonstrate that large-scale patterns and rates of evolution are not compatible with those hypothesized on the basis of extrapolation from modern populations and species [29, 33, 14, 30]. Evolution observed in modern populations and species might, in some cases, appear as slow, continuous changes toward occupying much of the available morphospace. Long-term, large-scale evolution however is characterized by the rapid divergence of lineages near their origin, followed by long periods during which the basic body plans and ways of life are retained. Intermediate forms, as well as the continual divergence of major lineages into the morphospace between distinct adaptive types are often missing [11];

2 Plate tectonics indicate that changes in the position and configuration of the continents, and their influence on climate and the capacity for dispersal, have been major driving forces in the evolution of life [20, 2];

3 Studies of genomics and molecular development show that long term evolution is not the result of the selection of alternative alleles controlling specific traits, or the progressive accumulation of new mutations in an additive fashion, as proposed by quantitative genetics [37, 1]. Instead development involves a hierarchy of genetic control, including the precise timing and positioning of regulatory genes, and a cascade of downstream genes, commonly producing pleiotrophic results. In spite of these advances in knowledge, quantitative genetics continues to treat polygenic traits in a statistical manner, as if they resulted from the additive effects of a large number of essentially equivalent genes [4];

4 An array of theoretical and experimental evidence produced by the recent progresses in genomics (mostly through large scale genome

sequencing projects) have forced us to look critically at the "central dogma" of genetics (that the causal flow from code (DNA) to message (RNA) to assembly of substance (polypeptide) is unidirectional, with one item of code -a gene- ultimately making one item of substance -a polypeptide-, and the set of polypeptides making a body), and to the relationships between genetic and organismal complexity [47].

While many of the shortcomings of the Modern Synthesis, mostly connected to its overtly reductionistic nature (i.e. [50, 7, 8]), had already been identified, and widely debated in the literature - including the lack of attention paid to biological hierarchies [10, 11]; the importance of adaptive mutational systems [16] and epigenetic inheritance systems [28]; the developmental constrains caused by morphogenetic fields, and the phenomena of self-organization and the emergent properties of complex systems [24, 18, 27]; the importance of the interaction between organisms and their environment [32]; the metazoan-centric perspective of the Synthesis which leads to biased assessments when it comes to judging the importance of taxon-specific evolutionary processes [34]; and the flawed assumption that macroevolution can be equated with "microevolution writ large" [12] - a further point, little acknowledged until recently, is that the Modern Synthesis was a virtually ahistorical endeavor. A phylogenetic tree, that is, an hypothesis of evolutionary relationships among species, was the only illustration in The Origin of Species [6], but phylogenies, and discussions about their importance in evolutionary studies, are virtually absent in the most significant publications of the "founding fathers" of the modern synthesis [9, 26, 35, 45]. The difficulty of dealing with problems such as the definition and identification of taxonomic units, be these species or higher taxa, the lack of a rigorous method for phylogenetic inference and the lack of interest of many systematists in evolutionary processes contributed to systematics being confined to a marginal role during the development of the Synthesis. Phylogenetic systematics [22, 23, 15], however, finally provided biologists with a rigorous methodology to investigate phylogenetic relationships, and study both patterns and processes (i.e. mechanisms) of evolution [21]. It is now almost universally accepted that phylogenetic trees provide the pattern from which explanations of processes responsible for evolution must be derived [39, 41, 3, 44].

Phylogenies are now key components in the study of: relationships among taxa; speciation; evolution of any character; community evolution (i.e. organism-organism associations) and biogeography (i.e. organism-areas associations). In spite of this situation a large component of evolutionary work is still conducted with no regard for biological history. It is apparent that, in order to overcome the limitations of neo-Darwinism and obtain a new evolutionary synthesis, we need to be able to integrate the findings of the various biological disciplines within a rigorous phylogenetic framework [4]. Due to the space constraints of this article I will concentrate my attention on two areas of research firmly rooted in phylogenetic framework, evolutionary developmental biology and historical biogeography, and discuss the contributions that these two disciplines can provide toward the attainment of a more holistic evolutionary synthesis.

2. Historical biogeography

Biogeography is the discipline concerned with the history of organism's distribution and the historical and ecological factors that produced it. Biogeography is an old branch of biology, practiced systematically since the 19th century (e.g. [48]), and it can, for the sake of simplicity, be divided in two subdisciplines: ecological biogeography and historical biogeography. Ecological biogeography is the discipline concerned with the identification and investigation of the ecological processes that occur over short temporal and small spatial scales, while historical biogeography, on the other hand, deals with evolutionary processes that occur over the long temp and on a large spatial scale. Biogeography has been plagued for most of its history by the use of narrative methods, in which only the presence/absence of species from an area is recorded, without search for evidences of historical relationships between taxa and areas. This procedure does not allow for rigorous inferences to be made, especially in the case of disjunct distributions of species. These patterns can be caused by different processes, such as: a common ancestor occurring in one area and later dispersing into another (dispersal), or a population of the common ancestor that was originally widespread in an area that became fragmented and the descendents have survived in the fragments until now (vicariance). Organisms live in a world that is not only biologically dynamic, with phenomena such as colonization, speciation and extinction being all but rare, but also geophysically dynamic, and geological phenomena, such as volcanic, orogenic and tectonic (i.e. continental drift) activity can influence biological distributions at many geological scales, by creating new areas (for example islands), dividing pre-existing habitats or causing mixing of biotas when physical barriers are removed, such as when different plates collide with one another. On the geological time scales of evolution the physical development of our planet is a major engine of change. Changes in the position and configuration of continents influence the patterns of evolution in many different time scales.

While dispersal has been the dominant explanation of species distribution for centuries because of the idea of Earth's supposed stability - species were supposed to have centers of origin from which they disperse, an idea advocated by Darwin and Wallace in the 19th century, and more recently by many evolutionary taxonomists, including Mayr and Simpson - the discovery of plate tectonics led to strong criticism, and later abandonment of the concept of centers of origin by most systematists. The discovery of plate tectonics and the growth of phylogenetic systematics allowed the development of techniques with greater predictive power in attempts to explain the distribution of organisms, and eliminated the need to rely on ad hoc assumptions [25]. In fact, general patterns of geographical area (and biota) relationships can be elucidated by comparing area phylogenies derived from different groups occuring in a region. Once the relationships between geographical areas and the biotas that inhabit them are known, and are represented in the form of cladograms (branching diagrams), they can be used to make inferences about the modes and processes of speciation that have contributed to the origin of biological diversity in the groups of organisms under study [13]. When modern biogeographic methodologies are employed, very powerful hypotheses can be produced, including assessments of the roles of dispersal and vicariance in the origin of new species [31]; identification of the ways in which species manage to track favorable habitats in the context of continuously changing environments, a key contribution to the persistence of the species themselves and the establishment of punctuated patterns of speciation [11]; identification of the causes of extinction [36], and the discovery that ecological communities experience phylogenetic constraints, due to the evolutionary history of the taxa present in each community [49]. Ideas that had surged to the role of paradigm, such as the role of the Indonesian region as a center of speciation [2] can be overturned by careful historical biogeographic analyses [43], and historical studies of biogeography will be even more of a centerpiece of evolutionary studies in the future, due to the pressing needs in conservation biology that will require more of these rigorous analyses, in order to be able to identify the areas of priority where intervention is most needed.

3. Evolutionary developmental biology

In spite of the enormous accomplishments made in the past few years in evolutionary biology, especially with respect to molecular evolution we still don't have a clear understanding of the correspondence between genotype and phenoype, much as how changes in the genetic material are manifested in the morphology of organisms. Thus some fundamental questions of evolutionary biology still lie unanswered over 140 years after the formulation of the Darwin-Wallace theory of evolution: how do new morphological structures form? How do new organs and bauplans originate? Is there a relationship between genome size and morphological complexity? We believe that answers to these questions can now come from an integration of robust phylogenetic hypotheses and data obtained from the study of the fossil record in combination with the modern techniques of molecular developmental biology.

Evolutionary developmental biology is the new discipline that draws together molecular developmental biologists, systematists and paleontologists with the common goal of understanding the evolution of mechanisms responsible for controlling the development of living forms. Comparative embryology performed in a phylogenetic framework, together with advances in the understanding of the functioning of the genome, has so far led to the identification of genes that control major developmental processes-from the establishment of body plans; to the realization that formation of appendages and sense organs are shared across the animal kingdom and have arisen early in animal evolution; then when organs are lost (e.g. limb loss in snakes), it does not necessarily imply the loss of the developmental potential to form those organs; that life history stages (embryos, larvae, adults) can develop and evolve separately, and that such changes can provide opportunities to modify and modulate embryonic development, specialization or diversification of adult structure, and the evolution of novel structures; that homology is hierarchical, with homologous genes initiating development of structures that are not homologous (i.e. Pax-6 genes and the eyes of arthropods and vertebrate) and that homologous structures may develop by processes that are not homologous [19].

The major contribution of this new discipline has probably been the discovery that there are developmental pathways among metazoans, and presumably other organisms, that are evolutionarily ancient, commonly distributed and sometimes co-opted for purposes other than their original functions. Studies of the occurrence and expression of developmental regulatory genes in model organisms indicate that the continuing evolution of these developmental pathways creates the basis for developmental and morphological innovations, which, through their interaction with natural selection, result in the emergence of novel forms. Among these regulatory genes, the ones that have received the greatest attention have been the Hox genes. These regulatory genes are active during the early phases of metazoan development, and code for specific proteins that activate other genes, thus regulating a host of processes within cells [40]. Hox genes are uniquely arranged in a linear sequence along the chromosome, which correspond to both the linear and temporal sequence of their activation along the anterio-posterior axis of the embryo. The number of Hox genes - arranged in a cluster along a chromosome - is broadly comparable to the degree of complexity of the organism. There is now abundant proof that the evolution of metazoans has been strongly influenced by the evolution of Hox genes [5, 37, 1, 46, 4]. The origin of multicellularity and complex body plans among animals may have been due to the evolution of the Hox code near the end of the Precambrian, and once evolved, the subsequent duplication and divergent evolution of these genes in the various animal lineages provided the basis for the radiation of many, if not most, metazoan phyla.

The study of Hox and other regulatory genes has certainly contributed to renew the interest of developmental biologists in evolutionary questions [17], as well as to broaden the genetic and molecular perspective of evolutionary processes, from a very reductionistic one based on highly theoretical population models to a much more comprehensive view. However most of the advances in this area have been made possible by the availability of phylogenetic hypotheses, that allow researchers to map the origin of morphological and genetic characters onto phylogenetic tree produced through independent characters, and to study the coevolution of genotype and phenotype. Without robust phylogenies little progress could have been possible, due to the impossibility to identify the primitive and derived states of the various characters, and to determine the polarity (direction) of the evolutionary transformation (e.g. see [42]).

4. Conclusion

Large-scale evolutionary phenomena therefore cannot be understood solely on the basis of extrapolation from processes observed at the level of modern populations and species. Patterns and rates of evolution are much more varied than had been conceived by Darwin and the neo-Darwinists. Evolutionary theorists need to have not only a good understanding of the dynamics of modern populations, but will also have to find a way to incorporate this with the new concepts and information produced by molecular developmental biology, systematics, geology and the fossil record in order to develop a new evolutionary synthesis. Phylogenetic systematics has provided a sound evolutionary methodology for establishing relationships among taxa, as well as formulating and testing evolutionary hypotheses. This strong foundation has allowed for the development of two key disciplines, historical biogeography and evolutionary developmental biology (evo-devo), both of which will be key players in the attempt to obtain a more holistic theory of biological evolution than neo-Darwinism has provided. Historical biogeography, complemented by robust fossil data, demonstrates that physical aspects of the earth's history have had a significant, even if varied, impact on the history of life. Molecular and evolutionary developmental biology makes it possible to understand how major changes in the anatomy and physiology of organisms have occurred, and how such changes can be influenced by natural selection. A new, and more holistic, evolutionary synthesis will thus only be attainable within a robust phylogenetic framework.

References

[1] ARTHUR W., *The Origin of Animal Body Plans*, Cambridge University Press, Cambridge 1997.

[2] BRIGGS J.C., *Global Biogeography*, Elsevier, Amsterdam 1995.

[3] CARROLL R.L., *Patterns and Processes of Vertebrate Evolution*, Cambridge University Press, Cambridge 1997.

[4] CARROLL R.L., *Towards a new evolutionary synthesis*, Trends Ecol. Evol. **15** (2000), 27.

[5] CARROLL S., *Homeotic genes and the evolution of arthropods and chordates*, Nature **376** (1995), 479.

[6] DARWIN C.R., *On the Origin of Species by Means of Natural Selection*, John Murray, London 1859.

[7] DAWKINS R., *The Selfish Gene*, Oxford University Press, New York 1976.

[8] DAWKINS R., *The Extended Phenotype: The Gene as the Unit of Selection*, W.H. Freeman and Co., San Francisco 1982.

[9] DOBZHANSKY T., *Genetics and the Origin of Species*, Columbia University Press, New York 1937.

[10] ELDREDGE N., *Unfinished Synthesis*, Columbia University Press, New York 1985.

[11] ELDREDGE N., *Reinventing Darwin: the Great Evolutionary Debate*, John Wiley & Sons, New York 1995.

[12] ERWIN D.H., *Macroevolution is more than repeated rounds of microevolution*, Evol. & Dev. **2** (2000), 78.

[13] ERWIN D.H., ANSTEY R.L., *New Approaches to Speciation in the Fossil Record*, Columbia University Press, New York 1995.

[14] ERWIN D.H., WING S.L., *Deep Time: Paleobiology's Perspective*, The Paleontological Society, Lawrence, KS 2000.

[15] FARRIS J.S., *The logical basis of phylogenetic analysis*, in: "Advances in Cladistics", N.I. Platnick and V.A. Funk eds., Columbia University Press, New York 1983.

[16] FOSTER P.L., *Adaptive mutation: the uses of adversity*, Annu. Rev. Microbiol. **47** (1993), 467.

[17] GILBERT S.F., OPITZ J.M., RAFF R.A., *Resynthesizing evolutionary and developmental biology*, Dev. Biol. **173** (1996), 357.

[18] GOODWIN B., *How the Leopard Changed its Spots*, Charles Scribner & Sons, New York 1994.

[19] HALL B.K., *Guest editorial: Evo-devo or devo-evo - does it matter?*, Evol. & Dev. **2** (2000), 177.

[20] HALLAM A., *An Outline of Phanerozoic Biogeography*, Oxford University Press, Oxford 1994.

[21] HARVEY P.H., PAGEL M.D., *The Comparative Method in Evolutionary Biology*, Oxford University Press, Oxford 1991.

[22] HENNIG W., *Grundzuge einer Theorie der Phylogenetischen Systematik*, Deutscher Zentralverlag, Berlin 1950.

[23] HENNIG W., *Phylogenetic Systematics*, University of Illinois Press, Urbana 1966.

[24] HO M.W., SAUNDERS P., *Beyond Neo-Darwinism*, Academic Press, London 1984.

[25] HUMPHRIES C.J., PARENTI L.R., *Cladistic Biogeography*, Oxford University Press, Oxford 1999.

[26] HUXLEY J., *Evolution, the Modern Synthesis*, Allen & Unwin, London 1942.

[27] KAUFFMAN S.A., *At Home in the Universe*, Oxford University Press, Oxford 1995.

[28] JABLONKA E., LAMB M.J., *Epigenetic Inheritance and Evolution: The Lamarckian Dimension*, Oxford University Press, Oxford 1995.

[29] JABLONSKI D., ERWIN D.H., LIPPS J.H., *Evolutionary Paleobiology*, University of Chicago Press, Chicago 1996.

[30] JACKSON J.B., LIDGARD S., MCKINNEY F.K., *Evolutionary Patterns: Growth, Form and Tempo in the Fossil Record*, University of Chicago Press, Chicago 2001.

[31] LIEBERMAN B.S., *Paleobiogeography*, Kluwer, New York 2000.

[32] LOVELOCK J.E., *The Ages of Gaia: a Biography of our Living Earth Updated and Revised*, Norton and Company, New York 1995.

[33] MCKINNEY M.L., DRAKE J.A., *Biodiversity Dynamics: Turnover of Populations, Taxa, and Communities*, Columbia University Press 1998.

[34] MARGULIS L., SAGAN D., *Microcosmos: Four Billion Years of Microbial Evolution*, Touchstone Book, New York 1991.

[35] MAYR E., *Systematics and the Origin of Species*, Columbia University Press, New York 1942.

[36] NOVACEK M.J., WHEELER Q.D., *Extinction and Phylogeny*, Columbia University Press, New York 1992.

[37] RAFF R., *The Shape of Life*, University of Chicago Press, Chicago 1996.

[38] RIDLEY M., *Evolution*, Oxford University Press, Oxford 1997.

[39] ROSE M.R., LAUDER G.V., *Adaptation*, Academic Press, San Diego 1996.

[40] RUDDLE F.H., BARTELS J.L., BENTLEY K.L., KAPPEN C., MURTHA M.T., PENDLETON J.W., *Evolution of Hox genes*, Ann. Rev. Gen. **28** (1994), 423.

[41] SANDERSON M.J., HUFFORD L., *Homoplasy: The Recurrence of Similarity in Evolution*, Academic Press, San Diego 1996.

[42] SANTINI F., STELLWAG E., *Phylogeny, fossils and model systems for the study of Hox genes*, Mol. Phyl. Evol. (2002), in press.

[43] SANTINI F., WINTERBOTTOM R., *Historical Biogeography of Indo-Western Pacific Coral Reef Biota: Is the Indonesian Region a Center of Origin?*, J. Biogeog. (2002), in press.

[44] SCHLUTER D., *The Ecology of Adaptive Radiation*, Oxford University Press, Oxford 2000.

[45] SIMPSON G.G., *Tempo and Mode in Evolution*, Columbia University Press, New York 1944.

[46] STELLWAG E., *Hox gene duplication in fish*, Cell Dev. Biol. **10** (1999), 531.

[47] SZATHM ÝRY E., JORD ÝN F., PÝL C., *Can genes explain biological complexity*, Science **292** (2001), 1315.

[48] WALLACE A.R., *On the zoological geography of the Malay Archipelago*, J. Linn. Soc. London **4** (1860), 172.

[49] WEBB C.O., *Exploring the phylogenetic structure of ecological communities: an example for rain forest trees*, Am. Nat. **156** (2000), 145.

[50] WILLIAMS G.C., *Adaptation and Natural Selection*, Princeton University Press, Princeton 1966.

CELL SYSTEM COMPLEXITY AND BIOLOGICAL EVOLUTION

Michele Sarà
Dipartimento per lo studio del Territorio e delle sue Risorse,
Corso Europa 26, 16132 GENOVA - ITALY
sara@dipteris.unige.it

Abstract According to the recent knowledge on cell and genome functioning the whole cell system directed by genes, and not the genes alone, should be considered as the basic evolutionary level. Gene changes cannot be considered, as in the neo-darwinian orthodoxy, only as stochastic changes because are submitted to several endogenous and exogenous constraints. Moreover, natural selection works on phenotypes that don't exactly mirror genomic changes but are strongly influenced by the epigenetic cell processes. The cell system produces the proteome and the whole cell phenotype through a continuous molecular talking based on a bidirectional information flow linking genome and membrane receptors. However, in a pluricellular organism, the single cell systems shall be always considered inside the framework of tissue interactions.

The dynamic autopoietic processes based on the cell system complexity are characterized by robustness but also plasticity. Stresses, i.e. stimuli beyond the threshold, may provoke their breakdown but may be also the cause of incipient evolutionary changes. These become effective if may be transmitted through generations and are not eliminated by natural selection. Transmission may occur through epigenetic or epigenetically induced genetic mechanisms. As the epigenetic way seems presently too little stable to represent a general mechanism I consider an epigenetically induced genetic way as the most fruitful working hypothesis for linking the cell system autopoietic changes to the evolutionary novelties. It could be based on the genetic assimilation of plastic traits.

1. Introduction

The evolutionary course of life depends by the interaction of different hierarchical levels of biological organization (communities, populations, organisms, cells, macromolecules, etc.) [33], each one of great complexity. However, the cell level may be considered the basic level because gene reproduction, variation and expression, from which the heritable phenotypic traits arise, are cell system processes.

In the light of current molecular biology, the genomic regulatory structure that requires the integrated complexity of the cell system, and not the single genes as in the classical Neo-Darwinian paradigm, represents the material basis of the evolutionary mechanism. Moreover, on one hand, the production of genetic variation utilized in phenotypes construction cannot be considered as purely stochastic and, on the other, natural selection does not act on phenotypes that mirror exactly their genotypes. It acts on phenotypic traits that arise epigenetically from the holistic complexity of cell system and in multicellulars from the whole organism complexity. In addition, this integrated activity requires an external source of information that comes from other cells of the organism (in multicellulars) and from the environment. This information influences and may also modify the genotype through genetic assimilation [47, 48, 40] and therefore phenotypic variations induced by environmental stresses may be transferred through generations [34]. The evolutionary role of the environmental information that flows directly in cells and organisms, flanking natural selection and interacting with it, even if little acknowledged, is a major problem of contemporary biology [35, 36, 37, 38].

2. Cell complexity and phenotype construction

The production of cell phenotype traits in the process of eukaryotic cell differentiation requires schematically three different levels of molecular interactions, in turn holistically integrated. Firstly, there is the genomic regulatory process with interactions between different genes. These interactions do not follow only hierarchical cascades of bifurcating regulatory decisions but also complex networks with trans-holonic interconnections [27, 20]. This may lead to the synexpression groups of genes involved in the same biological process [30]. Also the leading role of homeotic genes may be seen in an interactive context, morphogenetic fields, that can alter their activity [1]. Secondly, there are the interactions between DNA, RNA, and proteins that occur in the fundamental process of gene expression, in the two phases of transcription and translation. Alternative splicing and posttranscriptional modification of RNA transcripts can generate many more proteins than the number encoded by genes [45]. A bidirectional information, not only from DNA to proteins but also from proteins to DNA, regulates transcription. This occurs through the transcription complexes, with particular DNA sites (enhancers) accessible to specific DNA-binding proteins, and the chromatin structure dynamics [11, 19]. Thirdly, there are the signaling networks [18] and the interconnected transduction pathways required by cell morphogenesis and behavior [23]. The interplay of signaling and transcription shows that the transcriptional state of the cell depends upon its prior history and the signals it receives. In turn, the signals produced by a cell and its ability to respond to signals are programmed in a large extent by its transcriptional state (Bronner-Fraser and Sternberg, 2000). A molecular cross-talk operates continuously between the cell membrane receptors and the genes. This indicates that not only the signaling pathways communicate with one another but the entire cell functions as a single unity and the different stimuli are integrated into a coherent response [32]. A cross-road of the functional and morphogenetic cell activities is represented by the transcription complexes into which proteins and genes come in contact regulating the fundamental process of gene expression. The new picture of a complex intracellular networking (Brugge and Mc Cormick, 1999), to which genes, proteins and signaling molecules cooperate, substitutes the old picture centered on genes and marks the passage from the genomic to the postgenomic era.

This picture becomes necessarily wider when we consider the development and functioning of a multicellular organism but basically requires the same interplay of an intrinsic genetic information within cells with signals transmitted between cells. Cell communication through intercellular signals to which participate hormones, growth factors, antibodies, viruses, etc. result in inductive interactions which control cell cycle and differentiation [12]. Genotype and phenotype cooperate intimately for the organism morphogenesis and behavior through functional modules that constitute the epigenotype. The new scenario produced by the recent approach of genetics and developmental biology on a molecular ground (EVO-DEVO) puts an end to the old phenotype-genotype dualism with the primacy of genotype on phenotype. Moreover, because the epigenetic processes are influenced by environmental stimuli new perspectives are open for a role of direct environmental influence on evolution.

3. Inherited changes in cell systems and their impact on evolution

Changes of the morphological and functional cell properties occur continuously during organism development and cell differentiation. These changes are due not only to mutations, e.g. stochastic errors in DNA replication leading to codon changes. The new scenario of complexity in cell activity shows that genomic changes due to transposons and other changes in epigenetic systems may be relevant and may have evolutionary consequences whether persistent through generations.

The insertion of transposons in DNA filaments affects the dynamics and fluidity of the genome in bacteria, plants and animals. Transposable elements are generally represented by retrotransposons, obtained through reverse transcription of RNA in DNA as in oncogenic viruses and in the HIV virus responsible of AIDS. In plants but also in several animals and in man their quantity is very high, for man about 35% of the whole genome. The foreign DNA, which produces often highly repeated sequences, has been initially considered of little genetic value, as a junk or parasitic DNA. But, gradually, its importance in DNA functioning has been acknowledged and presently it is clearly established for DNA regulation and expression. Transposons change the genomic organization being coopted in several structural and regulatory gene systems. They do not act autonomously but cooperate with the preexisting genes changing their regulative properties through position effects. Their behavior is then more as symbionts than of parasites. As observed by Girard and Freeling [9] transposon insertion may alter the gene expression level as well as the spatial pattern of expression of contiguous genes with qualitative and quantitative regulatory effects. Transposons may contain enhancers with tissue specificity, promoters of gene expression systems and, through trans-factors and epigenetic changes as DNA methylation, may act as gene suppressors and silencers. Then, transposons play an important role in genomic changes and reorganization [21]. Herbert and Rich [13] have observed that reverse transcription in eukaryotes can store advantageous ribotypes in DNA filaments and successively transfer them to germ cells with evident evolutionary effects. Transposon activity has been also associated to the paramutation phenomena. Paramutations, discovered in plants, do not respect the Mendelian law that says that in the course of generations alleles segregate unmodified. The precise molecular mechanism of paramutations is unknown but current opinion considers that they are induced by molecular interactions of transposons and genes producing allele activity or silencing through DNA methylation or changes in chromatin structure. But, a meiotic inheritance of epigenetic conditions, as in paramutations, has been observed also in mammals and in organisms that do not methylate their DNA as yeast and *Drosophila*.

Paramutations are only one aspect of the epigenetic inheritable states that may be very important for human pathology but also for the evolutionary theory. Epigenetic inheritance, e.g. epimutations, differs from gene mutations because do not require changes in DNA nucleotide sequences and then in codons and does not follow Mendelian rules. Transgene experimentation has shown that epigenetic effects frequently arise in DNA-DNA and RNA-RNA interactions. Wolffe and Matzke [49] suggest that these epigenetic effects are due to recognition of the nucleic acid sequence homologies. In the case of DNA the interaction can lead to gene silencing and chromatin modifications that may be heritable. But epigenetic control of gene expression is essential also for the normal organism development because it directs gene differentiation and prevents by tumor pathology and other serious diseases. The epigenetic states inherited through cell divisions have been called epigenetic inheritance systems (EIS) [16, 17]. They control cell differentiation during development and being generally erased during gamete maturation are not transmitted to the following generation. There is, however, growing evidence that some epigenetic states can be meiotically inherited. Genomic imprinting is a non-Mendelian phenomenon in which alleles are differently expressed according to the parental source. It is produced by DNA methylation transmitted through gametes but it must be reset at each generation. In the case of paramutations and epimutations a stable inheritance can sometimes arise [14]. Cubas et al. [4] have shown in *Linaria vulgaris* that epimutations due to DNA methylation can phenotypically revert through the somatic development and are less stable that gene mutations. But, long term effects can be also produced favoring stable changes in gene expression or DNA mutation and recombination processes. It is generally assumed that epimutations may have some relevance for plant evolution at short and long term. Plants are more favored than animals because epimutations may be produced in plant meristem and then transmitted to the following generations not only by germ cells but also by vegetative propagation. But, the soma/germ barrier postulated by Weissman is not stringent also for animals (Buss, 1987). Many animals have open germ cell lines. In animals, as vertebrates, with close germ cell lines the passage of epimutations may occur in the first phases of ontogenesis when the germ line is not yet specified [29]. Other mechanisms are the insertion of transposable elements into the germ cell genome [24] or through mutated DNA in somatic hypermutation and its feedback to the germline [43]. Stable epimutations have been found in *Drosophila* and in mammals. In mouse the fur color may be inherited through generations by methylation of the agouti gene [28].

The implication of the genetic system in the non-Mendelian inheritance represented by the epigenetic inheritance is more or less great. According to Landman [22] the epigenetic inheritance (acquired characters inheritance or IAC) may be distinguished in nucleic, epinucleic and extranucleic. In the nucleic mechanisms there is loss of non-essential

DNA or RNA or acquisition of new DNA through transposable elements. In the epinucleic mechanisms, to which epimutations belong, inheritance regards DNA conformational changes due to DNA methylation or acetylation of histones bound to the regulatory complexes of gene transcription. Another mechanism is the activation of alternative promoters of a same gene that modulate its expression in relation to the environmental conditions [39]. In the extranucleic mechanisms a morphogenetic pattern produced by a gene expression stabilized by cytoplasmic factors may be inherited. An example is the cortical inheritance of *Paramecium* and other Ciliates. The role of epigenetic inheritance in evolution has been emphasized by Landman [22] and Jablonka et al. [15] and has been after confirmed. Discussion regards chiefly the diffusion of the phenomenon and its role in speciation [46] and adaptation [38].

Prionic inheritance is a special type of non-Mendelian inheritance shown in mammals and yeast. The prion phenotype, normal or changed, may be transmitted from a cell generation to the following through a molecular phenetic way [31]. The role of prions in cell activity and inheritance is likely more diffuse than known. Recently, Serio and Lindquist [41] have found that the yeast PSI+ factor, an epigenetic modulator of the end of translation, follows a prion inheritance. PSI+ is the prion form of the nuclear protein Sup 35 p, a member of the terminal complex of translation. The prionic protein may exist at least in two distinct states each one associated with a different phenotype and function.

4. Environmental information; cell complexity and evolution

The main aspect of the new views on the complex eukaryotic cell activity through interacting molecular networks and cascades, involving all cell compartments from membrane to nucleus, is its linkage with the environmental information, external or mediated, in multicellulars, by the other organism cells. This information, processed by the cell membrane signaling system, is transferred through the transduction pathways to the gene expression systems where is linked with genetic information. A crucial problem is how much of this environmental information, needed by the cell physiology and morphogenesis, influence the organism evolutionary processes. The epigenetic systems are the main places in which the two information streams, downstream or genetic and upstream or environmental, interact. They may be considered as cell windows open to the environmental information. Epigenetic processes can influence evolutionary course in two ways.

Firstly, epigenetic processes can represent a constraint of natural selection. This depends by the fact that gene expression in the two different phases of transcription and translation is regulated by proteins and other factors, which induce activation and silencing of alleles, alternative splicing of transcripts with the production of several proteins from the same gene, etc. The result is that a genotype may produce different phenotypes in relation to different environmental conditions. As selection acts on phenotypes and not on genotypes, the selective regimes, and then the evolutionary course, may be strongly influenced by changes in phenotypes produced by epigenetic processes. Also the phenotypic plasticity, i.e. the different phenotypes that are produced by environmental cues in relation to the extent of the reaction norm of a genotype, may lower the intensity of selection on gene polymorphism, changing the selective regimes to which species and traits are submitted [44]. The cell molecular basis of plasticity is not hitherto clear, even if recently its importance has been focused and its role in evolution acknowledged [40]. In fact, phenotypic plasticity role increases in changing environments on the basis of its evolvable genotype.

A second and harder way by which environmental information can influence evolution does not work by changing the selective regimes but autonomously, flanking or cooperating with the work of natural selection. As this challenges the Neo-Darwinian paradigm that considers natural selection the only source of adaptations. the evolutionary role and importance of this way are highly controversial even if supported by a growing evidence. Direct environmental information may be introduced in evolution through the processes of adaptive mutability, stable epigenetic inheritance and genetic assimilation of phenotypic variations. Adaptive mutability, a phenomenon known for bacteria as *Escherichia coli* and for yeast (Cairns et al., 1988, [7, 6, 25], Bridges, 2001), is a complex mutation mechanism that, even if not directed, seems strongly influenced by environmental cues. Hypermutation, the increase of mutations in a specific locus generally induced by environmental stresses, plays an important role in this mechanism favoring the adaptation of the organism to environmental requests. Stable epimutations, environmentally influenced, exist, as discussed before, but seem to be rare events. On the contrary, genetic assimilation of phenotypic variability induced by environmental factors, hitherto understated, is likely the more important alternative way for organism adaptation to environment. Phenotype canalization followed by its genetic assimilation was experimentally demonstrated by Waddington ([47, 48], Waddington, 1957) half a century ago and furnished experimental evidence to what postulated by Baldwin [2] and known as Baldwin effect. A considerable bulk of data has after confirmed the validity of Waddington's observations: Among these the existence of phenocopies that mimic genocopies [50]. Recently, Rutherford and Lindquist [34] have described a molecular mechanism for the genetic assimilation of new traits under environmental stresses in *Drosophila*. In these flies a heat environmental stress, as well as mutations or chemicals, can inhibit the expression of the heat-shock protein Hsp90 that regulates several metabolic and transductional pathways assuring the cell homeostasis. The protein inhibition allows the emergence of a cryptic genetic variability that may be expressed in the following generations (capacitation process). The emerging variability regards traits as deformed eye, altered segmentation, reduced wings, deformed legs, etc., similar to the macromutations considered by Goldschmidt [10] as source of "hopeful" monsters that may become relevant also for macroevolution. After several fly generations, during which flies are submitted to natural selection, some mutants persist to develop even if Hsp90 is returned to its normal level and the environmental stress is ceased. The acquired traits have been then genetically assimilated and may become a permanent acquisition of the species. Genetic assimilation is filtered by natural selection but the expression of the new heritable traits is due directly to the phenotype canalization induced by environmental stimuli. A genetic assimilation of single parts of the reaction norm in phenotype plasticity and then of new adaptive traits produced by the environmental conversion or modulation of the phenotype [42] has been suggested by Schlichting and Pigliucci [40]. Genetic assimilation may explain the controversial extremely rapid Cichlid evolution in Victoria and Tanganyika lakes with more than 500 sympatric species originated in less than 12000 years [8]. The genetic assimilation of the very plastic mouth structures may have favored species isolation in sympatry by divergent feeding behaviors. A possible role of genetic assimilation may be also suggested for the results of plasticity experiments on an African Cichlid gills [3]. Other examples are the different morphologies of aquatic and terrestrial leaves of buttercups that may be found also in different species, the different flax phenotypes induced by fertilizers [5] and the hereditary callosities shown by dromedary legs in relation to posture. Recently, the adaptive radiation of the length of legs in different populations of the lizard *Anolis* adapted to different Caribbean habitats has been related to a genetic assimilation of the phenotypic plasticity, after amplified by natural selection [26].

5. Conclusions

With the beginning of this century and the so-called postgenomic era the main interest in molecular biology research has shifted from single gene identities and activities to their expression and to the context into which they operate. The proteome resulting from the context-dependent gene activities is now at the focus of the research. This new scenario emphasizes the cell complexity arising by the interaction of the different levels of cell signaling, transduction pathways, gene expression and gene regulation. Downstream genetic and upstream environmental informations interact in complex networks with a continuous cross-talk ensuring the integrated cell behavior. The interplay between the two information flows finds a main place in the cell epigenetic systems.

As epigenetic systems may influence the genome and may be also directly inheritable, cell heredity should be seen also in the light of epigenetics and not only of gene mutations. The impact of epigenetics and especially of the genetic assimilation of phenotypic plasticity on organism evolution is challenging for the evolutionary theory.

References

[1] ARANDA-ANZALDO A., *The Hox-gene research programme and the shortcomings of molecular preformationism*, Riv. Biol./Biol. Forum **93** (2000), 57–82.

[2] BALDWIN J.M., *A new factor in evolution*, Am. Nat. **30** (1896), 441–541.

[3] CHAPMAN L.J., GALIS F., SHINN J., *Phenotypic plasticity and the possible role of genetic assimilation: Hypoxia-induced trade-offs in the morphological traits of an African cichlid*, Ecol. Lett. **3** (2000), 387–393.

[4] CUBAS P., VINCENT C., COEN E., *An epigenetic mutation responsible for natural variation in floral simmetry*, Nature **401** (1999), 157–161.

[5] CULLIS C.A., *Control of variation in higher plants*, in: "Evolutionary processes and metaphors", M.W. Ho and S.W. Fox eds., J. Wiley, New York 1988.

[6] FOSTER P.L., *Mechanisms of stationary phase mutation: a decade of adaptive mutation*, Annu. Rev. Genet. **33** (1999), 57–88.

[7] FOSTER P.L., ROSCHE W.A., *Adaptive mutation in Escherichia coli strain FC40*, J. Genetics **78** (1999), 7–11.

[8] GALIS F., METZ J.A.J., *Why are there so many cichlid species?*, Tr. Ecol. Evol. **13** (1998), 1–2.

[9] GIRARD L., FREELING M., *Regulatory changes as a consequence of transposon insertion*, Dev. Genet. **25** (1999), 197–200.

[10] GOLDSCHMIDT R.B., *The material basis of evolution*, Yale University Press, New Haven 1940.

[11] GRUNSTEIN M., *Histones as regulators of genes*, Sci. Amer. **267** (1992), 68–74B.

[12] GUTKIND J.S. (ED.), *Signaling networks and cell cycle control: the molecular basis of cancer and other diseases*, Humana Press, Totowa, N.J. 2000.

[13] HERBERT A., RICH A., *RNA processing and the evolution of eukaryotes*, Nat. Genet. **21** (1999), 265–269.

[14] HOLLIDAY R., *Mechanisms for the control of gene activity during development*, Biol. Rev. **65** (1990), 431–471.

[15] JABLONKA E., LACHMANN M., LAMB M.J., *Evidence, mechanisms and models for the inheritance of acquired characters*, J. Theor. Biol. **158** (1992), 245–268.

[16] JABLONKA E., LAMB M.J., *The inheritance of acquired epigenetic variations*, J. Theor. Biol. **139** (1989), 69–83.

[17] JABLONKA E., LAMB M.J., *Epigenetic inheritance and evolution. The Lamarckian dimension*, Oxford University Press, Oxford 1995.

[18] JORDAN J.D., LANDAU E.M., IYENGAR R., *Signaling networks: the origins of cellular multitasking*, Cell **103** (2000), 193–200.

[19] KADONAGA J.T., GRUNSTEIN M., *Chromosomes and expression mechanisms: the packaging is the message*, Curr. Opin. Genet. Dev. **9** (1999), 129–131.

[20] KAUFFMAN S.A., *The origins of order, self-organization and selection in evolution*, Oxford University Press, New York 1993.

[21] KIDWELL M.G., LISCH D.R., *Transposable elements and host genome evolution*, Tr. Ecol. Evol. **15** (2000), 95–99.

[22] LANDMAN O.E., *The inheritance of acquired epigenetic variations*, Annu. Rev. Genet. **25** (1991), 1–20.

[23] LAUFFENBURGER D.A., *Cell signaling pathways as control modules: complexity for simplicity?*, Proc. Natl. Acad. Sci. U.S.A. **97** (2000), 5031–5033.

[24] LIDHOLM D.A., LOHE A.R., HATL D.R., *The transposable element mariner mediates germline transformation in Drosophila melanogaster*, Genetics **134** (1993), 859–868.

[25] LOMBARDO M.J., ROSENBERG S.M., *Hypermutation in stationary-phase Escherichia coli: tales from the lac operon*, J. Genet. **78** (1999), 13–20.

[26] LOSOS J.B., CREER D.A., GLOSSIP D., GOELLNER R., HAMPTON A., ROBERTS G., HASKELL N., TAYLOR P., ETTLING J., *Evolutionary implications of phenotypic plasticity in the hindlimb of the lizard Anolis sagrei*, Evolution **54** (2000), 301–305.

[27] MORATA G., STRUHL, *Fly fishing downstream*, Nature **348** (1990), 547–548.

[28] MORGAN H.D., SUTHERLAND H.G.E., MARTIN D.I.K., WHITELAW E., *Epigenetic inheritance in the mouse*, Nat. Genet. **23** (1999), 314–318.

[29] NIEUWKOOP P.D., SUTASURYA L.A., *Primordial germ cells in the invertebrates*, Cambridge University Press, Cambridge 1981.

[30] NYERS C., POLLET N., *Synexpression groups in eukaryotes*, Nature **402** (1999), 483–487.

[31] PAGEL M., KRAKAUER D.C., *Prions and the new molecular phenetics*, Tr. Ecol. Evol. **11** (1996), 487.

[32] PAWSON T., NASH P., *Protein-protein interactions define specificity in signal transduction*, Genes Dev. **14** (2000), 1027–1047.

[33] ROLLO C.D., *Phenotypes: their epigenetics, ecology and evolution*, Chapman and Hall, London 1994.

[34] RUTHERFORD S.L., LINDQUIST S., *Hsp90 as a capacitor for morphological evolution*, Nature **396** (1998), 336–342.

[35] SARÀ M., *Biological evolution: a holistic organism-centered approach*, Riv. Biol./ Biol. Forum **86** (1993), 347–359.

[36] SARÀ M., *A "sensitive" cell system*, Riv. Biol / Biol. Forum **89** (1996), 139–148.

[37] SARÀ M., *New perspectives on the role of constraints in evolution*, Riv. Biol./ Biol. Forum **92** (1999), 29–52.

[38] SARÀ M., *Lamarckian aspects in contemporary biology*, Epistemologia **22** (1999), 223–250.

[39] SCHIBLER U., SIERRA F., *Alternative promoters in developmental gene expression*, Annu. Rev. Genet. **21** (1987), 237–259.

[40] SCHLICHTING C.D., PIGLIUCCI M., *Phenotypic evolution. A reaction norm perspective*, Sinauer, Sunderland, Ma. 1998.

[41] SERIO T.R., LINDQUIST S.L., *PSI+: an epigenetic modulator of translation termination efficiency*, Annu. Rev. Cell. Dev. Biol. **15** (1999), 661–703.

[42] SMITH-GILL S.J., *Developmental plasticity: developmental conversion versus phenotypic modulation*, Am. Zool. **23** (1983), 47–55.

[43] STEELE E.J., BLANDERS R.V., *The reverse transcriptase model of somatic hypermutation*, Phil Trans. R. Soc. Lond. B **356** (2001), 61–66.

[44] SULTAN S.E., *Evolutionary implications of phenotypic plasticity in plants*, Evol. Biol. **21** (1987), 127–178.

[45] SZATHMARY E., JORDAN F., PAL C., *Can genes explain biological complexity?*, Science **292** (2001), 1315–1316.

[46] VARMUZA S., *Gametic imprinting as a speciation mechanism in mammals*, Theor. Biol. **164** (1993), 1–13.

[47] WADDINGTON C.H., *Canalization of development and the inheritance of acquired characters*, Nature **150** (1942), 563–565.

[48] WADDINGTON C.H., *Genetic assimilation of an acquired character*, Evolution **7** (1953), 118–126.

[49] WOLFFE A.P., MATZKE M.A., *Epigenetic regulation through repression*, Science **286** (1999), 481–486.

[50] ZUCKERKANDL E., VILLET R., *Concentration-affinity equivalence in gene regulation: Convergence of genetic and environmental effects*, Proc. Natl. Acad. Sci. USA **85** (1988), 4784–4787.

SELF-ORGANIZATION AND PREBIOTIC ENVIRONMENT

Silvano Traverso
Rivista di Biologia / Biology Forum,
Via Assarotti 31, 16122 GENOVA - ITALY
silvano.traverso@tilgher.it

Abstract The hypothesis that the first step towards life took place in an aqueous medium gives rise to a number of thermodynamical problems. Here it is suggested that the hydrophobic layer made of hydrocarbons and other hydrophobic molecules which presumably covered the surface of primordial seas would have provided an environment more suitable than water for the concentration and polymerisation of prebiotic molecules. Furthermore, it is stressed that the emergence of self-organised molecular systems, such as Kauffman's autocatalytic networks, requires an environment in which molecules can easily interact with each other. While water weakens electrostatic interactions, because of its high dielectric constant, a hydrophobic medium seems to be a more suitable environment for the development of self-organised catalytic systems.

1. Introduction

It is generally assumed that four thousand million years ago on the primitive Earth some self-organised processes took place which led to the origin of the first living systems. In order to become an organised system, molecules had to overcome many thermodynamical difficulties. To begin with, in order to build a molecular system, a restricted environment is needed, in which molecules can concentrate, polymerise and interact. It is currently estimated that amino acid concentrations in primordial oceans may have been equal to approximately 10^{-12} M, at most, and the estimates for nucleotide concentrations are even lower. At such low concentrations, random collisions of amino acids would have been extremely rare and polymerisation highly improbable. Alternative environments have been suggested – such as tidal pools, evaporating lagoons etc. – in which amino acids might have increased their concentrations. But in such environments any initial form of life would have been destroyed by the ultraviolet radiation coming from the sun, since the ozone layer did not exist in the primordial atmosphere.

In any case, a fundamental thermodynamical difficulty arises when assuming that the formation of the first polymers occurred in an aqueous medium: peptide and nucleotide polymers are unstable in water, being rapidly and inevitably destroyed by hydrolysis.

2. The hydrophobic superficial layer

It is possible that on the primitive Earth an alternative environment existed in which such thermodynamical difficulties would have not arisen: hydrocarbons and other organic compounds might have accumulated, giving rise to hydrophobic environments.

It has been suggested that a thick (1–10m) oil slick made of hydrocarbons and other hydrophobic molecules covered the surface of primordial seas [5, 3]. Such an oil slick would have derived from the polymerisation of methane in a reducing atmosphere. Although different models have been more recently proposed for the composition of the early atmosphere, there is no convincing evidence that the atmosphere prior to 3.9×10^9 years ago was not reducing. In prebiotic experiments carried out under reducing conditions hydrophobic material ('tar' or 'tholins') is often obtained.

Even in the absence of a reducing atmosphere, a large amount of complex organic compounds would have presumably been present on the early Earth, delivered by extraterrestrial infall [1, 2]. It has been estimated [2] that over a period of 100 million years 10^{16}–10^{18} kg of organic carbon would have been delivered to the early Earth (the total organic carbon in the present biosphere is 6×10^{14}). Much of this delivery would have probably been hydrophobic material. Indeed, the organic content of the Murchinson meteorite consisted mainly (90%) in a complex aromatic hydrocarbon polymer.

Hydrocarbons can also be formed in hydrothermal vents [4]. Whatever their origin (extraterrestrial infall or synthesis under reducing conditions) it is not unreasonable to hypothesise that non polar organic compounds might have formed a hydrophobic layer at the surface of primordial oceans. We [8] suggested that such a hydrophobic superficial layer might have provided an environment more suitable than water for the concentration and polymerisation of organic molecules fundamental to life.

In order for amino acids coming from the atmosphere (synthesised by Miller's reactions or carried by meteorites) to have diffused in the water, they would have had to have passed through the hydrophobic layer. Hydrophobic amino acids may have shown an energetic tendency to remain in the layer, thermodynamically 'trapped' in the hydrophobic medium. Thus, increasing their concentrations to relatively high values.

Molecules necessary for life would have found a much more suitable environment for accumulation in the hydrophobic layer rather than in water. For instance, the concentration of phosphate might have increased because the absence of Ca^{++} in the hydrophobic medium would have prevented it from precipitating as apatite. Because of their hydrophobicity, the concentration of nucleotide bases might have also increased in the hydrophobic layer. Incidentally, it is worth noting that pyrimidine prebiotic synthesis would have been highly improbable in water, while it may have been more likely in a hydrophobic medium [7].

Once organic molecules reached sufficiently high concentrations, they might have self-assembled into aggregates or polymerised. Indeed, the hydrophobic layer would have provided an environment protected from hydrolysis in which formation of aggregates and polymerisation reactions would have been energetically favoured. Furthermore, polymers in the layer may have been protected against the dangerous ultraviolet sun radiation, due to the presence in the layer of many organic molecules capable of absorbing UV rays.

3. Self-organization and interactions

However, the formation of aggregates or polymers is only the beginning. A dynamically organised molecular system has to arise. Great interest has been aroused by the discovery of self-organising physical and chemical systems, such as Bénard's system or Belousov-Zhabotinsky's reaction. These are steady state systems being kept far from thermodynamical equilibrium by external constraints. In such conditions, long range correlations emerge. It has been suggested that biological complexity may have arisen in a way similar to such phenomena, as a result of self-organising processes in non equilibrium conditions.

However, it should be stressed that in the above mentioned systems organisation is not due to internal short range interactions but to external non-equilibrium constraints: molecules do not have to interact with each other to show coherent collective behaviour [6].

Therefore, such phenomena do not seem to help in understanding the origin of one of the most fundamental properties of living systems: specificity. All biological functions, from enzymatic catalysis to genetic information processing, are based upon specific interactions between molecules. Presumably, interactions played a fundamental role even in the early prebiotic systems.

Current research on the origin of life follows two distinct, almost completely separate, paths. Physicists and mathematicians focus on self-organising properties of living systems, such as collective behaviours and long-range correlations, whereas chemists deal with molecular structures and specific short-range interactions. In order to understand the spontaneous origin of (pre)biological specificity, a kind of self-organisation would be needed in which systemic properties emerge as a result of chemical interactions.

An interesting approach is that of Kauffman's auto-catalytic networks. According to Kauffman, given a sufficiently diverse mixture of organic molecules, if each of them, besides being a reaction product, catalyses one reaction in the system, an organised state arises (through a sort of phase transition) in which any molecule indirectly catalyses its own synthesis. This kind of self-organisation, being based on molecular interactions and chemical diversity, would seem to build a bridge between the physico-mathematical and chemical approach to the origin of life.

Unfortunately, autocatalytic networks only seem to work on computer screens. As of yet, no laboratory experiment has been successful. Once again, the problem seems to be the dispersion and hydrolyzation of molecules by water. Moreover, any catalytic system has trouble in water since catalysis (in the real world as opposed to the virtual space of computers) is based on electrostatic interactions, and these are weakened by water's high dielectric constant (which is equal to 80).

In the prebiotic scenario, spontaneous formation of autocatalytic networks in water does not seem very plausible.

A hydrophobic medium, with a low dielectric constant, would have been a more suitable environment for catalysis, since electrostatic interactions between polar molecules would have been more effective. In the hydrophobic superficial layer, polar amino acids might have been occasionally incorporated into growing non polar peptidic chains. Thus, some hydrophobic peptides could have assumed catalytic properties. Such early catalysts could have been relatively efficient due to the low dielectric constant of the hydrophobic layer. Significantly, nowadays active sites of enzymes are typically hydrophobic except for a few polar amino acids which effectively interact with the substrate thanks to the absence of water within the site.

Thus, it seems reasonable to hypothesise that early catalysis and possibly early self-organised catalytic networks arose in the hydrophobic superficial layer.

It may be further speculated that, sooner or later, catalysts and their substrates ended by being included in the micellar structures which formed at the interface with water, giving rise to the compartmentalised stage of prebiotic evolution.

4. Conclusions

To conclude, whenever self-organising processes are hypothesised at the origin of life, the environment in which such processes are supposed to have taken place should be taken into consideration. Initial steps towards biological complexity, from polymerisation to the development of self-organised dynamical systems, would seem to have been easier in a hydrophobic medium rather than in an aqueous one. An interaction-based self-organisation – the kind of self-organisation more relevant to the origin of life – would have presumably encountered less thermodynamical difficulties in a hydrophobic medium, such as the hydrophobic superficial layer, rather than in the traditional primordial soup.

Acknowledgments

The content of this contribution is in large part based on research carried out in collaboration with Prof. Renzo Morchio.

References

[1] Anders E., *Pre-biotic Organic Matter from Comets and Asteroids*, Nature **342** (1989), 255–257.

[2] Chyba C.F., Sagan C., *Endogenous Production, Exogenous Delivery and Impact-Shock Synthesis of Organic Molecules: An Inventory for the Origin of Life*, Nature **355** (1992), 125–131.

[3] Holland H.D., *The Chemical Evolution of the Atmosphere and Oceans*, Princeton U.P., Princeton 1984.

[4] Holm N.G., Charlou J.L., *Initial Indications of Abiotic Formation of Hydrocarbons in the Rainbow Ultramafic Hydrothermal System Mid-Atlantic Ridge*, Earth Planet. Sci. Lett. **191** (2001), 1–8.

[5] Lasaga A.C., Holland H.D., Dwyer M.J., *Primordial Oil Slick*, Science **174** (1971), 53.

[6] Nicolis G., *Fisica dei sistemi lontani dall'equilibrio e auto-organizzazione*, in: "La Nuova Fisica", P. Davies ed., Bollati Boringhieri, Torino 1992.

[7] Morchio R., Redaelli A., Traverso S., *Proteins, Nucleic Acids and Genetic Codes*, Rivista di Biologia / Biology Forum **94** (2001), 37.

[8] Morchio R., Traverso S., *The Hydrophobic Superficial Layer: The Primordial Cradle of Life?*, Rivista di Biologia / Biology Forum **92** (1999), 105.

III

HISTORY AND PHILOSOPHY OF SCIENCES

JAMES AND FREUD ON PSYCHICAL DETERMINISM*

Paolo Casini
Dipartimento di Pianificazione Territoriale e Urbanistica,
Università "La Sapienza", ROMA - ITALY
Paolo.Casini@uniroma1.it

Abstract William James and Sigmund Freud shared some important views about psychology as a natural science: they were both trained in medical studies, and accepted the leading ideas of evolutionary physiology and biology; they both reacted against some aspects of German systematic psychology, and against the English associationist school as well. Both James and Freud approached the experimental study of the psyche from a deterministic point of view. The present paper stresses Freud's radical determinism, as stated in his early *Essay on psychology* and elsewhere, as opposed in principle to James' dichotomy between the requirements of cause-effect connection in the method of psycho-physiological research, and his "dilemma of determinism" concerning moral responsability and free-will.

1. William James meets Sigmund Freud

The father of psycho-analysis was deeply gratified by the friendly invitation of the American psychologist G. Stanley Hall at Clark University, Worcester, in september 1909: "In Europe I felt as though I were despised; but over there I found myself received by the foremost men as an equal" [4, XX, 52].

In his autobiographical account, mentioning his meetings with several colleagues and the general acceptance met by his lectures *Über die Psychoanalyse*, Sigmund Freud adds: "Another event of this time which made a lasting impression on me was a meeting with Willam James the philosopher" ([4, XX, 52]; cf. [15, I, 70]).

We may ask whether to call James a "philosopher" was not, on Freud's side, a kind of a *lapsus*. For only during the last decade, say between the

*A more detailed version of the present paper has been published in Italian in [3].

end of the XIXth century and the beginning of the XXth, the experimental psychologist had turned into the philosopher of Pragmatism, the celebrated author of *The Will to Believe* (1897), *The Varieties of Religious Experience* (1902), *Pragmatism* (1907). The psychologist's masterpiece was still *The Principles of Psychology* (1890), written during the Eighties, which gained soon a worldwide audience in the scientific milieu and were often quoted as an authoritative reference-book in the international psychological literature (see [17, II, 91–111]). However, in Freud's writings there is no trace of a direct knowledge of this work. Freud quoted only second-hand a remark of James concerning the erotic instincts and homosexuality in the ancient world as an outcome of habit[1]. Anyway, both Freud and James were physicians by training and psychologists by vocation. They agreed in believing that a new experimental science was developing both from the crisis of the German philosophical psychology, and from the abstractions of psycho-physics. The future of the new psychology laid on the ground of cerebral anatomy and physiology and on the main road of evolutionary biology[2]. Theodor Meynert, the author of the classic textbook *Psychiatrie*, published in Vienna in 1884 and soon translated into English, was one of Freud's teachers; James too referred to his book as an authority for cerebral anatomy and physiology[3]. Both fully accepted Darwin's theory of the descent of man, as carefully distinguished from Spencer's metaphysics of cosmic evolution and from his *Psychology*, which was severely criticized by James[4]. James' philosophical problems were deeply rooted in the British Empiricist school, but his thought diverged increasingly from the associationist tradition of Hume, John Stuart Mill, Alexander Bain. He was also influenced by the founder of the German psychology, Johann Friedrich Herbart, and by the psyco-physics of Hermann von Helmholtz, Wilhelm Wundt, Theodor Fechner.

Freud, also a physician by training, was not very given to reading books on psychology. His knowledge of Herbart and Fechner was, it seems, only indirect, acquired through the Herbartian compendium by G.A. Lindner, *Lehrbuch der empirischen Psychologie*, used in current teaching and through Meynert's work (cf. [15, I, 443–53]). But, as

[1] See James' remark concerning this "unnatural vice, of which the notion affects us with horror", in [12, II, 438]. Freud's source was Havelock Ellis, *Studies in the Psychology of Sex*; 1897 ff.; cf. [20, chap. II, 8].

[2] On Herbart, Fechner and Freud see [15, I, 445–47]; see also [16].

[3] James borrowed with a slight change the *Meynert scheme* concerning the brain's functions, cf. [12, I, 26 ff.].

[4] In [12]; see also *Remarks on Spencer's Definition of Mind as Correspondence* and *Herbert Spencer*, in [1, vol. V, 7–20; 107–122].

a medical student, he had followed for two years the seminars of the philosopher-psychologist Franz Brentano, who had even recommended him as translator into German of some of John Stuart Mill's essays (cf. [15, I, 87]). In his *Autobiografical Study* Freud mentions two capital discoveries of his student years: Darwin's theory of evolution and, at a turning-point of his career, the appeal of Goethe's fragment on *Die Natur*[5]. Let me mention only very briefly several other aspects of his formative years, which have been explored in detail by his biographers: his first researches in the histology of the nervous system in the Viennese physiology laboratory under the direction of Ernst Brücke, his studies on the anatomy of the brain with Theodor Meynert, the Parisian period devoted to the study of the theory and practice of neuro-pathology with Charcot, the beginnings of the method of hypnosis in psycho-therapy with Josef Breuer, to say nothing of some "philosophical" connections: not only as Brentano's seminaries, but the secret fascination of the works of Schopenhauer and Nietzsche (cf. [15, I, *passim*] and [20, 517–18]).

Before their meeting of 1909, both James e Freud attended a medical conference on hypnosis in Paris (1891). No attempt shall be made here to compare in detail the intellectual outlooks of the two men in those years. We may remark however that they shared a common scientific orientation. Both contributed to the crisis of positivistic psychology and to the rapid growth of research in the related fields of dynamic psychology and psycho-pathology; but, in spite of their reciprocal esteem, James and Freud had diametrically opposed views about two controversial topics which are commonplace today, but were crucial at that time: the problem of the unconscious, and the determinism-chance dilemma in mental processes. I shall try to focus on these two related aspects of one and the same debate.

In his Worcester lectures, which first launched psycho-analysis in the United States, Freud formulated a clear-cut and rigorous methodic principle concerning determinism. He summarized his own views on the interpretation of dreams and on the meaning of slips and jokes in the light of the unconscious mechanisms of censorship and repression: "As you see – he commented – psycho-analysts are marked by a particularly strict belief in the determination of mental life. For them there is nothing trivial, nothing arbitrary or haphazard. They expect in every case to find sufficient motives where, as a rule, no such expectation is

[5] "The theories of Darwin, which were then of topical interest, strongly attracted me, for they held out hopes of an extraordinary advance in our understanding of the world; and it was hearing Goethe's beautiful essay on Nature read aloud in a popular lecture by Professor Carl Brühl just before I left school that decided me to become a medical student" [4, GW, XX, 8].

raised. Indeed, they are prepared to find *several* motives for one and the same mental occurrence, whereas what seems to be our innate craving for causality declares itself satisfied with a *single* psychical cause" [5, XI, 38][6]. Elsewhere, alluding to the resistance his critics opposed against the theory of infantile sexuality, he affirmed: "From the intellectual point of view we must, I think, take into account two special obstacles to recognizing psycho-analytic trains of thought. In the first place, people are unaccustomed to reckoning with a strict and universal application of determinism to mental life; and in the second place, they are ignorant of the peculiarities which distinguish unconscious mental processes from the conscious ones" [5, XI, 52][7].

Even in his popular Worchester lectures Freud presented the determinism of mental phenomena and the unconscious as two strictly correlated concepts. His allegiance to the experimental method and a deep philosophical persuasion never made him question the hypothesis that a rigorous law of cause-effect connection regulates the whole activity of the human mind. The positivist faith in psychical determinism and psycho-physical parallelism, which was widely shared at the end of the XIXth century in positivist psychology, also oriented Freud's researches in the anatomy and physiology of the nervous system. The same faith inspired the medical theory and praxis of his *Studies on Hysteria* (1892–95), as well as the controversial draft *Project for a Scientific Psychology* (1895), and finally the beginnings of psycho-analytic theory. It is well known that in his *Psychopathology of Everyday Life* (1901) Freud considered psychical determinism as the hidden background of slips, jokes and *lapsus*. These are, in his view, minor mental phenomena, apparently escaping the continuous concatenation of psychical events, but in reality confirming the general rule of determinism, thanks to the unconscious mechanisms of repression. It is also possible to explain, as Freud remarked, how "the illusion of free will" is generated, but only in some choices of little importance; whereas in accomplishing the most crucial acts of the will we are fully conscious of the sentiment of an absolute, compact (*lückenlos*) determinism [7, VI, 240–54].

---

[6] We must add that Freud always shared the traditional determinism of German experimental psychology of writers such as Herbart, Helmholtz, Meynert (not to mention Nietzsche).

[7] Cf. also a similar statement in [6, XV, chap. 2]; see on this topic [20, 103–5].

2. James: the 'Stream of Consciousness' and the 'Dilemma of determinism'

James first discussed the problem of psychical determinism in 1884, when he was already at work on his *Principles of Psychology.* His approach was different from that of Freud. He started from a dualistic perspective, believing that where chemical and physiological processes are concerned, the causality connection is constantly working in the *continuum* of the brain processes. On this basis, he reshaped in his own way the old dilemma of free will. Bearing in mind the philosophical implications of Charles Renouvier's contingentism, James developed in his *Principles of Psychology* a subtle and articulate analysis of the stimuli acting on the human will, of the conflicts between desires and reason, of the several tensions and hesitations preceding every actual choice. His essential concern is about the intensity of the effort involved in the process. When our will chooses, we are conscious that its activity is working within the mind, and that its effort is more or less intense, according to the higher or lower degree of the resistance it meets. We may try to measure the intensity of this effort simply by introspection, or we may ask ourselves whether it be possible to quantify it, "the only questions being, is the effort where it exists a fixed function of the object, which the latter imposes on the thought? Or is it a such an independent 'variable' that with a constant object more or less of it may be made?". But as the effort passes through infinite degrees of intensity, we must conclude that it is of indeterminable extent. And "if it be really indeterminate, our future acts are ambiguous or unpredestinate: in common parlance, *our wills are free.* If the amount of effort be not indeterminate, but be related in a fixed manner to the objects themselves... then our wills are not free and all our acts are foreordained"[8].

However such a solution of the free will controversy is only of apparent simplicity, for we have not at our disposal, in this case, exact methods of measurements. If we analyze the tortuous movements of the will preceding every act of choice, even in the case of the simplest or effortless volitions, it is impossible to avoid the suspicion that each of these be in the last analysis, the mechanically determined result of a sequence of modifications within the physical mass of our brain: "The general continuity of things and the monistic conception of the world may lead one irresistibly to postulate that a little fact like effort form no real exception to the overwhelming reign of deterministic law" [12, II, 572].

[8] I summarize the section "The question of 'freewill' ", in [12, II, 571–79].

No philosophical doctrine standing behind the experimental method, nor brain physiology, nor even introspection, are able to explain in a satisfying way the complexities of the volition process. However, from the point of view of experimental method, the deterministic postulate of a rigorous cause-effect connection is unavoidable by definition. This means that the scientist's judgement must be suspended: "My own belief is that the question of free-will is insoluble on strictly psychologic grounds – James concludes – ... The present writer does for the alternative of freedom, but since the grounds of this opinion are ethical rather than psychological, he prefers to exclude them from the present book". James, in a footnote, adds these grounds "will be found indicated, in somewhat popular form, in a lecture on *The Dilemma of Determinism*, published in 1884" [12, II, 572][9].

This essay presupposes, in connection with the dilemma of determinism, some of James' theories concerning the body-mind problem, the stream of consciousness, the rejection of the unconscious, the unity of the mind functions, as stated later in the *Principles of Psychology*, and even anticipates by implication some philosophical reflections of the Pragmatist period. *The Dilemma of Determinism* is not really a "popular" paper, as James called it. On the contrary, he subtly argues for and against the notions of chance and free will, determinism and indeterminism, touching in the same time upon a series of moral, metaphysical and epistemological consequences. He does not intend to demonstrate positively that human will *is* really free, but only introduces and discusses critically the various issues of this controversial topic. He first allows that the causality principle as applied in the hard sciences "proceeds from our indomitable desire to cast the world into a more rational shape", and that "the world has shown itself, to a great extent, plastic to this demand of ours for rationality". But nobody knows to what extent it will show itself plastic in the future, i.e. will satisfy our causal principle. For the latter is in no way a Kantian a priori category in the empirist's view, who can only put the following question: "The principle of causality ... what is it but a postulate, an empty name covering simply a demand that the sequence of events shall some day manifest a deeper kind of belonging of one thing with another than the mere arbitrary juxtaposition which now phenomenally appears"[10].

[9]Cf. [13, 145–83]; also in [14]; [1, 114–40]. The authority on James and Pragmatism in Italy is Antonio Santucci; see his selection [18] and especially his comprehensive [19]. Cf. also two outlines: [11] and [2] (both with bibliography).

[10]W. James, *The Dilemma of Determinism*, in [14, 116].

It should be kept in mind that several philosophical authors, like Francis Herbert Bradley, Thomas Hill Green, Charles Renouvier, Alfred-Jules-Émile Fouillée, and Charles Sanders Peirce had recently reopened on a new ground the old controversy concerning free will, contingency, chance and determinism. In his own way James compared the traditional statements of classic, or *hard*, determinism, to those of a new kind of *soft* determinism. He did not doubt that determinism is pivotal for formulating rigorous scientific laws, but vindicated by contrast the full significance of the word and notion of *chance*. However, chance has not a positive, but only a negative content, for it only hints at the possibility of aleatory and unpredictable events.

Thus, within the proper sphere of experimental psychology, James' critical examination of the concept of chance leads, as a result, to a full acceptance of both indeterminism and free will. It is true that consciousness is subdue, on the one side, to the deterministic sequence of the various mechanisms of brain physiology; and, on the other side, to environment, biological evolution, social context and so on. Anyway the freedom of the will, being supported by such notions as chance and possibility, is admissible only in the sphere of moral choices and values. The most interesting part of James' paper is devoted to a subtle analysis and a passionate defence of the intriguing "dilemmas" of moral conscience[11]. On this point, stressing the link between free will and moral responsibility, James did not bother to exceed the self-imposed limits of experimental psychology. Since 1884 he endeavoured to check his dilemma against a moral and philosophical touchstone, already announcing some themes of Pragmatism and *will to believe*.

3. Freud in 1895: his *Project for a Scientific Psychology*

Freud, as I suggested above, is not likely to have read the essays of James the philosopher of Pragmatism, particularly those hinting at the freedom of the will as opposed to scientific determinism. Freud's faith in the absolute dogma of necessity, widely shared by late-positivistic epistemology, was apparently never shaken by doubt (cf. [15, I, 401]). Moreover, he was relatively indifferent to the problems of epistemology during the last decade of the XIXth century, when he was interested in the treatment of the "hysteric" disease by means of hypnotism, and

[11]It is worth remembering that in the years 1880–1885 James exchanged some remarkable letters concerning the dilemma of free will and necessity with the French philosopher Charles Renouvier and with his friend Shadworth Hogdson, the author of *Philosophy of Experience* (1885); see [17, I, 611–43, 670–710].

began to speculate on the sexual basis of this disease. During these early researches Freud wrote his tentative outline of the activities and functions of the brain from the point of view of mechanistic neurophysiology, known as *Project for a Scientific Psychology* [*Entwurf einer Psychologie*]. The manuscript of this unfinished work, which was composed hastily during the summer vacations of 1895, has been preserved among the papers of Wilhelm Fliess and appeared in print only in 1950 [8, I, 295–397][12].

This highly controversial essay has been variously commented upon by Freudian scholars and biographers, and is fundamental for everyone interested in investigating Freud's methodic shift from neurophysiology to psychoanalysis[13]. The Freud-Fliess correspondence of these months shows that the former's main concern was to detect the key to the mechanism of repression in the uncertain borderline between consciousness and the unconscious, as it seemed to result from the symptoms of obsessive neuroses and from the treatment of hysteria by means of hypnosis. To resolve the problem of repression Freud sketched his model of brain activity, which has been highly praised for being relatively modern and up-to-date, since it was based on the recent discovery of neurones, their morphology and their contact-barriers, to whom Charles S. Sherrington two years later was to give the name of synapses.

In his outline of "a psychology that shall be a natural science", Freud uses a mechanist terminology, a series of quantitative, physical and even electro-mechanical metaphors. In this hypothesis the neurones work as material particles transmitting, through charge and discharge processes, quantities of energy which are reducible to the axiomatic model of the laws of motion [8, I, 295–397]. Here the principle of "neuronal inertia" is fundamental as a kind of homoeostatic equilibrium, consisting in a state of relative rest, subjected to uninterrupted troubles, coming both from the five senses and from the endogenous stimuli of the brain apparatus. In Freud's hypothesis the neurones are of two different kinds, sensory or motor: "There are *permeable* neurones (offering no resistance and retaining nothing), which serve for perception, and *impermeable* ones (loaded with resistance...) which are the vehicles of memory and so probably of psychical processes in general. I shall call the former system of neurones φ, and the latter ψ". The φ neurones have just the passive role of receiving the stimuli; the ψ neurones are also transmitters of energy, both of endogenous and external origin. So the reflex theory

[12]The original draft was first published in [9, 371–466].

[13]For a full discussion see [20, *passim*]; cf. also the detailed summary of the text in [15, I, 439–78].

of psycho-physics is reinterpreted in dynamic terms: energetic charges are first accumulated within the elementary unit of the neuronale cell, then they transit and are discharged, producing alternatively pain and pleasure states. The contact-barriers function like conductors between the φ neurones and the ψ neurones. The latter only are able to store energy in the form of memory. Freud also speaks of a kind of alteration which is produced in the latter system of cells, bringing a degree of *Bahnung*, or "facilitation", in the transmission of the stored energy: "We can then say: memory is represented by the facilitations existing between the ψ neurones" [8, I, 300 ff.].

Freud's *Project* has been often interpreted in terms of an ingenious anticipation of cybernetics or of late XXth century discoveries in the neurosciences. The essay rather shows, in my opinion, Freud's deterministic faith at work, exploring a recently discovered continent and following – as it were – a circular path. Of course, the neurones are not observed in their actual functioning, like it is possible nowadays through the electronic microscope. They are only conjectural units, classified and endowed with *ad hoc* properties, in view of a detailed hypothesis concerning the structures and functions of the psyche. The hypothesis itself has a mechanical character, but is supported in a rather vague and problematic way, where the phylogenesis of the brain is concerned, by Darwin's theory of the descent of man. For the distinction between the two systems of neurones ϕ, ψ, is presented as an achievement of human evolution, and its evidence consists in the distinction between the central and peripheral nervous system. The ϕ system, which is supposed to be in contact with the environment, should be located in the grey matter of the spinal cord; the ψ system, which is supposed to regulate endogenous functions only, in the grey matter of the encephalon[14].

But, by definition, the main burden of Freud's "psychology" consisted in the effort of identifying the bridge between consciousness and the unconscious; or, if we adopt the terminology of the *Project*, of explaining how we know that there are unconscious neuronal processes of both ϕ and ψ type. This was indeed a complicated puzzle. Freud remarks clearly that psychology, as a natural science, should explain consciousness, but "since this consciousness knows nothing of what we have so far been assuming – quantities and neurones – it should explain this lack on knowledge to us as well... It follows from the postulate of consciousness providing neither complete nor trustworthy knowledge of the neuronal processes, that these are in the first instance to be regarded to their

[14]For the darwinian background see esp. [20, *passim*].

whole extent as unconscious and are to be inferred like other natural things. In that case, however, a place has to be found for the content of consciousness in our quantitative ψ processes. Consciousness gives us what are called *qualities* – sensations which are *different* in a great multiplicity of ways... It may be asked *how* qualities originate and *where* qualities originate" [8, I, 307–8]. In order to find a way out from this dead end, Freud introduced here an *ad hoc* postulate: a third system of neurones he called ω, whose state of excitation "gives rise to the various qualities – that is to say, *conscious sensations*" [8, I, 309]. This is clearly a circular argument: Freud begs the question, yet he is well aware of the fact that all his conjectures are "complicated and far from perspicuous... no attempt, of course, can be made to explain how it is that excitatory processes in the ω neurones bring consciousness along with them " [8, I, 311].

Freud never solved the puzzle. Commentators have discussed in detail whether the unfinished *Project* may be considered, on the one side, his extreme endeavour to keep psychology within the limits of a physico-mechanical physiology, respecting the requirements of the deterministic procedures of the hard sciences; or rather, on the other side, a decisive shift towards a soft and wholly alternative psychology. In fact Freud, by abandoning his draft in the hands of his friend Fliess and never asking it back, seemed to have forgotten it altogether. This strange behaviour has been interpreted as the evidence of a profound gap between Freud's physiological and psychopathological studies of the late Nineties and his next discoveries in psycho-analysis. But it has also been shown that the hypotheses, the terminology and the brain functions, as stated in the *Project*, have strict affinities not only with the language, the laws and the psychical structures underlying the theoretical background of the *Interpretation of Dreams* (1900), but also with the further developments of psycho-analytic theory and practice[15].

As we have seen, Freud's deterministic postulate also stood the test of the passage from quantity to quality, from the unconscious to consciousness. For instance, he gave for granted the continuity of brain processes and psychological phenomena like censorship and repression, which he considered responsible for jokes and dreams mechanisms. The fourth and lost section of the *Project* was devoted to these themes; but even in the preserved text Freud does not hesitate to apply a rigid causal connection to the several functions of conscience: judgment, cognition,

[15]This is one of the main theses of F.J. Sulloway's book.

memory, theory building and so on [8, *passim*]. The "qualitative" approach to consciousness seems to remain lip-service.

Many commentators have rightly remarked that there is really no theoretical gap between the *Project* of 1895 and subsequent Freud's work. In this sense, the draft fully shows the Achilles' heel of the founder and theorist of psycho-analysis, namely the dilemma of determinism he never resolved, nor even questioned critically, which nonetheless represents a permanent challenge within the frame of meta-psychology. Freud never gave up the hope of confirming his theories and conjectures thanks to new discoveries in brain physiology. But he also believed to have anticipated in some measure the future discoveries by his circular argument; i.e. by extending the deterministic principle from "quantity" to "quality", from brain physiology to the whole of psychical processes and to meta-psychology, including the acts of the will. However, a radical break was open in the brain-mind *continuum*, and this theoretical inconsistency permitted him to reintroduce surreptitious, unpredictable and aleatory events. For instance, the therapeutic technique of free association should be able to reconstruct in a *free way* the sequence of *deterministic* events of the past, proceeding back from the effects to the causes, until the original trauma and cause of the disease be found and removed.

This paradoxical juxtaposition of the deterministic principle upon aleatory events is clearly stated in the *Interpretation of dreams*, chapter VII, where Freud summarizes the tenets of his *Project*. As it is well known, he distinguishes in the oneiric processes some constant functions – condensation, displacement, censorship, deformation, regression – which contribute to fashioning the incoherent and deceitful images of dream by hiding to consciousness the repressed materials. In the light of the *Project*, repression is the key of this deterministic process: "I cannot help concluding – as Freud resumes in his *On dreams* – that there is a causal connection between the obscurity of the dream-content and the state of repression (inadmissibility to consciousness) of certain of the dream-thoughts, and that the dream had to be obscure so as not to betray the proscribed dream-thoughts" [10, V, 672]. The unconscious process of censorship, repression and deformation produced by dream hinders the access to consciousness. But how to prove the postulate that: a) this process be deterministic, and therefore b) its causes and effects be recognizable in detail? Nonetheless this postulate is the shaky foundation of the interpretive and therapeutical technique of the analyst. From this point of view the epistemological objections of critics like Karl R. Popper and Ernest Nagel, for whom the whole "scientific" corpus of psycho-analysis lays on unfalsifiable foundations, seem to be insuperable.

Freud was an admirer of Arrian's *Res gestae Alexandri Magni*: like Alexander, he cut perhaps too boldly the Gordian knot of the complexities of the body-mind problem. This original sin at the beginning of the successful career of psycho-analysis may explain, at least partly, the everlasting crisis of psycho-analytic theories and practices, and their difficult relationship with the progress of brain physiology in the late XXth century. The *Project for a scientific psychology* seems to be the relic of a pious illusion rather that an anticipation, for its rigid determinism is at odds with the complexities of the brain functions as revealed by recent research in the neuro-sciences.

William James looked at the psycho-analytic movement with great hopes even if, just after the Worcester meeting, he confessed in private letters to be bewildered about some aspects of Freud's personality and methods[16]. We may conclude these remarks by two simple questions. Was not the "dilemma of determinism", as James put it, more open to future research and more far-sighted than Freud's positivistic determinism? Is not the problem of free will – which represents perhaps the utmost degree of complexity in nature – insoluble from the point of view of the hard sciences, and destined to wrestle with the quicksands of the soft sciences?

References

[1] BURKHARDT F.H. ET AL. EDS., *Essays in Philosophy; The Works of William James.*

[2] CALCATERRA R.M., *Introduzione al pragmatismo americano*, Laterza, Roma-Bari 1997.

[3] CASINI P., Rivista di Filosofia **XCIII** (1) (2002), 1–23.

[4] FREUD S., *An Autobiographical Study*, in: "The Standard Edition of the Complete Psychological Works", English translation edited by Strachey J. et al., The Hogarth Press, London.

[5] FREUD S., *Five Lectures on Psycho-analysis*, in: "The Standard Edition of the Complete Psychological Works", English translation edited by Strachey J. et al., The Hogarth Press, London.

[6] FREUD S., *Introductory Lectures on Psycho-analysis*, in: "The Standard Edition of the Complete Psychological Works", English translation edited by Strachey J. et al., The Hogarth Press, London.

[16] "I hope that Freud and his pupils will push their ideas to their utmost limits, so that we may learn what they are. They can't fail to throw light on human nature; but I confess that he made on me personally the impression of a man obsessed by fixed ideas": quoted by R.B. Perry [17, II, 122]; cf. James' letter to Miss Calkins of september 19, 1909: "I strongly suspect that Freud, with his dream-theory, of being a regular *halluciné*" [17, II, 123].

[7] FREUD S., *Psychopathology of Everyday Life* (1901), in: "The Standard Edition of the Complete Psychological Works", English translation edited by Strachey J. et al., The Hogarth Press, London.

[8] FREUD S., *Project for a Scientific Psychology*, in: "The Standard Edition of the Complete Psychological Works", English translation edited by Strachey J. et al., The Hogarth Press, London.

[9] FREUD S., *Anfängen der Psychoanalyse*, M. Bonaparte, A. Freud and E. Kris eds., London, Imago Publishing Co., London 1950.

[10] FREUD S., *On Dreams*, in: "The Standard Edition of the Complete Psychological Works", English translation edited by Strachey J. et al., The Hogarth Press, London.

[11] GUARNIERI P., *Introduzione a James*, Laterza, Roma-Bari 1985.

[12] JAMES W., *Principles of Psychology*, Dover Publications Inc., New York 1950.

[13] JAMES W., *The Dilemma of Determinism*, in: "The Unitarian Review", XXII, Sep. 1884.

[14] JAMES W., *The Will to Believe and other Essays in Popular Philosophy*, Longmans Green and Co., New York-London-Bombay 1897.

[15] JONES E., *The Life and Work of Sigmund Freud*, Basic Books, New York 1953.

[16] VON KARPINSKA L., *Über die psychologischen Grundlagen des Freudismus*, "Internationale Zeitschrift für Psychoanalyse", II, 1914, 305.

[17] PERRY R.B., *The Thought and Character of William James, as Revealed in Unpublished Correspondence and Notes, together with his Published Writings*, Little, Brown, and Company, Boston; and H. Milford, London 1935.

[18] SANTUCCI A., *Il pragmatismo*, U.T.E.T., Torino 1970.

[19] SANTUCCI A., *Storia del pragmatismo*, Laterza, Roma-Bari 1992.

[20] SULLOWAY F.J., *Freud, Biologist of the Mind. Beyond the Psychoanalytical Legend*, Basic Books, New York 1979.

PROBABILISTIC ASPECTS IN GEORGE D. BIRKHOFF'S WORK

Luca Dell'Aglio
Dipartimento di Matematica,
Università della Calabria,
87036 ARCAVACATA DI RENDE (CS) - ITALY
dellaglio@unical.it

Abstract The aim of the present paper is to analyse different kinds of occurrences of the concept of probability in George D. Birkhoff's work. The occurrences refer to transversally theoretical ambits such as dynamical systems and ergodic theory, and lead directly to the discussion of some epistemological questions concerning the interpretation of physical phenomena, also in relation to the modern concept of 'deterministic chaos'.

1. Introduction

In the context of their modern development it is interesting to seek to understand how probabilistic laws tend to emerge historically, particularly for as far as the field of classical mechanics is concerned.

In this context, the work of George D. Birkhoff is of great interest for different related reasons. First of all because it is the work of one of the protagonists of the development of theoretical mechanics during the 20th century, in connection with the topics of the theory of dynamical systems. Secondly because this work was strongly influenced by the conceptions of Henri Poincaré, who has the merit of offering one of the most comprehensive approaches to the idea of probability in modern scientific thinking. Furthermore, Birkhoff's work tends to show traces of a deeper awareness of the importance of probabilistic laws, also in connection with the growing relationship between probability calculus and measure theory.

This paper is an attempt to analyse some aspects of Birkhoff's work and to describe the role played by probabilistic concepts in it. The interest in such a topic is supported by the fact that Birkhoff's work has apparently not been subjected to any close historical scrutiny, except for some reports produced on the occasion of his death [36, 52] and for some references to his results in the historical reconstruction of some specific aspects of modern celestial mechanics, dynamical systems theory and ergodic theory [1, 32, 31, 35].

2. Papers concerning probabilistic concepts in Birkhoff's work

In Birkhoff's work, there are different papers that, more or less directly, display links with probabilistic topics. First of all, some references to these topics may be found in the first phase of the scientific production of the American mathematician, namely in the works leading up to the publication, in 1927, of *Dynamical Systems* [10]. However, the link with probabilistic concepts was to become clearer in some papers after 1930 – particularly in those related to the ergodic theorem and to its physical interpretation, such as the essay *Probability and physical systems* [14] and the epistemological paper *Stability in causal systems* [20], written in collaboration with C.D. Lewis Jr. Furthermore, some of his research on mechanics, before and after the publication of *Dynamical Systems*, reveals links with the modern concept of 'deterministic chaos'. Finally, in his philosophical works, of which *Aesthetic Measure* [15] is the best known, a discussion of 'the principle of sufficient reason' is presented in one of the lectures that Birkhoff delivered at the Rice Institute in March 1940 [16].

3. The concept of probability in *Dynamical systems*

It is possible to find an explicit occurrence of the concept of probability in Birkhoff's work in Chapter 7 of *Dynamical Systems.* This occurrence is neither accidental nor marginal, since this chapter – whose contents and the very definition of probability given in it echo the previous papers [3] and [8] – is one of the more general points in the treatise by the American mathematician; at this point in fact he gives a definition of some basic concepts of the modern qualitative theory of dynamical systems, such as wandering and non wandering points, central and recurrent motions.

In this context we have the following definition of probability related to the behaviour of a trajectory: "Let us define the 'probability' that an arc PQ lies in a given region Σ as the ratio of the time interval for the part that does lie in Σ to the total interval" [10, 196]. Expressed in formal terms [37, 361], this amounts to stating that:

$$p(PQ, \Sigma) = \tau/T$$

where τ is the measure of the set

$$\{t \in [0, T] : f(P, t) \in \Sigma\}$$

that is

$$\tau = \int_0^T \varphi_\Sigma[f(P, t)]\, dt$$

$f(P, t) = P_t$ is the generic point of the trajectory and φ_Σ is the characteristic function of the set Σ.

Even if not explicitly stated by the American mathematician, this definition may be related to two other considerations of the concept of probability.

First of all, it may be related to the way in which the concept of probability was dealt with in the development of statistical mechanics in the second half of the 19th century. The concept of probability as 'time average' actually tends to emerge in the work of L. Boltzmann, from 1866–8 on [23, 42], and, on this basis, it was utilised within the development of statistical mechanics, finding a similar formulation in the work of, among the others, Paul Ehrenfest [25]. With hindsight, such a connection with the tradition of statistical mechanics is not accidental in the overall context of Birkhoff's work, since, as we will see in the following section, it is a sort of physical substratum to the consideration of ergodic theory by the American mathematician.

Furthermore, there is another significant reference to the definition of probability given by Birkhoff in *Dynamical Systems* which concerns the idea of probability that Poincaré discussed in his works on celestial mechanics [47] in relation to the well-known 'recurrence theorem'. These considerations – already to some extent present in the paper *Sur le problème des trois corps et les équations de la dynamique* [43] – are developed by the French mathematician in their most complete form in Chapter XXVI, 'Stabilité à la Poisson', in the third volume of *Les méthodes nouvelles de la mécanique céleste* [45]. In particular, in paragraph 296 of this chapter, explicitly entitled 'Probabilités', there is a definition of probability concerning the behaviour of a trajectory. In this context, in particular, Poincaré formulated the recurrence theorem in probabilistic terms, stating that the "probabilité pour qu'une molecule ne traverse $U_\circ$ qu'un nombre fini de fois est infiniment petite, si l'on admet que cette molécule est à l'intérieur de $U_\circ$ à l'origine du temps" [45, III, 151]. Likewise, Birkhoff introduced the concept of probability for the purpose of considering some theorems which have Poincaré's recurrence theorem as prototype. For example, in *Dynamical Systems* there is the following theorem of a 'probabilistic' nature, concerning the set M_1 of non-wandering points – that is (in contemporary terms) of those points P such that for any neighborhood U there exist arbitrary large values t such that $f(U,t) \cap U \neq \emptyset$: "the probability that an arc lies in a given neighborhood of M_1 approaches unity uniformly as the interval of time increases indefinitely" [10, 196].

It is worth noting that, in this manner, Birkhoff's researches are a direct prosecution of Poincaré's idea of considering physical properties with probability 1 explicitly; a feature that was first formalized in terms of measure theory by C. Carathéodory [22].

4. Birkhoff's approach to ergodic theory

Not surprisingly, the concept of probability in terms of 'time averages' characterises also Birkhoff's approach to the ergodic theorem. As is well known [19, 26, 28, 32, 34], the history of this theorem characterizes an embryonic phase in statistical mechanics in relation to Boltzman's well-known 'ergodic hypothesis'. After some criticism, this hypothesis gave way to different re-formulations and, in particular, to the so-called 'quasi-ergodic hypothesis' by Paul and Tatiana Ehrenfest, which, among other things, stated the existence of the limit of the ratio τ/T, for $T \to \infty$, on the basis of physical considerations alone [25, 32–33].

This question received a preliminary answer from the demonstration by J. von Neumann, in 1931, of the so-called 'Mean Ergodic Theorem' (MET), which states the existence of such a limit in terms of convergence in the mean. The events that immediately followed this result shed considerable light on Birkhoff's attitude towards the ergodic theorem. Soon after he was told about the MET by von Neumann in October 1931 [19, 281], Birkhoff re-elaborated it, obtaining an analogous theorem expressed, however, in terms of pointwise convergence; it is worth noting that this enabled the American mathematician to publish his result [13] before von Neumann [38].

In these historical events, it is possible to recognize a slightly different attitude towards the consideration of the ergodic theorem. In fact, although von Neumann himself had declared the necessity of an extension of the MET [19, 281], Birkhoff often underlined the importance of this result but also highlighted its partial limitations on a number of occasions. For example, in *Probability and physical systems*, we read: "Von Neumann's result establishes that a similar "mean ergodic theorem" holds in the highly technical sense of "convergence in the mean", although throwing no light on the existence of a time-probability along the individual motions" [14, 366]. It is worth noting that such a statement in no way signifies a clash of opinion between Birkhoff and von Neumann, but rather indicates the presence in the American mathematician's conceptions of an opposition – sharper than in von Neumann, and this is the difference, – among more or less significant formulations of the ergodic theorem. The point is that the really significant formulation of such a theorem for the American mathematician is a proposition in terms of the probability of trajectories, in accordance with what he expressed in his research on dynamical systems; this helps us to some extent to understand why Birkhoff did not hesitate to publish his version of the ergodic theorem before von Neumann.

In this regard, the way the ergodic theorem is dealt with in Birkhoff's papers is highly significant. For example, in the paper where this theorem is demonstrated by the American mathematician [13], it appears in the following meaningful form: "For any dynamical system of type (1) $[dx_i/dt = X_i(x_1, \ldots, x_n)\ i = 1, \ldots, n]$ there is a definite 'time probability' p that any moving point, excepting those of a set of measure 0, will lie in a region v; that is

$$\lim_{\pm\infty} \tau/\tau' = p \leq 1$$

will exist, where τ denotes total elapsed time measured from a fixed point and τ' the elapsed time in v" [13, 660].

From a methodological standpoint, the way in which the ergodic theorem is obtained by Birkhoff is also significant: his demonstration in fact is completely different from von Neumann's. In particular, it is achieved on the basis of a 'strong recurrence theorem', that is to say of an extension of Poincaré's recurrence theorem, to which the immediately preceding paper [12] was dedicated. In this manner, Birkhoff's demonstration displays a sort of continuity with some of his preceding research, insofar as it is a variation of a demonstration already researched in the study of dynamical systems [12, 651]; it actually refers to the asymptotic behaviour of the quantities $t_n(P)/n$ and $t'_n(P)/n$, where $t_n(P)$ and $t'_n(P)$ respectively indicate the time interval of the nth crossing of the trajectory for P with a surface of section S and the time interval that such a trajectory lies in a given region V of the phase space before of the nth crossing with S.

5. Probability and physical systems

The ergodic theorem plays a central role also in Birkhoff's conceptions concerning the relationships between probability and physics, present in some works from 1920 on, in particular in *Stability and periodicity in dynamics* [9] and in the cited papers *Probability and physical systems* and *Stability and causal systems*. In this last paper, for example, there is a rather significant discussion on the question of the choice between 'causal systems' – systems for which "the subsequent development (...) from a known initial condition, that is, a condition in which the measurable variables have known values, is uniquely determined by these values" [20, 304] – and 'non-causal systems'. Significantly, Birkhoff's position in this context is not in contrast with the consideration of probabilistic laws. For example, in Section 9 of this paper we read: "Whether or not the universe itself is one vast causal system is a question open to speculation. Is it theoretically possible, with the aid of an enormously complicated system of differential equations, to determine the whole course of destiny from a complete knowledge of the state of the universe at any given instant?" [20, 323]. On the basis of a realistic evaluation of the progress of physics in the first decades of the 20th century, the answer given by the authors to this question is very clear-cut: "It is scarcely necessary to say (...) that causal systems are not the only ones considered by modern physicists. Notably in the quantum mechanics, we have the 'principle of indeterminacy', and, in atomic theory generally, the theory of probability plays an important rôle. According to these theories the more classical causal theories of physical systems are true only in a statistical sense" [20, 323]. Furthermore, making explicit reference to the throwing of a coin, they state: "In order to treat any non-causal system on a mathematical basis, it seems necessary to introduce the notions of probability. Such notions involve the assumption that certain sequences of ratios approach limits. (...) The assumption that certain sequences converge, that is, that certain probabilities exist, is part of the physical law in question. There is no a priori reason why such a law should not be wholly adequate" [20, 323–4].

However,notwithstanding these considerations,the authors make some criticism of laws of a probabilistic nature. Besides the classical argument of the possibility of an interpretation in deterministic terms, – explicitly indicated as related to a "much more exact treatment" [20, 324] – some considerations are developed concerning the so-called 'closed variational dynamical systems'; it is here that the epistemological consideration of the ergodic theorem plays a role. On this subject, as in *Dynamical Systems*, the problem is related to the probability that a point P_t in the phase space belongs in a region V in an isoenergic manifold Ω; that is to say which is the limit of the quantity $T(\tau)/2\tau$, where, as usual, $T(\tau)$ is the time interval that the point P_t lies in V with t in the interval $(-\tau, \tau)$. Underlying any attempt to empirically determine such a limit is the hypothesis that it exists, as highlighted by the initial development of statistical mechanics. On the other hand, the fact that the existence of this limit is assured by the ergodic theorem in the case of 'closed variational dynamical systems' – and that it is equal to V/Ω under the so-called hypothesis of 'metrical transivity' – entails a sort of logical centrality of these systems compared with statistical mechanics; that is to say a sort of logical dependence of a non-causal theory from a causal one. This is the case of a logical dependence which, however, according to the authors, has no absolute validity owing to the presence of quantum mechanics. The result is an attempt to go beyond the contraposition between causal and non-causal theories, assuming an alternation between theories of a different nature: "One is now tempted to ask if there is a "causal" theory in back of the quantum theory. If so, is *it* the fundamental theory of the universe, or is there a still more fundamental non-causal system in back of it? May there not be an infinite number of such theories, alternatively causal and non-causal, each theory more "fundamental" than its predecessor?" [20, 324].

6. Birkhoff and the idea of 'deterministic chaos'

There is one further perspective, related to the modern concept of 'deterministic chaos', from which it is possible to examine the presence of elements of a probabilistic nature in Birkhoff's work. Actually, since Birkhoff is historically situated between Poincaré, who first introduced this concept [46], and Edward Lorenz, the American meteorologist who first highlighted its technical importance in the second half of the 20th century [33], one may imagine that Birkhoff had a significant role in this thinking process. However, the problem of the relationships between Birkhoff's research and the modern concept of 'deterministic chaos' is rather complex, owing to the possible presence of a sort of imbalance between the technical and the conceptual aspects of the question.

What cannot be denied is the presence in the American mathematician's work of the development of many of the technical aspects of the modern theory of 'chaos'. This is related mainly to the attention focused in Birkhoff's work, again following Poincaré's, on the study of homoclinic points, for example in the papers *On the periodic motions of dynamical systems* [11] and *Structure analysis of surface transformations* [21]. In these contexts, the so-called Smale-Birkhoff theorem – which is now considered as one of the basic conditions for the presence of chaotic behaviour – plays a central role. This is a theorem for which Birkhoff demonstrated a corollary [11, 376] and it may be assumed that he foresaw its validity in the case of dimension 2 [48, 64], [49, 775].

However, the point is that Birkhoff's research, as well as Poincaré's, is related to the consideration of conservative systems and not of dissipative ones: the latter is an area that was to be the main context of development of the modern concept of 'deterministic chaos' [51].

7. Conclusions

The foregoing considerations allow a sort of reconstruction to be made of the way in which probabilistic concepts tend to emerge in George D. Birkhoff's work. Referring to the tradition of statistical mechanics, the approach to the concept of probability of the American mathematician is based on Poincaré's work on celestial mechanics related to the properties of recurrence of a dynamical system; it is essentially along this path that Birkhoff, in the years that followed, arrived at his own formulation of the ergodic theorem and its use in the interpretation of physical phenomena.

It should be noted that these considerations are of interest not only from a historical point of view, as is shown by the fact that Birkhoff's approach to the concept of probability in terms of 'time averages' has recently undergone significant development, both technical – concerning the study of the probabilistic behaviour of a trajectory [2, 30] – and epistemological – in relation to an interpretation of probability of a frequentist nature *via* ergodic theory [41].

References

[1] BARROW-GREEN J., *Poincaré and the three body problem*, American Mathematical Society, Providence 1997.

[2] BOYARSKY A., GORA P., *Laws of Chaos. Invariant Measure and Dynamical Systems in One Dimension*, Birkhäuser, Boston 1997.

[3] BIRKHOFF G.D., *Quelques théorèmes sur le mouvement des systèmes dynamiques*, Bull. Soc. Mat. de France **40** (1912), 305–323.

[4] BIRKHOFF G.D., *The restricted problem of three bodies*, Palermo Rend. **39** (1915), 265–334.

[5] BIRKHOFF G.D., *Dynamical systems with two degrees of freedom*, Nat. Acad. Proc. **3** (1916), 314–316.

[6] BIRKHOFF G.D., *Dynamical systems with two degrees of freedom*, Trans. Amer. Mat. Soc. **18** (1917), 199–300.

[7] BIRKHOFF G.D., *Surface transformations and their dynamical applications*, Acta Math. **43** (1920), 1–119.

[8] BIRKHOFF G.D., *Über gewisse Zentralbewengungen dynamischer Systeme*, Gesell. Wisse, Göttingen 1926, 81–92.

[9] BIRKHOFF G.D., *Stabilità e periodicità nella dinamica*, Period. Matem. **6** (4) (1926), 262–271.

[10] BIRKHOFF G.D., *Dynamical Systems*, Amer. Math. Soc., New York 1927.

[11] BIRKHOFF G.D., *On the Periodic Motions of Dynamical Systems*, Acta Mat. **50** (1927), 359–379.

[12] BIRKHOFF G.D., *Proof of a recurrence theorem for strongly transitive systems*, Proc. Nat. Acad. Science **17** (1931), 650–655.

[13] BIRKHOFF G.D., *Proof of the ergodic theorem*, Proc. Nat. Acad. Science **17** (1931), 656–660.

[14] BIRKHOFF G.D., *Probability and physical systems*, Bull. Amer. Math. Soc. **38** (1932), 361–379.

[15] BIRKHOFF G.D., *Aesthetic Measure*, Harvard University Press, Cambridge (MA) 1933.

[16] BIRKHOFF G.D., *Three public lectures on scientific subjects*, Rice Institute Pamphlet **28** (1941); reprinted in Birkhoff G.D., *Collected Mathematical Papers*, vol. III, American Mathematical Society (1950), 755–838.

[17] BIRKHOFF G.D., *What is the ergodic theorem?*, Amer. Math. Mont. **49** (1942), 222–226.

[18] BIRKHOFF G.D., *Collected Mathematical Works of George David Birkhoff*, Providence R.I. 1950.

[19] BIRKHOFF G.D., KOOPMAN B.O., *Recent contributions to the ergodic theory*, Proc. Nat. Acad. Science **18** (1932), 279–282.

[20] BIRKHOFF G.D., LEWIS C.D. JR., *Stability in causal systems*, Philosophy of Science **2** (1935), 304–333.

[21] BIRKHOFF G.D., SMITH P.A., *Structure analysis of surface transformations*, Journ. de Math. **7** (1928), 345–379.

[22] CARATHÉODORY C., *Über den Wiederkehrsatz von Poincaré*, Berl. Ber. (1919), 580–584.

[23] CERCIGNANI C., *Ludwig Boltzmann e la meccanica statistica*, La goliardica pavese, Pavia 1997.

[24] DEVANEY R.L., *An Introduction to Chaotic Dynamical Systems*, Addison Wesley, New York 1988.

[25] EHRENFEST P., EHRENFEST T., 1911. *Begriffliche Grundlagen der statistischen Auffassung in der Mechanik*, Encyklopädie der mathematischen Wissenschaften, vol. 4, part 32 (1911); english translation *The Conceptual Foundations of the Statistical Approach in Mechanics*, by Michael J. Moravcsik, Cornell University Press 1959.

[26] FARQUHAR I.E., *Ergodic Theory in Statistical Mechanics*, Wiley-Interscience, New York 1964.

[27] GUCKENHEIMER J., HOLMES P., *Nonlinear Oscillations, Dynamical Systems, and Bifurcations of Vector Fields*, Springer, New York-Berlin-Heidelberg 1983.

[28] KHINCHIN A.I., *Mathematical Foundations of Statistical Mechanics*, Dover, New York 1949.

[29] KOOPMAN B.O., *Birkhoff on Dynamical Systems*, Bull. Amer. Math. Soc. **36** (1930), 162–166.

[30] LASOTA A., MACKEY M.C., *Probabilistic Properties of Deterministic Systems*, Cambridge University Press, Cambridge 1985.

[31] LEFSCHETZ S., *Geometric Differential Equations: Recent Past and Proximate Future*, in: "Differential Equations and Dynamical Systems", J.K. Hale and J.P. La Salle eds., Academic Press, New York 1967, 1–14.

[32] LO BELLO A., *On the Origin and History of Ergodic Theory*, Boll. St. Sc. Mat. III (1) (1983), 37–75.

[33] LORENZ E.N., *Deterministic nonperiodic flow*, J. Atmos. Sci. **20** (1963), 130–141.

[34] MACKEY G.W., *Von Neumann and the early days of ergodic theory*, in: "Proceedings of Symposia in Pure Mathematics" 50 (1990), 25–38; reprinted in Mackey G.W., *The Scope and the History of Commutative and Noncommutative Harmonic Analysis*, American Mathematical Society (1991), 311–324.

[35] MIRA C., *Some Historical Aspects Concerning the Theory of Dynamic Systems*, in: "Dynamical systems: a renewal of mechanism. Centennial of George David Birkhoff", S. Diner, D. Fargue and G. Lochak eds., World scientific, Singapore 1986, 250–261.

[36] MORSE M., *George David Birkhoff and his mathematical work*, Bull. Amer. Math. Soc. **52** (1946), 357–391.

[37] NEMYTSKII V.V., STEPANOV V.V., *Qualitative Theory of Differential Equations*, Princeton University Press, Princeton 1960.

[38] VON NEUMANN J., *Proof of the quasi-ergodic hypothesis*, Proc. Nat. Acad. Science **18** (1932), 70–82.

[39] NITECKI Z., *Differentiable Dynamics*, The MIT Press, Cambridge (MA) 1971.

[40] PERCIVAL I.C., *Chaos in Hamiltonian systems*, in: "Dynamical Chaos", M.V. Berry, I.C. Percival and N.O. Weiss eds., The Royal Society, London 1987, 131–144.

[41] VON PLATO J., *Probability in Dynamical Systems*, in: "Logic, Methodology and Philosophy of Science VII", J.E. Fenstad et al. eds., North-Holland, Amsterdam 1989, 427–443.

[42] VON PLATO J., *Creating Modern Probability*, Cambridge University Press, Cambridge 1994.

[43] POINCARÉ H., *Sur les équations de la dynamique et les problème de trois corps*, Acta Math. **13** (1890), 1–270.

[44] POINCARÉ H., *Calcul des probabilités*, Gauthiers-Villars, Paris 1896 (revised edition 1912).

[45] POINCARÉ H., *Les méthodes nouvelles de la mécanique céleste*, Gauthiers-Villars, Paris 1892–99.

[46] POINCARÉ H., *Science et Méthode*, Flammarion, Paris 1908.

[47] SHEYNIN O.B., *Henri Poincaré's Work on Probability*, Arch. Hist. Ex. Sc. **42** (2) (1991), 137–171.

[48] SMALE S., *Diffeomorphisms with Many Periodic Points*, in: "Differential and Combinatorial Topology", S.S. Cairns ed., Princeton University Press, Princeton 1965.

[49] SMALE S., *Differentiable Dynamical Systems*, Bull. Amer. Math. Soc. **73** (1967), 747–817; reprinted in Smale S., *The Mathematics of Time*, Springer, New York-Heidelberg-Berlin, 1–82.

[50] SMALE S., *On the problem of reviving the ergodic hypotheses of Boltzmann and Birkhoff*, in: "International Conference on Non-Linear Dynamics", New York Academy of Science 1979; reprinted in Smale S., *The Mathematics of Time*, Springer, New York-Heidelberg-Berlin, 137–144.

[51] STEWART B., THOMPSON M., *Nonlinear Dynamics and Chaos. Geometrical Methods for Engineers and Scientists*, Wiley, New York 1986.

[52] VEBLEN O., *George David Birkhoff (1884–1944)*, in: "Year Book of The American Philosophical Society", 1946, 279–285; reprinted in Birkhoff G.D., *Collected Mathematical Works of George David Birkhoff*, 1950, xv–xxi.

THE METAMORPHOSIS OF HOLISM

Elena Gagliasso
Università degli Studi di Roma "La Sapienza",
Dipartimento di Studi Filosofici ed Epistemologici,
"Villa Mirafiori", Via Carlo Fea 2, 00199 ROMA - ITALY
elega@tiscalinet.it

Abstract This article explores anti-reductionist tendencies and the variety of their theories and metodologies. Anti-reductionism is seen principally as a critique of reductionistic methodologies: from vitalism (XVIII century), to Naturphilosophie (XIX century), to systemic theories ([5]), or from autopoiesis (Maturana, Varela, 1980), to auto-organizing theories ([18]). These different areas are examined from historical and theoretical points of view, taking into account ontological and epistemological aspects. The final part focuses on key aspects of contemporary organicism and holism.

From the conceptual point of view, while the definitions of the diverse types of reductionism in biology are sufficiently shared by the scientific community [28, 31, 30], this doesn't happen for the composite anti-reductionist front, that isn't neither coherent nor united, nor definable in a unitary way.

Reductionism has represented the main road of the scientific knowledge for Biology, in spite of the fact that, above all in the German area, a great part of the research was founded – until the first half of the century – on a vitalist thematic structure, that was common to embryologists, physiologists and morphologists [20, 26].

Anti-reductionism presented itself essentially as criticism. The cone of powerful but circumscribed light of the production of theories and the reductionist biological methodologies, i.e. drawn from methodologies characteristic of the exact sciences (therefore extradisciplinary) or from sectors of Biology that are more formalizable and predictive than Biology itself, (therefore intradisciplinary), has pushed to the margin the peculiar aspects of the living world that were not saturated by these methodologies.

We can therefore talk of a true constellation of anti-reductionisms, a spectre that goes from the various forms of the XIX century sensist vitalism to the present systemic and self-organizational theories. In the first case, from the sensists to C. Bernard's vitalist mechanisism[1], it was above all a matter of metaphysics and panpsychism. Finalism and vitalism were connected in order to vindicate the specificity of the living world, passing in the substantialistic finalism of the *Naturphilosophie* of the following century, and after that period in the realm of substantialistic vitalism of the German embryologists, whose most important theoretician was H. Driesch (Driesch, 1909).

Subsequently, and in clear metaphysical discontinuity with vitalism, the researches based on the importance of the intra– and extra–organismic relationships, characteristic of the organicism of the XX century (Haldane, 1924; [40]; Waddington, 1968) and the researches on the Systems (von Bertalanffy, 1953), set up a scientific apparatus for holism[2], often integrating metaphysics and methodologies, and producing a paradigm that has been a source of research and critical reflection.

The retroactive *loops* that co-determine the links of the possible states between organisms and environments, or the complex epigenetic phases that connect the expression of the genome with the differentiation process and development, were studied in this context. The study of the *autopoiesis* systems (or self-organizational systems) in the ambit of the paradigms of complexity has indirectly contributed to moderate the most recent reductionism, that is more faceted and aware of the role of the systemic retroactivities and of the emergent properties, according to their levels of relationship. The definition of supervenience properties (Rosenberg, 1984; Stanzione, 2000), was elaborated in order to keep a distance from the vitalist implications of the term "emergence" that is now broadly used by the evolutionist biologists (Mayr, 1990).

Between soft reductionism, (informational and probabilistic reductionism), and the holistic and organicistic position, centered on relationships and on processes in time, the controversy continues, even if between these positions there is a smaller distance than the distance that once opposed the mechanists and the classical vitalists. For the

[1]M. Di Giandomenico (ed.), *Claude Barnard. Scienza, filosofia, letteratura*, Bertani, Verona 1982.

[2]The term ***holism***, created in the Twenties by J.C. Smuts, decreed the end of the conceptions based on matter, casuality an mechanics, not only for the living world but also in Physics. As it happened with sensism in the XVIII century, there is a new attempt to fill the gap between inorganic and living matter on a frontier between Science and Philosophy. This was Smuts' intent in his work ***Holism and Evolution*** (1926). Smuts' globalizing dimension was an attempt of a scientific re-enchantement of the world, that implied the co-presence of life in mind and matter [27].

contemporary anti-reductionists, that are aware not only of the *olos*, i.e. the integrated organism, but also of the manifestations of historical time in Biology, the present statistic reductionism, "weakened" in comparison with the original deterministic-laplacian pretensions, still fails to understand the singularity of the living world and his fundamental distinctions from the world of the inorganic.

According to organicism, cybernetic reductionism continues, even if used in Ecology, to undervalue the ecological thought, and doesn't completely understand the specificity of the organism in its dialectical relationship with the environment. According to radical holism, every form of analytical reasoning is a type a monocausal reductionism, an updating of the old Cartesian games.

A difference of perspective between the two areas is persistent, notwithstanding the fact that many methodologies are now common. This situation is linked to the *choice of the meanings* to give to the phenomena, to the role of the context in which they take place and to the interactive dynamicses between the organisms. Differences are traceable also in the sense that organicism and reductionism attribute to shared concepts: for example, the concept of emergence, that is used by both areas, but with different meanings, that are distinguishable only when the term is analysed in a discursive context. In fact is the discursive context, that continues to be the true discriminant between the organicistic and the reductionist perspective, in a period in which their methodologies are superimposing.

Another radical discard, underlined by the divergent uses of the same concept, that divides the two approaches concerns the positioning of the observer: "if, to study Mathematics, is sometimes enough to be angels, to study Biology we sometimes need to feel ourselves as animals" (Canguilhem, 1969, p. 38). When, in accordance with the reductionist intent, biology assumes a predictive direction, the eye of the observer has to be, so to speak, "angelic", not involved in his biological-material subjectivity, that it shares with the rest of the living, that vast community of which the subject, as individual of a given species, is a part. But the same 'angelic' position was expressed also by certain positions of holism of the first half of the XX century. This idea is inherent, for example, to the claim of the autonomy of biology proposed by the German philosopher of Biology E. Ungerer, in defence of the concept of abstract totality[3]. This is also true for L. von Bertalanffy's *Theory of*

[3]Among the German post-kantian scientists and philosophers, the biologist and epistemologyst E. Ungerer attempted a "logical" approach to biology, analysing the fundamental concepts of this science through the well-known contrast between vitalism and mechanicism

Systems that, from a holistic choice, has elaborated a true mathematical formalization, applicable to biology, but transversally also for the other sciences[4] that study sets and not analytical investigations, and in which the properties of the whole determine the state of the parts.

Vitalism, holism, organicism are traditionally set in epochal succession, but, from the theoretical point of view a distinction between ontological and epistemological level is necessary [6].

At the ontological level, for vitalism, that represents more a metaphysics than a methodology of research, the living world is radically separated from the inorganic for its peculiar vital properties. The objective consists in the comprehension and explanation of the nature of what is defined "Life", looking for this "nature" in properties that are extraneous to matter and identifying that "quid" (that for many vitalists cannot be analysed[5]), that distinguishes "life" from the "nonlife".

The vitalist program has worked through heterogeneous modalities along three vectors: substantialistic, energetic and finalist.

In the first case, there was the hypothesis of an unique *sui generis* substance, the *protoplasm*, extraneous to the inorganic matter, subsequently defined *colloidal state*, a true state of matter[6]. The second vector was freely inspired by the energetic theories of Physics, that hypothesized the existence of a vital force, peculiar to and distinctive of life[7]. *Nisus formativus* according to Blumenbach, *Lebenscraft*[8] according to Mueller,

---

and their overcoming. Ungerer's proposal laid the foundation of the organicistic position, grounded on two principles: the over-determination of the whole, linked to the concept of totality end the re-definition of finalism. Ungerer said that "in the science of organic Nature there are no multiple goals, but only one goal, i.e. the conservation of the totality of an object in becoming [33].

[4]The Austrian biologist and epistemologyst L. von Bertalanffy focused his attention on these themes for the first half of the century (*The Critical Theory of Morphogenesis* in 1928, *Theoretical Biology* in 1932, *The Biological Image of the World* in 1949 and *Outlines of the General Theory of Systems* in 1959). The attempt to find an axiomatic criterion for Biology, avoiding both reductionism and vitalism, was then transformed in the utopia of the unification of knowledge on a systemic ground. Even though the attempt was a failure, it was a valuable euristic hypothesis in several fields, in psychology and cybernetics.

[5]This is one of the classical definitions of the concept of life in [7]. Concerning the transformations of the concept of life see the essay [15].

[6]The concept of *protoplasm*, that now indicates the material of the cell, was a term that was originally introduced in biology indicating "the material substratus from which life arises", P. Duris, G. Gohau, 1999.

[7]The passage from a substantialistic vitalism to a vitalism inspired by an energetic force is gradual, and mediated by Physics. Between substance and force there was the concept of vital fluid, with the theories of the German vitalist of the end of the XVIII century Blumenbach (F. Jacob, 1971, cit.). However, "many forms of vitalism are absolutely legitimate extensions of the Cartesian program in mechanistic biology, realised with newtonian tools" (McLaughin, 1991, quoted in Mayr, 1998, p. 225).

[8]About the ideal, cultural and political consequences of the *Lebenscraft*, in Nazi Germany see [17].

entelechy, and *élan vital* according to Bergson, are some of the denominations assumed by this concept. Finally there was a connection between the energetic hypothesis and the evolutionary finalism. For its resistance to the Darwinian connection between variation and selection, every finalism really has a cosmic background and often doesn't speak only of biology. Lamarckian finalism, and the various forms of cosmological and evolutionary finalism, that were constantly proposed until Theilard de Chardin, were all considered from this point of view[9].

The ontological weakness of vitalism is well synthesized by Mayr: "The term Life is a simple reification of the vital process, which doesn't exist as an entity in itself. It's possible to study the process of Life scientifically, but not the abstract concept of Life" (Mayr, 1998, p. 2).

The area of metaphysical mixtures is a peculiar aspect of vitalism, not only in its debuts, but also in its last expressions. The sensist philosophers of the XVIII century, in particular, represent an interesting example of such a metaphysical mixture. The *Philosophie de la Vie* is anti-mechanistic, coherently with the anti-cartesianism that unites all the movement of the *Ideologues*, and at the same time is far from any transcendence principle, and irreducible to any physicalism. The materialistic philosophy that subtends, is empiricist and phenomenical, and avoids to postulate inverifiable principles, but gives to the single atoms the attributes of living matter. Buffon's *organic molecules* are in contrast with the atoms of the brute matter, while according to P.L. Maupertuis the elementary atom, for the living as well as for the inorganic, is identically pervaded by a vital psyche [4]. In Buffon the living molecule is connected again to Leibniz's monads while in Maupertuis there is a non-dualistic reality pervaded by a material continuity that is integrated and sentient: "we must turn to some principle of intelligence (inherent in matter), to something similar to what we call desire, aversion, memory" (Maupertuis, 1768 in [4, p. 147]). One of the the most radical representative of metaphysical reductionism of the period, J.O. La Mettrie, commented: "They have spiritualized matter rather than materialized the soul" (La Mettrie, p. 175; [4, p. 70]).

However, vitalism, in the form of finalist principles of the living, runs like a carsic river through Biology and above all, through all the *biophilosophies* that, from the beginning of the XX century, postulate not forces and specific substances, but specific explanatory principles for Biology. *Finalism* or *teleology* are recurrent themes in the claim of autonomy on behalf of Biology until the first half of the XX century, until

[9]Concerning the recent discussions of lamarckian finalism see B. Continenza, 1999.

the moment of the decoding of the Dna and of the rebirth of the Darwinian thought, that connects historical dimension, casual variation and selection. The *Naturphilosophie* of the XIX century is finalist and all the German embryologists between the XIX and the XX century. Finalism is a central feature of all Bergson's philosophy of the living and the philosophy of physicists of the first part of the XX century such as Elsasser and Polanyi[10].

The character of the German vitalisms, defined as teleomechanical [20] is much more heterogeneous. A true rebirth of Spinozism follows the naturalism of Schelling, Herder and Gaste, proposing again the idea of the *Deus sive Natura.* It is a naturalism extraneous to the "geometric method", and connected to Leibniz's notion of "force" that influences the experimental methodologies of physiologists and embryologists: in their mechanical analysis of the physiological processes the vitalist wedge is inserted only for the embryonic processes with a finalist development.

In the case of holism, the living world consists in a reality that can be understood only in its global and inseparable unity: "the whole is greater than the sum of the parts" [5]. More than the reductionist method, is the analytical method, the target of criticism, because according to holism, reality acquires a meaning only if analysed from a systemic point of view. The *Theory of Systems* has the objective to found the study of the living as a system, i.e. as "a complex of elements in interaction".

In organicism, that from this point of view differs from holism, the concept of globality itself leads to another result: the particular properties are given by the relationships between the parts and the characteristics considered as defining of any living entity (an organ, an organism, an organic and environmental system) and are determined by all the relationships that interconnect the diverse entities and that transform both the entities and the relations.

Holism and organicism have often been considered by radical reductionism as nothing but a transformation of vitalism deprived of its transcendent elements. The unity of the whole that characterises systemic holism[11] is therefore in polemic with the characteristics, not of the formal and nomological reductionism or with the axiomatization, but with the "sectorial" principle of the factual reductionism: the analysis and the reduction to infinitesimal parts, to be studied separately and to be

[10]For a critical analysis of Elsasser's and Polanyi's work: [25, 10].

[11]The neo-kantians in Germany connected the philosophical reflection on the autonomy of Biology with the researches of their contemporary vitalist and mechanistic embryologists. W. Koehler, father of the *Gestalt* theory, studied the problems of the relationship between Physics and Biology through structural criteria, centered on unity and on the self-regulative transformations of the *form.*

reassembled subsequently, as components of the living machine, doesn't allow the explanation of the most specific phenomena of Life.

P. Weiss, another important exponent of holism (in controversy with the mechanicism proposed by the physiologist J. Loeb[12]), in an article meaningfully entitled "1+1 does not equals 2" enunciated one of the basic principles of emergence: the fact that in the analysis "toward the lower part" the research "loses in perspective what gains in precision", while instead, in a synoptic picture that starts from the above, new and remarkablecharacters could appear [40].

Holism, at its birth, is also a metaphysics that acts unitarily not only concerning ontology (i.e; "the whole" of living reality) but also at the epistemic level of scientific theories. The "whole", in fact, acts as a unifying principle of the laws in their complex, in every discipline, through stylistic analogies (not forms of reductionism) with new basic principles of quantic and relativistic physics.

This type of new disciplinary connection went through interesting metamorphosis: on one hand it has created hierarchical connections, used to make coherent the passages that, from quantic physics, will arrive to the connectionistic theories of the human mind, and trying to justify a principle of theoretical unity, (almost demonstrative of a pre-ordained intelligibility of the world) called *anthropic principle*; on the other hand, it has underlined the criterion of multiple connection, marked by the self-referential loops: a common condition of the theories inspired by the paradigm of complexity.

Complexity is the unifying principle of many disciplines toward the end of the XX century, a vast theme that sometimes is a source of misunderstandings for his vague borders and the abuses that have been made. The mathematical use of algorithms in order to understand in various fields the autocostitutive dynamics of the systems was the objective of the paradigm of Complexity, that for its characteristics of globality and possibility of formalization should have pulverized the old dichotomies (Reductionism *versus* holism) giving to the latter an axiomatized structure, defeating on the ground of the formal unification the practice of reductionism.

[12]The direct controversy is really between the physiologist E. Jennings and J. Loeb concerning the analytical or synthetical interpretation of the regulative principles of the "system of actions" in the inferior organisms. Simple machines or complex unified systems? A reconstruction of the debate is R. Cordeschi's essay [11].

The theme of Complexity is present like a unifying theme in many disciplinary sectors[13], it is fundamental in the explanations of the living[14] and it is frequently alluded to as a *passe-par-tout, "mot valise"* [32], by the diffused culture. The theme of Complexity shows in its applications the same metaphysical temptations of reductionism: to become a global interpretative key and a dominant paradigm. In the meantime, while dealing with global phenomena, since its beginning, Complexity has set itself as a real criticism of reductionism, from a factual and formal point of view [12].

The whole and its internal relations are at the center of the interest of some lines of research that are inspired by organized Complexity from the *autopoiesis* theories by H. Maturana and F. Varela to the theory of the self-organization by S. Kauffman. As it happened for von Bertalanffy's systemic holism, the evolutionist dimension seems, (above all in the phase of formation of such theories) set aside: the organized structure and its characteristics of recursivity have rules that, even if influenced by time, doesn't seem liable to the unpredictability of its contingency.

If holism and organicism are connected by their common aversion toward analysis they have to be separated as historical roots of the theories of Complexity: the algorithmic "extremes" of Complexity are nearer to a formalized holism, while its explanations of the self-costitutive mechanisms seem to be closer to a concrete form of the usage of models within organicism [3].

Organicism, more than indistinct totalities,[15]studies the ways through which organization creates the order of the living and promotes emergent properties. In organicism there's no cosmic unity, but a unity of organisms, a unity of correlations of their internal parts. There are also the correlations between organisms and their environments, the biological niches and the correlations between the environments and the geoclymatic global system. Organicism is tied to the specificity and to the

[13]In the theories on the complex systems we find: the Physics of Chaos, the biophysics studies on the origin of the living matter, the researches on the "structural coupling" and on the organisational shut-down of systems (Maturana, Varela, 1986), the research concerning ontogenesis and the stabilization of the cellular and neural structures, all the researches on *Artificial Life* of the Santa Fe' School [19] and the applications in Economy and Management techniques [13]. There is a vast epistemological and scientific production concerning these new areas of Complexity. For a critical analysis of the passage from the theories of deterministic chaos to the sciences of Complexity: M. Cini, 1994, chapter n. 3: "Il linguaggio della complessità", [3].

[14]For the fundamental role of this concept in contemporary Biology: S. Forestiero, 2000 *Enciclopedia Treccani* (Roma), sub vocem *Complessità*.

[15]The totalities referred to by the traditional sentence "the whole is greater than the parts", considered by Ageno "irritating and a bit idiotic" [2, p. 100].

autonomy of the sciences of the living world, not to a cosmic coherence that goes from the galaxies to the brains, as in holism.

This allows to face and to underline, also on the epistemic level, the particular criterions of explanation that accompany the formulation of the autonomy of Biology: the abstract *olos*, conceived as inseparable unity, is replaced by the dynamics of the diverse relational keys, each with its own ambit of research and disciplinary languages.

On the epistemological level, the strategies of organicism don't exclude the analytical procedure and the knowledge produced experimentally, but they integrate them in a more complex methodology. This fact allows to distinguish with greater clarity between *analysis* and *reduction*: two procedures that are often homologated, but whose possible overlap consists only in the fact that reductionism is accompanied by the analytical practice, while the contrary is not always true.

References

[1] AA. VV., *La parte del tutto, Saggi sull'Olismo*, Casamassima, Udine 1992.

[2] AGENO M., *Le radici della biologia*, Feltrinelli, Milano 1986.

[3] ATLAN H., *Complessità, disordine e autocreazione del significato*, in: "La sfida della complessità", Bocchi G., Ceruti M. eds., Feltrinelli, Milano 1985.

[4] BARSANTI G., *Dalla storia naturale alla storia della natura*, Feltrinelli, Milano 1979.

[5] BERTALANFFY VON L., *General System Theory*, Borntraeger, Berlin 1950.

[6] BERGANDI D., *L'idée d'émergence*, Sciences et Avenir (hors-série) (2000), 70–75.

[7] BICHAT X., *Recherches physiologiques sur la vie et sur la mort*, Gabon-Brosson, Paris.

[8] CANGUILHEM G., *La conaissance de la vie*, Vrin, Paris 1971.

[9] CERUTI M., LASZLO E., *Physis: abitare la terra*, Feltrinelli, Milano 1988.

[10] CORBELLINI G., *La grammatica del vivente*, Laterza, Bari-Roma 1999.

[11] CORDESCHI R., *Il riduzionismo e l'analogia organismo-macchina: il caso di J. Loeb e H. Jennings*, in: AA. VV., op. cit. (1992), 61–90.

[12] DUPUY J.P., *Ordres et Ordres. Enquête sur nouveau paradigme*, Seuill, Paris 1982.

[13] DUPUY J.P., DUMOUCHEL J.P. (A CURA DI), *L'auto-organisation. De la physique au politique*, Seuil, Paris 1983.

[14] DURIS P., GOHAU G., *Histoire des sciences de la vie*, P.U.F., Paris 1997.

[15] FANTINI B., GRMEK M.D., *Le definizioni di vita e di morte nella biologia e nella medicina contemporanee*, in: "Bioetica", A. di Meo, C. Mancina eds., Laterza, Roma-Bari 1989.

[16] GAGLIASSO E., *Verso un'epistemologia del mondo vivente*, Guerini Scientifica, Milano 2001.

[17] HERF J., *Reactionary Modernism*, Cambridge University Press, New York 1984.

[18] KAUFFMAN S., *At Home in the Universe*, Viking, London 1995.

[19] KAUFFMAN S., *The Origin of Order*, Dordrecht, London 1997.

[20] LENOIR T., *The Strategy of Life: Teology and Mechanics in Nineteen Century German Biology*, Dordrecht-Riedl, Boston 1982.

[21] MATURANA H., VARELA F., *El arbol del conocimiento*, Editorial Universitaria, Santiago 1985.

[22] MAYR E., *The Growth of biological Thought*, Belknap Press, London 1982.

[23] MAYR E., *A long Argument. Charles Darwin and the Genesis of modern evolutionary Thought*, Allen Lane, London 1991.

[24] MAYR E., *This is Biology - The Science of the Living World*, McGraw-Hill, New York-Milano-Sydney 1997.

[25] MONOD J., *Le hasard et la nécessité*, Seuil, Paris 1970.

[26] ORSUCCI A., *Dalla biologia cellulare alle scienze dello spirito*, Il Mulino, Bologna 1992.

[27] POGLIANO C., *Mistero o semplicità: la difficile scelta*, in: AA. VV., op. cit. (1992).

[28] ROSE S., *Lifelines*, Allen Lane, Penguin Press, London 1997.

[29] ROSE S., ROSE H., *Alas, Poor Darwin*, Cape, London 2000.

[30] ROSENBERG A., *A Field Guide to a Recent Species of Naturalism*, British Journal for Philosophy of Science **47** (1996), 1–29.

[31] SOBER E. (A CURA DI), *Conceptual Issues in evolutionary Biology*, MIT, Cambridge 1984.

[32] STENGERS I. (A CURA DI), *D'une science à l'autre*, Editions du Seuil, Paris 1987.

[33] UNGERER E., *Outlines of the Formation of Biological Theories*, 1919.

[34] WADDINGTON C.H., *Strategy of the Genes*, Allen and Unwin, London 1957.

[35] WADDINGTON C.H. (A CURA DI), *Towards a theoretical Biology*, vol. 4, Edinburgh University Press, Edinburgh 1972.

[36] WADDINGTON C.H., *Paradigm for an evolutionary Process*, in: C.H. Waddington (a cura di), op. cit. (1972), vol. 2, 106–128; (ed. or., 1969).

[37] WADDINGTON C.H., *Tools for Thought*, Edinburgh University Press, Edinburgh 1977.

[38] WADDINGTON C.H, *The Evolution of an Evolutionist*, Edinburgh University Press, Edinburgh 1975.

[39] WADDINGTON C.H., *Genetic Assimilation*, in: C.H. Waddington, op. cit. (1979), 59–92; (ed. or., 1961).

[40] WEISS P., $1 + 1 \neq 2$ *(When One Plus One Does Not Equal Two*, in: "The Neurosciences. A Study Program", C. Quarton et al. eds., Rockfeller University Press, New York 1967, 801–821.

EARLY APPROACHES TO THE MANAGEMENT OF COMPLEXITY IN ENGINEERING SYSTEMS

Ana Millán Gasca
Dipartimento di Matematica,
Università di Roma "La Sapienza",
Piazzale A. Moro 2, 00185 ROMA - ITALY

Abstract The evolution of large artificial systems essentially underpins modern systems thinking and the ideas on complexity. A key contribution of industrial engineering to this intellectual process is the consideration of organizational problems and the development of the corresponding management technologies, eventually based on mathematical models. This kind of problems lie at the crossroads of engineering and the human and social sciences. From the first attempts to quantify labour and production, to 20th century operations research, an interdisciplinary evolution took place regarding ideas such as optimization, decision making, and control. Discussions on the use of quantitative and mathematical instruments in this context played an important role.

1. Introduction

The large technological systems developed after World War II in the United States are some of the best examples of complex systems. They consisted of military project systems developed for waging the Cold War, such as the air defense system SAGE (Semiautomatic Ground Environment) or the Atlas Project for the production of an intercontinental ballistic missile. Their complexity had to do with the large number of their components, their heterogeneity and the multiple interactions among them; a contribution was made by the high degree of automation, the use of a wide range of technologies (telecommunications, aeronautics, computers), the new forms of man-machine interaction and the presence of an important organizational component to manage the project. The creation of these systems provided the crucible for the development of a engineering systems approach representing one of the main contributions made by US culture to the 20th century.

The complexity of 20th century technological systems represents a radical challenge to one of the main objectives of engineering design, namely to obtain behaviours to assign to the artificial world. The response to the challenge led to the emergence, during the second half of the century, of a 'new engineering', the socalled systems engineering; however, this was not so much a truly new branch of engineering science as an innovative theoretical and methodological approach to technological knowledge. The principal components consisted of automation and control theory, computer science, as well as organization and management, which could also be formulated in mathematical terms. Use was therefore made of resources already available in previous technological contexts, which were resumed in the fifties as part of a new way of viewing technological conditions as 'systems' and not just groups of individual machines or plants in a new global approach to such systems. As Thomas P. Hughes and Agatha C. Hughes wrote [12, p. 2]:

> Practitioners and proponents [of the systems approach] embrace a holistic vision. They focus on the interconnections among subsystems and components, taking a special note of the interfaces among the various parts. What is significant is that systems builders include heterogeneous components, such as mechanical, electrical, and organizational parts, in a single system. Organizational parts might be managerial structures, such as military command, or political entities, such as a government bureau. Organizational components not only interact with technical ones but often reflect their characteristics. For instance, a management organization for presiding over the development of an intercontinental missile system might be divided into divisions that mirror the parts of the missile being designed.

This "systems approach" was later extended to include civilian contexts, ranging from town planning to industrial engineering, from the bearing structures of public administrations to the service networks, each now viewed as examples of a system. One fundamental aspect of the action carried out by technicians and experts in their design or reengineering is related to management and organization, in close association with hardware aspects, automation and with network building; information technology made a fundamental contribution in this connection. In the latter part of the 20th century, this broad field of practice and of research and development activity, marked by the working methods and analytical approaches typical of the engineer, spread from the United States to all the industrialized countries. The knowledge, experience, and challenges of the great modern artificial systems essentially underpin the perception of the complexity emerging at that time from a broad cultural perspective. The systems engineering approach made an essential contribution to 20th century systems thinking, defined as a global approach to the study of physical, biological, technological and social systems [3, 7]. One of the key concepts in this new unifying approach is organization.

2. Mathematics, engineering, and organization

Engineers have a longstanding familiarity with organization problems related to the performance of the activities for which they are responsible. These problems have been a fundamental component of their work ever since the Industrial Revolution. The 19^{h} engineer was actually on many occasions more a manager than a "mechanic", both as regards his activities in manufacturing industries and in his role of civil servant. In management, engineers relied essentially on their decision-making and operational capacity, guided mainly by their experience in one or more sectors of activity. However, considerable thinking was devoted to the division of labour as an organizational principle, based on Adam Smith's presentation of this issue (he used pin manufacturing as an example), and there were several attempts to 'rationalize' labour by providing quantitative and mathematical tools that could be used to guide the organizational decision-making and to analyse the activity processes and sequencing. From the 18th century fortification builders to the first engineering companies, and to the overland and river transport network, technological development has been accompanied by the creation of plants and infrastructures that at the same time raise closely interrelated problems of organization and cost-benefit economic problems. The important contributions made to political economy by engineers throughout the 19th century were often based on this kind of experience in the field.

The need to provide a scientific and mathematical basis for engineering expertise which gradually emerged during the 18th century was one of the most important cultural consequences of the spread of Newton's work, who was considered by the leaders of Enlightenment as the symbol of the triumph of reason. Also the project to extend the model of a mathematicized science beyond the physical sciences was accepted with great naturalness by engineers. Underlying the attempt to construct a "mathématique sociale" is the idea of a social reform that would be made possible and effective by identifying the aspects that lent themselves to mathematical treatment, in the first instance demography and economics. Likewise, from the point of view of authors like Charles A. Coulomb (1799) and Jules Dupuit (1844) in France, Charles Babbage

(1832) in the UK, or Wilhelm Launhart (1885) in Germany, mathematicization was the key to providing a rigorous and rational definition of the organization of labour and of activities suitable for attaining the required objectives in terms of efficiency, available resources, cost control and so on. But also in this context, a strong resistance arose during the 19th century, of the same nature as that against the project to mathematicize economics, that is, linked to the reluctance to apply the cold and mechanical techniques of mathematics to problems involving the sphere of human freedom, of the "moral and political sciences". This accounts for the failure of these studies to develop during the 19th century in spite of the existence of real problems of growing complexity, and despite the above-mentioned contributions and others as well which, even if isolated and addressing problems of management in different contexts, nevertheless already contained several important theoretical and methodological elements. A central mathematical tool had in fact been identified, the "théorie de maximis et minimis" as Coulomb called it, that is, the methods of optimization; and Babbage had identified in a very general fashion the problem of managing activities.

3. Industrial engineering and operations research

The main problem addressed by Babbage were the factories, the growing complexity of which was due also to the automating of the work and the interaction between men and machines, and which were forced to operate in an increasingly competitive environment. While the introduction of machines was the most conspicuous aspect of the ongoing transformations, there also existed a set of 'immaterial' problems linked to the industrial process with which anyone concerned with manufacturing had to cope; the question was to "arrange the whole system of his factory in such a manner, that the article he sells to the public may be produced at as small a cost as possible" [1, (1st ed. 1832) p. 86]. In the early 20th century these problems emerged dramatically also because of the drift of the manufacturing world towards mass production, of the importance of a correct management of the 'human factor' in the factories and of the growing complexity of the relations with customers, suppliers and competitors. The responses were manifold, and had variably sized quantitative/mathematical components.

The pioneers of modern industrial engineering, like Frederick W. Taylor (1911) in the United States and Henri Fayol (1916) in France, concentrated on the processes, on the organization structures and on the materials flows and chains of command. Quantification was reduced essentially to the idea of measurement, geared to the need to link up organizational action to an empirical knowledge of the true facts regarding labour and activity in the factory. However, in around 1930 the first studies were carried out on applying the new statistical methods of quality control and, starting in the late '30s, mathematical programming began to be developed. This consisted of a set of tools to be applied to various problems of managing productive activities, logistics and transport in the military and industrial fields. This marked the beginning of an extraordinary blossoming of mathematical methods of management, of allocation of resources and planning, to explain which a set of concomitant factors is needed. Development actually took place in an impelling and even disordered fashion, especially in view of the fact that in this sector an independent discovery was made, first by Leonid Kantorovic in the Soviet Union, and a few years later by George Dantzig in the United States, of basic linear programming techniques; and also inside the United States, many researchers in the postwar years discovered they were working independently on topics that were theoretically linked together.

The factors that can offer a historical explanation of this development, which still remains to be studied in detail, include a fresh interest in the idea of the mathematicization of phenomena lying outside the field of the physical sciences, that is, phenomena related to the living world, society, the economy, man. This idea, originating out of the Enlightenment period, although rejected by many, had continued to recruit convinced supporters like Léon Walras (for economics) and Vito Volterra (for biology). It is present in John von Neumann's attempt to create innovative mathematical tools for the human sciences. It underlies the birth of what could be defined a scientific research and action programme, namely, operational research. Nowadays, the term operational research refers to a properly institutionalized branch of learning, often considered as a part of applied mathematics, dealing with problems of management and organization. Originally, however, in England in the years leading up to World War II, operational research was still something that had to be developed on the basis of a two-fold conviction with an Enlightenment flavour – the importance of the role of scientists in the defence and at the service of the State, and the usefulness of the mathematical approach regarding the efficient utilization of resources. This explains why the meaning and content of operational research underwent such rapid changes after the war.

The profound changes experienced by mathematics at the turn of the 19th and in the early 20th century contributed to creating the new tools that the supporters of mathématique sociale did not possess. Walras, and even Volterra himself, who nevertheless used the most sophisticated mathematical tools with great success, had nevertheless remained anchored to the model of classical mathematical physics. The axiomatic method, the pre-eminence of algebra over geometric intuition, the development of mathematical logic, of discrete mathematics and of probability, underpin the operation carried out by Von Neumann in creating mathematical tools for economics (with game theory) or for new sectors such as computer science and the cognitive sciences (with the theory of automata). And these new developments of mathematics form also the basis of the new techniques of operational research, such as linear programming and network flows analysis, which renewed the classical tools of optimization, the new probabilistic techniques (queuing theory, inventory control) and so on. Furthermore, the application of these techniques is marked by the spread of mathematical modelling: research is focused on individual aspects of a phenomenon, and a mathematical schema is constructed which lays no claim to providing causal explanations following the classical mechanic model, but describes these aspects for the purpose of obtaining predictions or to control their behaviour. The use of mathematics thus satisfies a criterion of functionality without any general epistemological implications regarding the nature of the phenomena. The 'weakening' made it possible to bypass objections that in the past had stood in the way of achieving the old enlightenment project of constructing a mathématique sociale.

At the same time, the mathematical analogy characteristic of making models (a model is not anchored conceptually to the phenomenon actually giving rise to it, and can thus be transferred and used whenever considered useful for a phenomenon even in a completely different sphere) and the exploration of the purely mathematical equivalence of the models had a dual multiplying and unifying effect: they represented the source of many ideas and transferred results, but also made it possible to establish links among results and theories developed in apparently different contexts. This mingling later made it difficult to perform a historical reconstruction of the conceptual pathways followed, among systems engineering, management sciences and operations research, control engineering and economics. Military applications, the source of substantial investments and also of ideal motivation, initially during World War II and then during the Cold War, acted as a catalyst to this development in the United States during the second half of the 20th century. However, there was also an interaction with civilian applications, particularly as far as the design and management of production systems was concerned. The high degree of mathematicization of the new systems management technologies represented a profound change compared with the case by case approach to organizational problems that had prevailed in the past, and thus opened up a new stage in the development of industrial engineering.

4. Systems, organization, control

System, organization, organism. These terms have a dual meaning according to whether they are used in biology or in the technological and engineering fields. In early reflections on the SAGE project, dating back to 1950, an explicit use was made of a conception of systems as an organism as well as of the related concepts of function and purpose in order to describe the aspects of complexity. The biological metaphor subsequently continued to exert a considerable influence, as in conceptions of the enterprise as a living system. However in the later phases of the large-scale US defence projects, as well as in systems science in the Soviet Union, a mathematical approach to the problem of organization was to prevail. In this approach a link was again envisaged between the capacity to manage and act upon reality and the mathematical description. In the second half of the 20th century the idea of organization was given a mathematical content. However, again with the help of mathematics, organization was conceived of in a completely general fashion, as envisioned by Aleksandr Bodganov at the beginning of the century. In 1921, in a speech delivered to the First all-Russian initiative conference on the scientific organization of labour and production, Bodganov presented his ideas on the construction of a universal science of organization, tektology. These ideas are an example of the holistic tendencies that prevailed in Russian culture at the time. He actually considered organization as a "universal process at work in the organic and inorganic worlds, in production, in social interaction, and in creative activity" [9, p. 48]. In *Tecktologiya: Vseobshachaya Organizatsionnaya Nauka* (1925, 3rd ed.) (cit. in [29, p. 238]) he claimed that:

> "structural relations may be generalized to the same degree of schematic formality as the relations of magnitudes in mathematics, and on this basis organizational tasks can be resolved by methods analogous to those of mathematics. Moreover, I see quantitative relationship as a special type of structural relationship, and mathematics itself as a branch of the general organizational science which for special reasons developed earlier. This explains the gigantic practical power of mathematics as an instrument for the organization of life".

The modern mathematical formulation of organizational technologies, the heir to an old 19th century aspiration, thus introduces the need for control into this context. This idea had already been present in the early presentations of Dantzig's mathematical model for programming interdependent activities: "its purpose is the close control of an organization" [30, p. 15]. In modern engineering systems, the prescriptive approach typical of technological thinking is formulated precisely in terms of 'control', and the idea of control over the organizational components of the system represents the equivalent of the automated control technologies governing the material components of the system. From the management point of view, the complexity of such systems thus corresponds to the difficulty of controlling them, namely, the difficulty of obtaining a prescribed behaviour of a given system. This approach to complexity typical of modern technological thinking, conveyed by the strong cultural impact of the power of the large modern engineering systems, has substantially boosted the methodological role of mathematics in 20th century systems thinking.

References

[1] BABBAGE CH., *The economy of machines and manufactures*, 1832, in: "The works of Charles Babbage", vol. 8, Campbell-Kelly M. ed., New York University Press, New York 1989.

[2] BAYART D., CRÉPEL P., *Statistical control of manufacture*, in: "Companion Encyclopaedia of the History and Philosophy of the Mathematical Sciences", vol. II, I. Grattan Guinness ed., Routledge, London 1994, 1386–1391.

[3] BERTALANFFY L. VON, *General system theory*, George Braziller, New York 1968.

[4] BIGGART J., DUDLEY P., KING F. EDS., *Alexander Bogdanov and the origins of systems thinking in Russia*, Aldershot, Ashgate 1998.

[5] BRENTJES S., *Zur Herausbildung der lineare Optimierung*, in: "Ökonomie und Optimierung", Lassmann W. and Schilar H. eds., Akademie Verlag, Berlin 1985, 298–330.

[6] DE WITT D., *The shaping of automation. A historical analysis of the interaction between technology and organization 1950–1958*, Uitgeverij Verloren, Hilversum 1994.

[7] BUCKLEY W., *Sociology and modern systems theory*, Prentice-Hall, Englewood Cliffs, NJ 1967.

[8] FORD L.R., FULKERSON D.R., *Flows in networks*, Princeton University Press, Princeton, NJ 1962.

[9] GLOVELI G., *Bogdanov as Scientist and Utopian*, in: "Bodganov and his work", Biggart J., Gloveli G., Yassour A. eds., Aldershot, Ashgate 1998, 40–59.

[10] HOUNSHELL D., *The cold war at Rand, and the generation of knowledge*, Historical Studies on the Physical and Biological Sciences **27** (1997), 237–267.

[11] HUGHES TH.P., *Rescuing Prometheus*, Pantheon Books, New York 1998.

[12] HUGHES TH.P., HUGHES A. EDS., *Systems, experts, and computers: The systems approach in management and engineering, World War I and after*, MIT Press, Cambridge MA 2000.

[13] ISRAEL G., INGRAO B., *The invisible hand*, MIT Press, Cambridge MA 1990.

[14] ISRAEL G., *La mathématisation du réel. Essai sur la modélisation mathématique*, Le Seuil, Paris 1996.

[15] ISRAEL G., *The two faces of mathematical modelling: objectivism vs subjectivism, simplicity vs complexity*, in: "The application of mathematics to the sciences of nature", P. Cerrai, P. Freguglia and C. Pellegrini eds., Kluwer Academic/Plenum Publishers, New York 2002, 233–243.

[16] ISRAEL G., MILLÁN GASCA A., *Il mondo come gioco matematico. John von Neumann, scienziato del Novecento*, La Nuova Italia Scientifica, Roma 1995.

[17] ISRAEL G., MILLÁN GASCA A., *The biology of numbers*, Birkhäuser, Basel 2002.

[18] LENSTRA J.K., RINNOOY KAN A.H.G., SCHRIJVER A. EDS., *History of mathematical programming. A collection of personal reminiscences*, CWI/North Holland, Amsterdam 1991.

[19] LEVIN M.R. ED., *Cultures of control*, Harwood, Amsterdam 1999.

[20] KOOPMANS T. ET AL., *Activity analysis of production and allocation*, Yale University Press, New Haven-London 1951.

[21] KLEIN J., *Controlling gunfires, inventories, and expectations with the exponentially weighted moving average*, Mary Baldwin College, unpublished 1999.

[22] LUCERTINI M., *Facing complexity in the management of organized systems*, in: "Mathematical models and technological methods in modern engineering systems", Lucertini M., Millán Gasca A. and Nicolò F. eds., Birkhäuser, Basel, forthcoming.

[23] LUNDGREEN P., *Engineering Education in Europe and the U.S.A. (1750-1930): The Rise to Dominance of School Culture and the Engineering Professions*, Annals of Science **47** (1990), 33–75.

[24] MILLÁN GASCA A., *Organization and mathematics: a look into the prehistory of industrial engineering*, in: "Mathematical models and technological methods in modern engineering systems", Lucertini M., Millán Gasca A. and Nicolò F. eds., Birkhäuser, Basel, forthcoming.

[25] MIROWSKI P., *Economics meets operations research in mid-century*, Social Studies of Science **29** (1999), 685–718.

[26] RIDER R., *Operational research*, in: "Companion Encyclopaedia of the History and Philosophy of the Mathematical Sciences", vol. I, I. Grattan Guinness ed., Routledge, London 1994, 837–842.

[27] RIDER R., *Operations research and game theory: early connections*, in: "Toward a history of game theory", Weintraub E.R. ed., Duke University Press, Durham, NC 1992, 225–239.

[28] SIMON H., *The sciences of the artificial*, MIT Press, Cambridge, MA 1969.

[29] URMANTSEV Y., *Tektology and GST: A comparative analysis*, in: Biggart et al. 1998, 237–253.

[30] WOOD M.K., DANTZIG G., *The programming of interdependent activities: General discussion*, in: Koopmans T. et al. eds. 1951, 15–18.

THE DIGNITY OF THE NATURAL SCIENCES

Pietro Omodeo
Dipartimento di Biologia Evolutiva,
Università di Siena
rota@unisi.it

> "I do not esteem a scholar of one science more than that of another, if both work with passion and common sense. All sciences further each other."
> Montesquieu, Pensées.

Abstract In this paper the author denounces the criticism, often very harsh, levelled against the natural sciences (biology included) which are deemed to be *soft science*, while physics and mathematics are judged as *hard science* since they produce predictable outcomes.

Such criticism has emerged from the scientific epistemology expressed by Bertrand Russel and K. Popper. Russel limited himself to declaring the superiority of physics methodology which was optimally to be adopted by all scientists. He also foresaw, however, the dangers that biology would present as a prediction science and warned against any biological programmes aimed at improving the human race. Popper, instead, implicitly stated that only physics was entitled to be considered a proper science and declared that most biological principles and theories were merely erroneous conjectures.

In any case, it may be convened that the peculiar nature of naturalistic epistemology is a result of the peculiar nature of its subject matter, as was implicit in the bygone distinction between *ex act* and *natural* sciences. With reference to biology in particular, it is the universal kinship of all living beings that allows even the most advanced theories to be set forth through generalization while at the same time refusing the universal validity of any principle (those descending from logical necessity excepted). Instead, provisional principles remain indefinitely open to modifications, improvement and even rejection.

For precisely this reason any programme for human biological amelioration through selection or genetic manipulation is bound to lead to diseaster, as predicted by B. Russel and illustrated by Aldous Huxley in *The brave new world.*

1. Introduction

1. As of late, the relationship between physicists and natural scientists, never one of great mutual respect, has taken a turn for the worse. Proof of this conflict has arisen repeatedly, but became particularly alarming when a highly renowned physicist stubbornly insisted on expressing his disdain for biology and his assuredness of the superiority of his scientific methods. What is more, he even expressed these views in the daily press and on television! He maintains that "evolution must be explained in mathematical terms, just like the 'first-order' scientific theories that I deal with. It is against the precepts of Galileo's scientific enlightenment to claim that a phenomenon lacking mathematical rigour to the point that it does not possess even minimal (third-order) scientific credibility may be applied to the human species, as evolutionary fanatics insist. If Galilean science had developed with the same criteria that defend evolutionary fanatics, we would be back in the stone age" ("La Sicilia", 11.II.1999).

He went on to say that "Any theory I might hope to propose, if I am unable to express it through a single or multiple equations, must be considered totally lacking in scientific value. When I say lacking, I intend that it is devoid of even minimal (third-order) scientific credibility. Such credibility normally regards phenomena which occur only once and therefore cannot be reproduced. Nevertheless, even such unique events are expressed in rigorously mathematical terms. For example, the origin of the universe may be explained through equations, even though it is impossible to replicate. This is not true of the origin of life. No equations exist. Only theories. The same is true of the biological evolution of the human species. It lacks any expression in mathematical terms" ("La Sicilia", 7.IV.1999).

A month later he continued: "The phenomena at the basis of the origin of life and the biological evolution of the human species occur only once in history. But this research fails to achieve even third-order credibility, since – as I stated earlier – only formulas, not theories, belong to the realm of Galilean science. Even today, this is the greatest achievement of reason over immanentism. One thing is certain: if I had not assumed the intellectual humility and mathematical rigour of Galilean science, during my forty years of research I would never have discovered antimatter, or the actual energy in interactions between quarks and gluons, or that protons resist breakage, nor would I have been able to invent the technology necessary for discovering the Third Column in the structure of Creation or understand the origin of the quark mixture in mesonic matter. These were all fundamental steps with first-order scientific credibility leading up to the unification of the Three Forces and the Three Columns at the outer limits of the universe.

Physicists need no help from biologists to formulate their theories".

We might simply shrug off such considerations as simply the wild musing of some harmless extremist. Unfortunately this is not the case, since Professor Antonino Zichichi holds a highly-reputed teaching post at the University of Bologna, has acted as president of the Italian Institute of Nuclear Physics, controls a vast sum of research funds, enjoys significant political support and is currently the president of the World Scientists' Federation, from which he might be expected to expel all natural scientists for their lack of Galilean methodology and general unworthiness.

Such a petty conflict may appear to originate from a deprived, small-town mentality, even if the anglo-saxon culture, with its term "soft science" has shown itself to be equally vulnerable. Just observe the reactions of "hard" scientists when addressing the realm of what they consider to be their "soft", incompetent counterparts. F. Hoyle is fierce and relentless, while S.A. Kauffman is eloquently indignated throughout the thousands of pages in which he deplores the inaccuracy of biologists who fail to make use of mathematics in their work.

2. Unfortunately the twentieth century witnessed a blurring of the distinction between the "exact" sciences and the "natural" sciences which had been blithely accepted as fact during the previous two hundred years. It was understood that "exact" scientists were actually involved with legitimate abstractions which as such were capable of producing mathematically exact results, while natural scientists worked with magmatic matter, which was much more unwieldy. This distinction did not imply that natural science was less dignified or that physicians and mathematicians were more noble scientists. On the contrary, many scholars such as Descartes, Johann Bernoulli, Alfonso Borelli, de Maupertuis, Galvani and Helmholz explored both realms. However, the Vienna Circle later coined the terms "standard science" and "nonstandard science" which implied greater and lesser worth (or to be more benevolent, greater or lesser "maturity"). With time a certain bad taste facilitated an even greater demarcation, with the appearance of the terms "hard science" and "soft science", in which the former is practiced by physicists while the latter is practiced by naturalists, geologists, ethologists, biologists, meteorologists and possibly even by astronomers. Geographers are excluded from both fields since they are not considered scientists at all.

The historical distinction was abandoned but its modern counterpart, while supported by few physicists[1], is dangerous for two reasons. First of all, it threatens to create a schism within the scientific community, just as the schism between humanism and science is starting to heal. Secondly, many biologists have been pressured through criticism and allurement into accepting this new creed, and thus desperately declare that their science is also one of predictable outcomes, and that as such they aim to improve plants, animals and even man himself through entirely predictably methodology. In this way they propose what Aldous Huxley had foreseen fifty years earlier: "The true revolution will not effect the outside world, but rather the hearts and minds of human beings". Such ambitions are not only dangerous, but also ridiculous.

2. Who is responsible for our current outlook on science?

3. At this point we must attempt to understand who is responsible for this outlook and its origin. Undoubtedly its most extraordinary, chaotic and contradictory contribution was made by Bertrand Russell. The most popular work by this brilliant, combative figure is *The Scientific Outlook* (1931) in which he blends intellectual audacity with paradox and a wealth of undue approximation.

"The scientific method as we know it was integrally formulated by Galileo, and to a lesser extent by Kepler [...] they performed research on particular phenomena in order to draw up quantitatively exact laws which could be used to predict other phenomena". This is what we read on page 16, and we are tempted to reply: the Galilean revolution may only be appreciated within the context of the new methodology imposed by the Renaissance naturalists, with their need for precision and accuracy[2]. Keplero would never have been able to formulate his laws if he hadn't had Tycho Brahe's precise observations and measurements of planet movement as his disposal.

On the heart of the matter, the *Scientific outlook* affirms: "textbooks make use of deductive methods while laboratories use inductive methods, and the only science that has thus far reached this perfection is physics" (p. 40). He then goes on "Thanks to the simplicity of its content, physics has reached a higher level of development than any other science. I consider beyond doubt the fact that all sciences aspire to this ideal" and adds: "In its highest form, science consists of a series of hierarchically

[1]See [3] and [1]. Marcello Cini also expresses his disagreement in various works.
[2]See [10], chapter XII. For a competent, balanced view of Galileo, see [4].

arranged theories in which the lowest positions are held by particular phenomena and the highest positions are held by general laws that apply to the entire universe" (p. 56).

The hierarchy of scientific methods suggested by Windelband immediately comes to mind, as does the routine sermon preaching the natural hierarchy of sciences with physics at the lead. We soon realise, however, that Russell's intent is different and that he presumes the existence of "a double logical relationship, with one directed upward and the other downward; the upward relationship proceeds via induction, while the downward relationship proceeds via deduction" (p. 39). As the author observes, this method applies to all of the sciences, and certain procedures it implies deriving from physics are not necessarily obbligatory: "Mathematics is a powerful tool, and scientists are naturally eager to use it [...] but a law may be perfectly scientific without being quantitative. Pavlov's laws regarding conditioning may act as an example [...]. Therefore, when dealing with a subject such as animal behaviour, for the moment we are obliged to accept qualitative laws which are no less scientific than if they were quantitative" and adds: "One advantage offered by quantitative precision when it may be attained is that it strengthens inductive argumentation" (pp. 44–45). This clarification is precious because, in contrast with the demands currently made by many, it recognises the pre-eminence of science's ability to supply knowledge over its ability to predict. In fact, the predictive role of the basic sciences, which may be associated with applied science and technological progress, is discussed with rare acumen in the second section, which also underlines the contribution of biology to man's well-being. In the third section Russell foresees a future in which biology and sociology provide the means for "scientifically improving" mankind and society. Russell's predictions regarding the use of biology and sociology to plan the future have no mathematical context; he maintains that only diseasters may be predicted on the basis of historical experience.

These predictions were immediately embraced by a number of authors including Aldous Huxley, who made use of them in his *Brave New World.*

4. A different and less encouraging scientific outlook is presented by young Popper in *Logic der Forschung* (1934). In this work Popper fights against the use of inductive reasoning in science and maintains that a true scientist must always begin with a theory, possibly substantiating it through observation, and mathematically formulate the conclusions he has reached through deductive means.

This premise, supported by circumstantial argumentation, stirred an uproar and brought about many effective protests[3], so there is no need to address the subject to add another voice to the choir. Further on, however, it will be useful to consider the so-called "secular" consequences of this thesis. For the moment, we should note that Popper, instead of identifying physics as the science to be emulated by the others as Russell does, instead takes physics to represent science in general and thus considers it exclusively. He never displays interest or understanding of the problems concerning the natural sciences (chemistry included) for example, nor does he cite the opinions of naturalists[4], be they right or wrong. The natural sciences are not to be found in any of the appendixes repeatedly added to the text, nor even in *Science and Philosophy*, a work that contains five essays by Popper published between 1956 and 1967 with the promising subtitle *Problems and Objectives of Science.*

Although it must be recognised that any field chosen by a scientist is legitimate, it is less justifiable to devote an entire career to convincing trusting readers that physics is the only worthy science. Simply due to ignorance of the other sciences.

Popper encountered his first conflict over biology in about 1947, thirteen years after his virulent approach with epistemology, when he discussed with Schrödinger the ideas presented by the latter in *What is Life?* (1946). Popper limited himself to identifying a weak point in the theories of his fellow colleague and citizen[5]. In 1952 he reviewed a volume on the origins and boundaries of the humanistic disciplines and the natural sciences. He found the work unconvincing, but finally was given the opportunity to express the underlying message of all his previous works: "I believe that natural science – in contrast with pure mathematics – is not true *scientia* or *episteme*, not because it is *techne*, but because it belongs to *doxa*"[6]. Translating from his technical terminology, this means that he considered the natural sciences to present erroneous conjectures or prejudices rather than science.

It is surprising that he was able to reach such a precise and definitive conclusion without ever having examined or discussed any natural science theory or principle.

In 1955 however he was forced to consider the theory of biological evolution within the context of his distinction among science/pseudo-science/metaphysics. He concluded that, since the theory could not be

[3] On this subject, see [12, 5, 6]; also see [7].

[4] He briefly mentions Linneo and Pasteur and in another note ironically cites J.B.S. Haldane, having encountered an article by him published in "Nature".

[5] This anecdote is told in *La ricerca non ha mai fine*, §30.

[6] Citation from *Conjectures and refutations*, 1969, final chapter.

entirely disproven, it was to be considered within the realm of metaphysics[7]. According to the author, this did not imply a definitive relegation, since metaphysical speculation could be elevated to the level of true science through continued research.

In 1961 Popper participated in a conference entitled "Evolution and the Tree of Knowledge", during which he haughtily offered superfluous advice on how biologists should carry out their work and his own personal theory of evolution, which his followers had discouraged him from publishing. During this presentation he confused the biological selection of breeders with the interaction that takes place between any activity and its milieu, and thus attempted to explain all human behaviour through evolutionary paradigms in which he recognised a dualism corresponding to Descarte's body/soul dualism....

His final encounter with biology took place years later, when Popper again met the Australian neurophysicist Eccles. That meeting led to a big joint project[8] in which Popper avoided questions of biological epistemology and limited himself to discussing matters concerning perception. The collaboration between the two great scholars was not particularly productive: the eagerly-awaited research by Eccles on the workings of the brain were poorly contextualized, while his mystical forays were interpreted within the theosophic (or anthroposophic) view of the "three worlds" so dear to Free-mason esoterism.

5. Before proceeding any further, it would be wise to compare these two thinkers who have acted as the prime contributors to our current view of scientific epistemology. Although they began from entirely different premises and reached different conclusions, their effects were much the same.

Russell loved teaching by means of paradoxes, a method which is effective because it challenges pupils to compete with their master and to apply themselves in searching for elements that may have been ignored or distorted in order to formulate ideas at odds with intuition or common knowledge. The limitation of paradoxical teaching is that its success depends on the master's agreement when a pupil identifies the weak point in a theory or at least on admission that the paradox was created to explore new solutions. If the master remains unmoved from his original statement or the pupil passively accepts the paradox as an intangible truth, the result is failure. It is also understood that such an approach

[7]In: *The demarcation between science and metaphysics*, 1955. In reality, Popper did not develop this belief, and in 1978 "Nature" published a short article in which he recognised the scientific quality of evolutionism. On this subject, see also [2].

[8]K. Popper, J. Eccles, *The self and its brain*, 1977.

may only be effective if there is direct dialogue or if the paradox is presented within the context of a challenge.

Russell presented his paradoxes in his writings, sometimes as a challenge, and sometimes with nonchalance, but was always ready to abandon his extreme positions. As such his books are enlivened with a playful sense of variety that makes them both enjoyable and worthy of meditation.

Popper used precisely the opposite technique: he proposed rules for theorising and taught how to expand one's knowledge, but with the indifference of a bureaucrat disinterested in his jurisdiction, declares: "these are the rules: without them philosophy and science are impossible".

Whatever their intentions, both authors have negatively influenced the epistemology of the natural sciences. For example, some of Russell's followers adopted the theories proposed by their master that they judged most convenient and most suited to their personal ends, without wondering whether they were proposed as paradoxes. As for Popper's multitude of devotees, out of hypocrisy or intellectual sluggishness some imitated their master's denial of the value of the natural sciences, along with its results and errors.

This incomprehension has reached a crisis point that requires urgent intervention.

3. How can reciprocal understanding be achieved?

6. It is commonly agreed that the best way to resolve intellectual controversy is through historical investigation, and this is therefore the key I will use to explain how the foremost exponents of the natural sciences worked in the past and today, with emphasis on biology. However, I first consider it necessary to seek common ground with reference to the value of direct observation, which Popper considered useless, and of discovery.

It is clear that observation is the basis for human knowledge and is also necessary even for the most humble of functions. However, philosophy has always understandably tended to emphasize intellectual developments rather than their underlying acquired knowledge. Therefore the concept of 'scientific discovery' is perceived to require the application of logical or algebraic procedures in order to be fully understood, or to involve questions regarding the inherent structure of matter, and so on. As such, certain philosophers consider geographical discoveries, the discovery of a powerful drug or the discovery of a new plant or animal species to be nothing more than acknowledgement of isolated facts, which thus lack scientific relevance.

In science the situation is considerably different. Many physicists consider the discovery of a supernova or a planet in another solar system to be an exceptional event. This cannot be denied: the discovery of sunspots, the 'Medici planets' and lunar mountains brought on the demise of Ptolemaic astronomy and changed our perception of the world. But at the same time, numerous physiscists and mathematicians deny the scientific value of the discovery of new continents on our planet and even maintain – Russell first and foremost – that geography is not a science at all.

This wide range of opinions has led to conflicting interpretations. The use of a network of orthogonal coordinates in geographical maps, such as the projection invented by Mercator, makes it possible to easily identify single points on the earth's surface. Since such a projection is also isogonic, it also enables us to calculate the shortest route between two ports.

Fifty years later, Descartes used the same orthogonal coordinates to formulate the equations that express the aforesaid points and lines. So why is Mercator's method considered a simple technical device, while the history of science dedicates considerable attention to Descartes' simple algebraic formulas that produce the same results? Could it be because the points dealt with by Descartes lack real dimensions? Or because the routes traced by a navigator represent mere segments while Descartes' lines stretch 'from the plus infinite to minus infinite'? Yet, on a map the two lines appear to be identical.

The only way to resolve such schizophrenia is through consensus, and there are few logical solutions. I believe we must also recognise discoveries as scientific knowledge, especially if they are identified with an aim and serve as a potential basis for a conceptual structure, that is, a scientific theory.

4. The methods used by naturalists and the predictive power of the natural sciences

7. Let us now discuss how naturalists work. Since Renaissance times they have collected homogeneous data, described them and then formulated generalisations involving as-yet unknown items and phenomena. Following some initial hesitation, expressed by the great scholar K. Gesner [10, chapter 3], they renounced formulating universal truths, instead advancing affirmations left open to subsequent verification and formulating principles without any premise of definitiveness.

Harvey reminds us that the audacity underlying eventual generalisations was borne of faith in nature itself, which proceeds with coherent patterns and tends to use constant means to achieve constant ends, as Aristotle himself taught [8, chapter 1, p. 4]. It was also borne of the vague, commonly-held notion that, since all living beings are in some way related, they must all be similarly equipped, since they are all born, nourished, grow, reproduce and eventually die.

As we know, faith was placed in the practice of generalisation by F. Bacon, who recongnised it as methodologically sound; such a procedure subsequently led to the formulation of countless principles in the fields of medicine, zoology and botany.

Certain realms of biology never go beyond the descriptive stage. This is due to contingent causes or the fact that these days it is simply impossible to proceed further, although in most cases some general principles are formulated. These principles might take the form of axioms or assumptions made in order to interpret and associate observed data or summary intuitions. They may be expressed as aphorisms (frequently used in the 1600s and 1700s), propositions or theorems, and they are sometimes tacitly accepted without explanation in treatises, given their obvious nature. As time passes and terminology is coined, these principles are often expressed in new terms, remaining somewhat transitory for even extended periods, in spite of the fact that they deal with fundamental aspects of life itself. They nonetheless serve as indispensable instruments in research and teaching.

Biological principles, through evolution and enrichment thanks to ongoing research, eventually acquire universal value.

The principles underlying the origin of living organisms represent a particularly paradigmatic case in this sense.

Approximately mid-way into the 1600s, Francesco Redi affirmed and demonstrated, in contrast with the ancient belief widely held at the time, that decaying matter does not give rise to vermin or insects, which instead hatch from eggs (*generatio ex putri non datur*). About a decade later, Malpighi's acute observations extended this concept to parasites, and he concluded that they too hatch from eggs and are not generated by decaying plant or animal fluids (*generatio aequivoca non datur*). In approximately 1750 Carlo Linneo generalised this theory by stating that all living beings are generated from eggs (*omne vivum ex ovo*). But this supposition proved inadequate when a better understanding of unicellular organisms was attained (since they produce no eggs) and when it was discovered that all living organisms are composed of one or more cells. Thus a century later R. Virchow and R. Remak formulated the aphorism *omnis cellula e cellula* (all cells are generated by cells). But later even this aphorism proved unsatisfactory and was thus replaced by *omne vivum e vivo* (all living organisms are generated by living organisms).

Even in this form, after 250 years of observation and research, this principle has been accepted with reservations, and for the duration of the twentieth century, and indeed even today, biologists and biophysicists have sought to reconstruct the process that generated the first living cells from organic non-living matter.

In any case, each step in this ongoing intellectual exercise has brought about real progress in the fields in which each affirmation has been proved experimentally valid, from protection against damage by insects and saprophytes to the prevention and cure of parasitic disease, the sterilisation and conservation of tinned foods (N. Appert, 1795), the pastorization of drinks and liquid foods (L. Pasteur, 1864) and asepsis and antisepsis in surgical practice (J. Lister, 1867).

8. The fact that the origins of biology lie in pure empiricism, that it has progressed by means of successive approximations and that its very principles are of a temporary nature have not won it favour with modern physics epistemologists, as we have seen. But this does not diminish the many noteworthy successes it has enjoyed, especially during the second half of the 1800s and again of the 1900s.

Most criticism has been unjust, and certain scholars of biological logic and theory have rightly protested. It must be noted that even in its simplicity, biological observation has profoundly influenced human culture (an example which equally applies to other fields of knowledge), while through association, its principles governing living organisms in general have led to discoveries of immense worth. One remarkable case regards precisely the definition of living organisms, which still has not been satisfactorily formulated (see [11]). Another more renowned case concerns the theory of evolution, which following a century of debate has come to occupy a central role in modern scientific culture. I will not recount these vicissitudes in detail (see [9]), but would simply like to emphasize that the current version of evolutionary theory is so solid and universal that it may be compared to the theories typically employed by physicists and technologists. Nevertheless it has been attacked on the basis of the presumed existence of decisive exceptions to its application. These are not actual exceptions however, but simply cases of erroneous formal application of some principle (such as that regarding genetic polymorphism, for example) or of insufficient understanding of mutations and the mechanisms employed by selection agents. Much remains to be learned on these matters, but this in no way weakens the overall structure of the theory.

9. Another argument regularly employed to stigmatize the low standards of biological research makes reference to its lack of predictive value, which is attributed to the inadequate application of mathematical mod-

els. This observation, repeated with obsessive obstinance, is totally unfounded. There are numerous cases in the fields of physiology, ethology and genetics in which a subject isolated from its environmental conditions may offer opportunities to make precise calculations and satisfactory predictions. This is what makes medical and veterinary prognosis feasible. This claim is particularly applicable to the mechanics of movement, the functioning of the sensory organs and biochemical and biocybernetic mechanisms. Basically, whenever experimentation may be performed under stable conditions and a suitable mathematical model may be found, it is applied.

In contrast, calculations and monitoring are made impossible in other fields by the combined effect of external factors, such as with the inheritance of one or more polygenic characteristics. Some morphogenetic phenomena are even influenced by seemingly inconsequential erratic factors that lead to unique patterns such as those visible in zebra stripes and finger and palm prints. Nonetheless, even in such cases we are able to produce reliable basic models that may be usefully described and classified.

In other cases, and the majority at that, predictability applies to such a short term since certain measurements employed evolve over time. In this limitation biology differs from astronomy and astrophysics only in scale: while a geneticist refrains from predicting the characteristics of a population beyond more that ten generations due to the great number of variables involved, no planetologist would dare to predict the exact time and place of a solar eclipse a million years in the future, since the duration of solar days and the distance from the moon vary, as does the diameter of the sun. Just as no botanist would ever precisely predict the date of extinction of a species he considers to be inevitable, no astrophysicist would set an exact date for observing the explosion of a supernova, even if he is assured of its imminence.

In spite of the obvious weakness of the allegations brought against biological research methodology, certain biologists have become convinced of their presumed inferiority and have invented mathematical and physical models with the aim of improving standards. General application of these models has been counter-productive, however, and it is becoming clear that aspiration to the absolute truths proposed by a handful of practitioners threatens to do irreparable damage.

Fortunately, recent research into the notion of complex systems and chaos theory has abated past polemics and steered most of the scientific community toward less intransigent positions.

5. The universal family of living organisms, generalisations and the comparative method

10. Let us now consider the object of biology: the living organisms whose peculiarities necessarily influence biological methodology. As has been apparent since the remote past, all living organisms demonstrate fundamental common characteristics: they are nourished, grow, reproduce and eventually die. As regards animals in particular, we may add that they are sensitive, responsive to stimuli, capable of movement and divided into genders.

This led to the belief in a sort of parenthood of animal species, and even in the possible parenthood of all living species. This ancient belief was almost definitively confirmed only in the second half of the 1900s by molecular biology research, but had already been encouraging generalisation and the formulation of inductive theories in biology for over two millennia. These theories made it possible to outline a general framework for the science of living beings.

At first these theories were accepted to be universal and considered useful in formulating syllogisms and making accurate predictions. However, during the Renaissance it became evident that their usefulness was limited by their inevitably transitory aspects. Over time it was increasingly clear that the transitoriness of these theories was precisely what made the method so valid and that generalisation was more fruitful and productive when it involved the basic characteristics of living organisms, and especially of cells.

For example, when the genetic code was deciphered and the pattern of correspondence between RNA, the twenty protein amino acids and the stop signals was mapped out, it was immediately affirmed and accepted that this pattern was the same in all living beings. And yet, experimentation had only been performed on less than one millionth of all possible cases. In fact, exceptions later emerged involving ciliated protozoans and mitochondria (organelles in eukaryotic cells) but the theory was not invalidated. It was simply added that at some point in evolutionary history these improbable mutations had been perpetuated.

Furthermore, when electronic microscopy and biochemical research confirmed that certain cells are surrounded by a cellular (or plasmatic) membrane consisting of a double lipidic layer with polarised groups facing either outside or toward the centre of the cell itself, the principle that such a membrane surrounds all cells, including those of microorganisms, fungi, plants and animals was readily accepted. Research had been done on only a tiny portion of all possible cases, but the theory was immediately accepted and biophysicists proceeded to carry out excellent research on just such cell membranes. Only once bacteria that develop and reproduce at a temperature of 100°C were discovered did it become possible to observe that their cell membranes are different and that this generalisation was not universally valid. This had been forseeable in light of certain physical and chemical considerations, but what had not been anticipated was that some bacteria flourish at such extreme temperatures. In any case the theory continues to be respected, and at most an appendix is added with little real relevance for theoreticians.

Descriptions of eukaryotic cells have affirmed that they are furnished with special organelles - mitochondria - which produce adenotriphosphate (ATP) or equivalent molecules by means of respiration. It was later confirmed that certain ameobae in asphyctic environments and certain endoparasitic organisms dwelling inside host cells actually lack these indispensable organelles. Nevertheless the generalisation remains valid, with the added clarification that cells which have evolved to survive in environments devoid of oxygen may lose their mitochondria and that parasitic cells survive ably without mitochondria when ATP is supplied by the host cell.

Examples of faulty generalisations abound, especially if one goes back in time: all mammals were said to bear live offspring until monotremes were found to lay eggs; all birds were said to brood over their eggs until the megapodids of New Guinea were found to let them incubate under decaying plant matter. Other generalisations have been repeatedly confirmed over time, such as the fact that all mammals nourish their offspring with milk and that their red blood cells are devoid of a nucleus.

11. *The parenthood of all living organisms on this planet* justifies the application of the comparative method as originally adopted in Aristotelian philosophy and philology. Common structures bear witness to a shared geneology, while individual differences are proof of the subsequent occurance and persistance of variations. This holds true at all levels of biological research, whether regarding the morphology of spermatozoa or the nervous system of vertebrates. Cladistics is a discipline which uses (sometimes less than satisfactory) algebraic calculations to construct geneological trees based on whether the ancestral characteristics (plesiomorphies) and derived characteristics (apomorphies) of a group of taxa are shared or exclusive. The comparative method is indispensable in phylogenic and systemic research and is applied uniquely in biology and philology.

Lastly, this method has been of great use in resolving numerous doubts in physiology and ethology. In fact, it has facilitated the identification of the essential elements which always exist when studying organs and complex behaviour, distinguishing these from the secondary elements that disappear or evolve as the living conditions of the species vary.

6. The importance of single events

12. Shortly before the second World War, Zimmer and Delbrück, two German physicists working with the Soviet geneticist Timofeef-Ressowsky, planned a research programme aimed at calculating the dimensions of a gene. They perceived this particle as a potential target for high-energy photons whose penetration power and mutagenic potential were already known. Each successful collision was revealed by the appearance of a mutation in the descendents of the bombarded subject. The density and scattering of the projectiles launched, together with the frequency of successful collisions calculated according to the frequency of the resulting mutations, were applied in balistic equations to calculate the dimensions of the target and therefore of the gene itself.

Not many years later, knowledge obtained through the wartime use of atomic weapons revealed that the results which had been achieved were correct, but only by chance. This is because photons do not affect the target directly, but rather by means of the superoxides and peroxides produced during the reaction.

Even if the 'target theory' failed to offer definitive results, I gladly mention it for two reasons: because it shows how productive collaboration among scientists from different cultures can be (Delbrück was to become a renowned geneticist); and because, as Erwin Schrödinger demonstrated, it underlines an important aspect of all living organisms. That is, while numerous principles of physics are substantiated by the fact that their phenomena involve huge quantities of particles that statistically define a trend, in biology the situation is different, since reproduction implicates that *important biological processes may be triggered by a single event.* For example, the mutation caused by a single high-energy photon reveals a submicroscopic event at a macroscopic level through a chain of biochemical and morphogenetic processes. The same may be said of the "founding effect" and subsequent "genetic drift": the offspring of a single or small number of individuals, due to chance sorting, display certain characteristics that distinguish them from the original forebears.

The importance of single, often unrepeatable events above all justifies the historical aspect of biology. This is equally true at the level of molecular biology, a science which would appear to be governed by deterministic laws.

7. Other basic aspects of living organisms and their methodological implications

13. Since ancient times we have attempted to understand the makeup of living beings according to the principles of physics and chemistry. However, since the basic tenets of these sciences may be expressed as rudimentary patterns, the attempts have been less than satisfactory. Only the wildest of imaginations could claim to be able to interpret the phenomenon of life through the four principles of Medieval physics (heat, cold, dryness and humidity) or the four elements of alchemy (earth, water, air and fire).

Nevertheless, back then students already had clear ideas regarding the fields of mechanics and hydraulics, and these sciences had progressed along with the methodology employed in constructing buildings and boats, acqueducts and fountains, winches and wheels. Thus it comes as no surprise that since antiquity, but especially during the Renaissance, many naturalists and physicians offered mechanical solutions for the questions that puzzled them. For example, Cardano and Rondelet were able to use hydrodynamic considerations to correctly explain why female dolphins and seals, which produce milk for their young, do not have protruding breasts, just as their male counterparts bear their testicles internally. Gesner, while inspecting an image of a sea monster with prominent ears, was prompted to liquidate it as an invention. During that same period, certain questions regarding the dynamics of flight and blood circulation were resolved correctly and satisfactorily.

It thus follows logically that from Descartes onwards, animals and even man came to be considered as machines in the sense that mechanical principles could be usefully applied to them and certain aspects of their behaviour could be compared to the workings of a machine, although this did not mean that living beings were no more than machines.

Defining exactly what a machine is even presents certain difficulties, just as with defining what a living being is. In any case, a machine must possess the ability to work, otherwise it should not be called a machine, just as a living being must carry out certain functions in order to be known as such.

It should also be noted that *all living beings are endowed with multiple, refined mechanisms of operation and self-monitoring, and possess a clear structure at all levels of observation: macroscopic, microscopic, submicroscopic, molecular and submolecular.* As the resolution with which we are able to observe cells increases, we discover ever-more-minute functional structures. At the microscopic level we may observe chloroplasts and mitochondria, veritable cellular generators, and golgiosomes, assembling devices for polysaccharides and conjugated proteins. At the submicroscopic level we observe ribosomes, where polypeptide chains are constructed, and the granules of neurotransmitters. At the molecular level we observe an endless series of enzymes, transmitters of oxygen. At the submolecular level we observe the clefts of enzymes that recognise and retain substrates, allosteric sites, the coenzymes that supply the right amount of energy to the right destination, the hinge in the hemoglobin tetramerous molecule that assures the Bohr effect, and so on.

14. This minute structural organisation in living organisms which all specialists are familiar with is extremely relevant from an epistemologic perspective, but it has not been adequately investigated. This may be a distant consequence of one of the greatest illusions in the history of biology, which may be usefully recalled and highlighted in the hope of limiting damage and preventing the repetition of similar events in the future, rather than in order to attribute fault. At the beginning of the 1900s it was believed that 'protoplasma', which was considered to be 'living matter', had a colloidal structure, and that therefore it was necessary to reveal the properties of colloids before its secrets could be understood. Back then this simple assumption was revered as an oracle by biologists and chemists the world over, and even by scientists in other fields such as Bertrand Russell. At least two journals were founded to exclusively explore matters of colloidal chemistry.

As research on the substructure of enzymes progressed, it became clear that no 'living matter' actually existed, and that the attribute of 'living' depended on aspects of structural organisation and corresponding self-controlling mechanisms. This illusion faded away in silence and dozens of works were left to collect dust on library shelves. Unfortunately, not everyone became aware of this radical development, and efforts were made to substitute the mythical properties of colloids with the even simpler, mythical properties of 'dissipating structures'.

15. Mention has been made of the fact that every living organism is endowed with various self-monitoring devices. These have been investigated by physiologists and biophysicists for over a century. At the physiological and macroscopic level, these control devices regulate the flow of energy, material and information. In animals such flows are governed by the nervous system, while in plants they are governed by mechanical or physico-chemical devices not unlike those observed in technology preceding the electronic age. In the isolated, self-sufficient cells of microorganisms just as in those of pluricellular organisms, the self-controlling devices are regulated by a rapid, precise molecular apparatus.

Self-monitoring guarantees the system's reliability and therefore its resistance over time in a variety of environmental situations. If these self-monitoring mechanisms go awry, the organism begins to die. In highly evolved animals it has long been clear that death coincides with an interruption of breathing, heart beat and reaction to stimuli (such as today's encephalograms). In microorganisms death is more difficult to diagnose, but when the enzymes regulating cell division and DNA duplication and repair malfunction, the result is genetic death – technically, an interruption of reproduction.

This quick overview of a phenomenon which has recently received less attention than it is due should emphasize that *self-monitoring is a necessary requirement for all living organisms* and that it has no known or conceivable equivalent in the realm of imaginable physics. It also clarifies how this basic characteristic of living organisms is best studied through models used in cybernetics and advanced technologies.

16. Living organisms display characteristics not to be found in the subjects investigated in other branches of science. We may presume that these characteristics appeared during various stages of their evolution. Reproduction, for example, takes place in even the most elementary forms of life, as do many of the self-monitoring mechanisms discussed earlier. Mobility appears at more advanced stages, when eukaryotic organisms generate a system of contractile proteins. Behavioural memory is the result of the development of a neuronal network in animals, while immunological memory develops later, with the ability to produce antibodies, and so on.

The acquisition of new characteristics through evolutionary events is known as 'emergentism'.

The emergence of new characteristics in living organisms has repeatedly been contested. The first critic of this theory I have found was Tommaso Campanella, who polemicised against the implications of Epicurean and Lucretian evolutionism. De Maupertuis was better able to support his criticism, but found himself obliged to confer certain properties that naturalists attribute exclusively to living organisms (such as aversion and memory) even to atoms. More recent disbelievers have included Thomas A. Edison and Bernhard Rensch.

The body of contestation against emergentism remains inconsistent, however, since it tends to attribute certain properties to particles governed by laws of physics and atoms governed by laws of chemistry which these particles and atoms never actually display. Furthermore, emergentism provides precious interprative tools for philogenetic and morphogenetic research as well as for classification in general, in addition to an aspect of biology that we might describe as evolution through complexity.

8. Conclusion

17. It is beyond doubt that in biology, just as in all other natural sciences, it would be inconceivable to formulate interpretations or theories that conflict with the laws of physics or chemistry. It is equally true that each new development in the fields of physics, chemistry, technology and even certain branches of mathematics has promoted parallel developments in pure or applied biology. It is furthermore doubtless that the formidable progress made in the field of physiology during the second half of the 1800s was made possible by the developments in chemistry and thermodynamics attained in large part by scholars of biology. The refinement of optical and electronic microscopy contributed decisively to progress in the fields of microbiology, histology and cytology. The refinement of X-ray crystallography and defractometry made it possible to reveal the structure of DNA, with all the consequences this involved.

These premises, taken together with the pretense of interpreting the entire realm of biological phenomena on the basis of physical and chemical interpretations, have led to the recovery of two bygone hypotheses. The first, attributed to Laplace, states that "if knowledge is possessed of the position and trajectory of each particle of matter in the universe at a given moment, it is possible to reconstruct all past and future events." This ultramechanistic thesis, though refuted by present-day physics theories, made a comeback during the second half of the 1900s when worded as such: "once the chemical and physical properties of all the elements comprising living matter have been revealed, it will be possible to deduce even the most minute details of all biological and psychological phenomena". This is the essence of absolute biological determinism, known as reductionism.

The second thesis which was recovered and logically connected and subordinated to the first, states that: "no characteristic is exclusive of living organisms, since they are all already inherent to some form of inorganic matter (and the particles composing it)". This thesis is labelled 'antiemergentist' since it denies that traits may emerge as a result of biological or individual evolution. An obvious implication of antiemergentism is that in biology as in psychology, any historical or simply probabilistic interpretation is made in vain.

The truth is that biology is a science of vast proportions and as such, at the chemical, biochemical and mechanical levels any hypothesis must be formulated in 'mechanistic' terms, which are actually deterministic.

But as the level progressively increases to comprise structures and functions of growing complexity, it is no longer possible to hypothesize in deterministic terms. Not only, but once comparative questions such as: "How does this joint differ in man from its equivalent in chimpanzees?" or "Why is the hemoglobin of a horse different from that of a donkey?" are considered, then the answer necessarily implies evolutionary and therefore historical considerations, even when investigating purely mechanical or biochemical matters.

Prediction is possible in historical terms, but as Russell taught, it is necessarily always negative in nature, such as in: "All wars bring harm upon not only losers but also winners". Evolutionary prediction furnishes similar conjectures, two of which should be kept in mind by those hoping to plan the biological future of humanity: "Systematic abuse of the natural wealth of this planet can only lead to catastrophe" and furthermore, "Programmed modification of the genetic makeup of man, given the unknown future that awaits us, is madness that can only produce unthinkable horror".

References

[1] AGENO M., *Le radici della Biologia*, Feltrinelli, Milano 1986.

[2] ANTISERI D., *Teoria della razionalità e scienze sociali*, Borla, Roma 1989.

[3] ANTONIETTI M., *Tra armonia e conflitto da Kepler a Kauffman*, in: "La matematizzazione della biologia", P. Cerrai and P. Freguglia eds., Quattro Venti, Urbino 1999.

[4] BELLONE E., *Galilei e la scienza*, in: "Quaderno filosofico" 10-11, 1984.

[5] FESTA R., *Induzione, probabilità e verisimilitudine*, in: "Introduzione alla filosofia della scienza", Giulio Giorello ed., Bompiani, Milano 1994.

[6] GIANNOLI G.I., GIANQUINTO A., *Introduzione alle metodologie della scienza*, Bagatto Libri, Roma 1992.

[7] MAYR E., *Il modello biologico*, McGraw Hill, Milano - New York 1999.

[8] OMODEO P., *Creazionismo ed evoluzionismo*, Laterza, Bari 1984.

[9] OMODEO P., *Fondamenti della teoria dell'evoluzione*, in: "La nuova critica" 75-76, 1985.

[10] OMODEO P., *Alle origini delle scienze naturali*, Rubbettino, Cosenza 2001.

[11] RIZZOTTI M., *Defining life, the central problem in theoretical biology*, University of Padova, Padova 1996.

[12] ROSSI P., *I ragni e le formiche, un'apologia della storia della scienza*, Il Mulino, Bologna 1986.

HOLISM: SOME HISTORICAL ASPECTS

Silvana Procacci
Dipartimento di Scienze filosofiche,
Università degli Studi di Perugia,
Via dell'Aquilone 8, 06123 PERUGIA - ITALY
procacci_s@yahoo.com, ipazia@unipg.it

Abstract A favourite thesis of the holistic and organismic biology is that the parts of organic wholes exhibit patterns of behaviour that do not show and are not predictable outside these wholes. The main task of holistic and organicistic movement is to defend the autonomy of biological science in comparison with the sciences of the inorganic phenomena. The second meaning of Holism is therefore to isolate a sense in which we can demonstrate a causal power of the whole on to the parts. Another important point is the active role of the organism in the adaptive process. Through the analysis of the conception of one of the pioneers of holistic doctrine, J.C. Smuts, we can understand the importance of Holism for the history of biology.

1. Introduction

During the first years of the twentieth century, the erosion of the classical mechanical paradigm started with Quantum Mechanics and Relativistic Theory has opened again the ancient debate between Holism and Reductionism. Near to the new frontiers of the physics, the sphere of the complex arises. Before physicists, in the early years of the twentieth century self-organisation and pattern formation were studied only by biologists, and only very recently they are becoming appreciated by physicists and chemists. Partly this is because of longstanding emphasis that physicists have given to linear systems: on the contrary, self-organisation is essentially non-linear in nature (see [6]).

Many philosophers and scientists, especially in the first half of the last century, have taken into consideration the following problems: 1) to affirm the independence and autonomy of biological science in comparison with physical and chemical sciences; 2) to discuss the legacy of the use of the method of analysis to study a living organism, since the analytic-reductionistic method is considered inadequate to investigate the nature; 3) to explain the marvellous and incessant increasing of evolutive phenomena, which show a directionality from simplex to complex, from inorganic to organic, from matter to mind. For these reasons, during this period many doctrines – like the Organicism and the Holism – developed taking into account these problems against reductionistic or meristic approach. Taking occasion to this *querelle*, the present contribution intends to clarify if not the origins – immemorial – at least about the author and the date that gave to Holism the notoriety.

2. The beginning: J.C. Smuts and his time

What is Holism? First of all, we can begin with the clarification of the term "Holism", a word derived from Greek "holon", "whole". Very roughly, it is the idea of global dependence. During the last century, it is a very used word and, sometimes used to excess. In linguistic and semantics, Holism is the doctrine that one can not grasp the meaning of a word in isolation before one understands the whole language to which the word belongs. In the philosophy of science, Holism means that every theory must address itself to *all* the evidence, facing the "tribunal of experience as wholes". In the social science, methodological holists claim that social phenomena may be studied at their own autonomous, macroscopic level of analysis, against methodological individualism, that affirms the social events are mere aggregates of actions and relations of the individual. Another holistic view is expressed by the *entangled* state in Quantum Mechanics. We can speak of Holism even for the Mach's principle in physics. Mach maintained that inertial mass of a body was determined by the distribution of matter in the universe as a whole.

Nevertheless this doctrine is very ancient and widespread in many authors, the word "Holism" was introduced for the first time in 1926, when Smuts publishes his *Holism and Evolution*, a book that deals with some of the problems which fall within the debatable borderland between science and philosophy[1]. As the same author says, this work regards the point of contact between the two disciplines with the purpose to describe the tendency towards complexity in nature. Generally, Holism in biology is expressed by the assertion that "the whole is more than the sum of its parts". Historically, this notion is awarded in particularly to a living organism to express the fact that if the parts of the organism are taken to pieces, it "is destroyed and cannot be reconstituted by again putting together the severed parts" [19, p. 101].

[1]For a monograph about this author and the main bibliography see [14].

The holistic movement represents a very difficult field for the historiography of science because there is not a specific and well consolidate tradition of research developed during the time neither an organic scientific community liable of a consistency research program. Holism expresses rather a few requirements that are sometime repeated in a stereotyped formula, without a real and appropriate philosophical investigation. From a historiographycal point of view, Holism in biology has been compared to many forms of Emergentism and Organicism, when it is not undervalued as a simple reproduction of the vitalistic theory. In general, the holistic idea has been used as alternative to the mechanistic method. Nevertheless it is not insignificant that the starting point of Smuts's holistic conception is not isolated, but corresponds to a common background of a lot of scientists and philosophers of the same period, expressed in the doctrine of "Emergence" and of "Organicism".

The concept of "Emergence" has been introduced by C. Lloyd Morgan, S. Alexander and C. Broad to characterise the appearance, during the evolutive process, of properties that are not predictable observing the properties of the previous evolutive phase (see [12, 13, 1, 4]). The natural reality is characterised as a wholeness progressive evolution from simplex to complex structures: every level of these hierarchical structures has properties irreducible to the properties of the lower. A second concept adopted by the holistic approach is the theory of "Organicism" (organismalism), introduced by W. Ritter in 1919 to characterise the impossibility to prescind by the organism in its wholeness to explain the behaviour of the elements[2]. But the most famous representative of Organicism is L. von Bertalanffy, who has the merit to show the organicistic current as alternative to the Vitalism and the Mechanism. In particular the organicistic doctrine points out three important characteristics of biological phenomena: 1) the wholeness of the organism (organic regulation); 2) the organisation (with proper laws for every level); 3) the evolving state (typical of the open system)[3]. Since the parts of an organic wholeness have behaviours consequent not only from the relations

---

[2]See [15, 17]. For an history of the main important representatives of this doctrine, see [10]. Even the biologist J. Scott Haldane [9] asserts that the properties of the parts are determined by the whole. The main problem of the organicism was the explanation of the individual form and its genesis. Already during the Romanticism the poet Coleridge discriminates the organic form from the mechanic one according to five rules: 1) the origin of the whole goes before the differentiation of the parts; 2) the whole is, from a logical and ontological point of view, the first, while the parts derived from it; 3) the form shows the process from it is borne; 4) the organism absorbs many components and transforms them; 5) the external aspect of the organism depends from internal processes: see [16].

[3]Bertalanffy in [2] defends the doctrinal autonomy from the organismic biologist's standpoint. See even [3].

of the parts of a same level but even of the lower and the higher, the organicists affirm that the properties of the whole determine the properties of the parts.

Many of the issues expressed above are resumed and investigated by J.C. Smuts, who distinguishes wholeness in the strict sense for mere aggregates, mechanical system, and chemical compounds. In a true whole, thinks Smuts, the parts lose forever their prior identity, while in aggregates, mechanical system, and chemical compounds the identity of the parts of elements is not lost but is always recoverable. In spite of this, there are certain entities, such as biochemical systems, which appear to have an intermediate status. These systems form the vast ladder of life. At the bottom of the ladder, mechanistic features predominate, while at the top there are holistic features. True wholes, free of any admixture of mechanism are exemplified in minds or psychic structures, which first appear among higher organism, and in particular in the human personality, "the supreme embodiment of Holism". Smuts sometimes speaks of atoms and molecules as wholes because even these structures are complex systems.

3. His cosmic vision

Smuts's cosmic vision regards the increasing of complexity and the real value of evolutionary process from matter to life and mind. He wants guaranties the novelties of the development. Both matter and life consist of unit structures whose ordered grouping produces natural wholes which we call bodies and organisms. He sees this character of "wholeness" as fundamental in the universe: wholes are not mere artificial construction of human thought but the signs of a real great evolutionary process. Their appearance was explaining by postulating a primordial whole-making, or "holistic" factor in the universe. Therefore, "Holism" is the term coined to describe this fundamental tendency in the universe to create wholes. Every organism, but even atom, is a whole, with specific internal organisation and self-direction. The conception of wholes therefore covers a much wider field than that of life, traceable from the inorganic form of nature to human associations like State, and to the creations of the human spirit in all its greatest activities. All the great main types of existence are due to Holism. In fact, long before organisms or minds arose, the holistic factor is producing elementary wholes of a purely physical kind. Later, through a series of creative leaps, it becomes more fully embodied in biological structure, the more perfect of which are minds and persons[4]. In the sphere of spiritual values Holism finds its clearest embodiment. So a whole is not an artificial construction, but an organic unity in which "all the parts appear in a subtle indefinable way to subserve and carry out the main purpose or idea" [19, p. 98]. This characteristic shows the existence of a tendency or energy in the universe to produce wholes. Evolution is properly the progressive development of the resulting wholes at all stages, from the inorganic to the most organised phenomena like human mind and spiritual value.

Smuts conceives Holism as a *real factor*, a natural principle that acts against the second principle of thermodynamic, creating order and complexity. Through all the stages of natural evolution "we see the ever-deepening nature of the whole as a specific structural synthesis of parts with inner activities of its own which co-operate and function in harmony, either naturally or instinctively or consciously. The parts so co-operate and co-function towards a definite inherent inner end or purpose [...]. We thus arrive at the conception of a universe which is not a collection of accidents externally put together like an artificial patchwork, but which is synthetic, structural, active, vital and creative in increasing measure all through, the progressive development of which is shaped by one unique holistic activity operative from the humblest inorganic beginnings to the most exalted creations and ideals of the human and of the universal Spirit. [...] Holism in all its endless forms is the principle which works up the raw material or unorganised energy units of the world, utilises, assimilates and organises them, endows them with specific structure and character and individuality, and finally with personality, and creates beauty and truth and value from them" [19, p. 107].

The challenge that we can see in this conception is to isolate a sense in which the *causal powers* of the whole can not be predicted on the basis of the causal powers of the parts. Causally interactive ensembles will be considered by the holist to display regularities of interaction that cannot be obtained by any method of superposition or aggregation of the causal interactions of their constituents.

Another important aspect of the Smuts's conception is to emphasize the active role of the organism in the choice of the adaptive process.

---

[4]In the field of mental activity we can define human personality as the highest and most complete whole in the universe. In fact human personality is a creative synthesis in which the earlier material elements, organic and psychical, are incorporated in a new complex whole. As Smuts underlines, "human personality takes up into itself all that has gone before in the cosmic evolution of Holism. It is not only mental or spiritual but also organic and material [...]; the structures of matter, life and mind are inseparably blended in it, and it is more than any of them" [19, p. 264]. Naturally, the importance of mental or spiritual factors must not blind us to the essential of physical factors: they are not independent and disembodied mind and disminded body are both impossible concepts.

The variation is not single, but behind it there is the whole force of the organism, that acts as a single adaptation complex, with constraints that, arising from internal organic competitions, regulate and address the morpho-functional conditions. He shows the limits of a neodarwinian point of view that reduced all the features and properties of the organism to the external factors like natural selection and blind and mechanical factors like the variations due to chance. Smuts traces the outlines of the concept of adaptations as the result of interactions among the genotype, phenotype and environment. The organism is not rigidly determined by genotype or environment, but it is the real subject of evolution[5].

4. Is Holism a useful scientific approach or is a simple philosophical vision?

Although the conception of Smuts's Holism is very fascinating to describe the increasing of order, it is clearly metaphysical, because it is not expressed in a formalised natural law and it can not be tested with actual facts and experiments. Nevertheless, Smuts – as many others defenders of the organicistic doctrine – shows the weakness or rather the one-sidedness and partialness of the scientific vision. In particular he supports to the analytic investigation the synthetic one, observing that "no description of the parts is a complete description of the whole object; it is only a partial description, and falls short of a true and full account in proportion as the object partakes of the character of the whole. [...] We may say generally that wherever an object shows structure or organisation (as very object does) a full description of it would involve at the very least an account of this structure or organisation as a whole, in addition to its detailed mechanisms and functions" [19, p. 322].

But, we can replay, do not we consider the science as studying the interrelations among the elements, searching to discover the structure of the whole object[6]? And even if we can investigate the relations among the parts, as the science really does, is it not impossible to know an object exhaustively, as notes Kant? Thirdly, how we can demonstrate that the emergence of a characteristic is an ontological trait inherent in some phenomena, rather than a sign of the scope of our knowledge at a

[5]This perspective will advance important researches, perhaps few known but not less significant for the theoretical biology. See [18].

[6]In other words, if Holism means that the parts have properties within the whole that they lack outside it, then it is true but hardly supports the doctrine of an unbridgeable gulf. In fact, a description of a part's relation to a whole *is* the specification of a property of the part. Thus, Holism in this sense is true of any whole whatever. Indeed, organismic biologists sometimes give an exact description of a physical field when describing the features of an organic whole. See, for example, [7].

given time or the expression of a determined theory available today but not tomorrow[7]?

These theoretical objections are very heavy, and carry us to the traditional philosophical problem of the relation between the whole and the parts. Smuts faces this thesis in the terms of the organisation and wholeness. From Gothe to Goodwin and Webster, the main risk about the riductionistic approach is the lost of the *morfé* and the overall meaning of the organism (see, for example, [5, 8]). To prescind from the epistemic character of the definition of "whole" and "part", Smuts has opened the door to the study of the complex systems and certainly, although biological research was not his main occupation, he has focused a lot of aspects that will be investigated afterwards.

References

[1] ALEXANDER S., *Space, Time and Deity*, Macmillan & Co. Limited, London 1920.

[2] BERTALANFFY L. VON, *Problems of Life*, New York 1952.

[3] BERTALANFFY L. VON, *General System Theory*, George Braziller, New York 1969.

[4] BROAD C.D., *The Mind and its Place in Nature*, Routledge & Kegan Paul, London 1925.

[5] CONTINENZA B., GAGLIASSO E., *Giochi aperti in biologia*, Franco Angeli, Milano.

[6] DAVIES P., *The New Physics*, Cambridge University Press, Cambridge 1989.

[7] DÜRKEN B., *Entwicklungsbiologie und Ganzheit*, Verlag von C.B. Teubner, Leipzig un Berlin.

[8] GOODWIN B.C., WEBSTER G.C., *Il problema della forma in biologia*, Roma 1988; tr. it. Armando.

[9] HALDANE J.S., *Mechanism, Life and Personality. An Examination of the Mechanistic Theory of Life and Mind*, 2nd ed., Hodder & Stoughton, London 1923.

[10] HARAWAY D.J., *Crystals, Fabrics and Fields. Metaphors of Organicism in Twentieth-Century Developmental Biology*, Yale University Press, New Haven 1976.

[11] HARAWAY D.J., *General System Theory*, George Braziller, New York 1969.

[12] MORGAN C. LLOYD, *Emergent Evolution*, Williams & Norgate, London 1923.

[13] MORGAN C. LLOYD, *Review of Holism and Evolution*, Journal of Philosophical Studies **2** (5) (1927), 93–99.

[14] PROCACCI S., *Alle radici dell'olismo. La filosofia della natura in J.C. Smuts*, ESI, Napoli 2001.

[7] For a major analysis of the critics against Holism see [14], in particular chap. IV.

[15] RITTER W.E., *The Unity of the Organism or the Organismal Conception of Life*, R.G. Badger, Boston 1919.

[16] RITTERBUSH P., *The Art of Organic Forms*, Smithsonian Institute Press, Washington 1968.

[17] RUSSELL E.S., *Form and function. A Contribution to the History of Animal Morphology*, John Murray, London 1916.

[18] SARÀ M., *The Problem of Adaptations: An Holistic Approach*, Rivista di Biologia **82** (1) (1989), 75–101.

[19] SMUTS J.C., *Holism and Evolution*, Greenwood Press Publishers, Westport, Connecticut 1973.

TOWARDS A HISTORY OF COMPLEXITY

A comparison between Europe and China

Tito M. Tonietti
Dipartimento di Matematica,
Università di Pisa,
Via Filippo Buonarroti 2, 56127 PISA - ITALY
tonietti@dm.unipi.it

> When intelligence is wide-ranging, it connects; when limited, it divides.
> Zhuang-zǐ.

Abstract The scientific revolution that took place in Europe during the XVII century was mainly based on simple systems. Complex systems, previously considered important, were put aside. In China, on the contrary, complex systems remained at the centre of interest, thus giving rise to a "very organic world-view".

1. What meaning can be given to complexity

Nowadays, complexity is a frequent topic of discussion. Explanations of the meaning of this term are less common. Some try to offer precise definitions, such as algorithmic complexity, whereas others prefer to use the common meaning. At this very meeting in Arcidosso, Marcello Buiatti and Vieri Benci appear to be examples of the first type, and R. Esposito and A. Politi of the second. The latter even appears to use the word in the sense of "simplicity". I will choose an intermediate position, taking the word 'complex' to refer to an event (a phenomenon, a system, a model) which cannot be modified in any of its aspects without some consequent change in the 'essential' characteristics of its behaviour (evolution, dynamics). This is clearly an extremely general meaning, according to which 'almost' all the world is complex. And this is undoubtedly true. Indeed, the surprising thing is that cases exist which are simple, or rather capable of simplification.

In Europe, the development of the scientific tradition was made possible by the isolation of certain phenomena which could be reduced to other, simpler forms. Galileo Galilei believed it was necessary to take away "the hindrances of matter". Only thus was he able to imagine (thanks to mathematics and the reconstruction of the phenomenon in a laboratory) an isochronous pendulum and the uniform acceleration of a falling body. His ability (and the secret of his success in the following centuries) lay in the demonstration that the basic characteristics of the system under examination could be found in a simplified model. Thus, the construction of a simpler model must consider irrelevant the inevitable differences in the desired characteristics, compared with the given phenomenon to be studied. [8, v. II p. 260]

This is the heart of the matter, because judgements of importance and relevance necessarily become subjective, they become value judgements, and end up by depending partly on historical contexts. Galileo's famous saying about the need to ignore the packing in calculating the cost of wool could be taken for granted in the XVII century. But it would create problems for economists nowadays. In the global market, the soul of the product has become the box that contains it, because without this, advertising would not be possible. And thus our media and our rubbish dumps are full of packages.

Together with some highly successful simple models, however, whole classes of similarities began to appear, which were presented (not always rightly) as heirs to the success of the original models. In the end, in a scientific community by now triumphant, and so sure of itself that it could see no limits to its powers, someone arrived at the hasty conclusion that the whole world could be reduced (sooner or later) to one simple system, if necessary with some simple subsystems. But simple systems remain the exception, even if it is a particularly interesting one, and usually systems prove to be complex. In the past, followers of Plato and Pythagoras long believed that the world could be reduced to proportions involving whole numbers. Subsequently people understood that in the *continuum* of real numbers, whole numbers and rational numbers represented a small minority, which was an exception with respect to the far more 'numerous' irrational numbers. It may even be shown that a relationship exists between the two questions, allowing us to make a more precise definition of our starting notion of complexity. And if we choose to proceed in this direction, we may begin to discuss stability, structural stability, (geometric) generic properties, and the other notions that are equally variously connected with our complexity. [16, 26]

Here, however, I prefer to choose, from the history of science, a few cases that appear to be complex, and to compare them with the better known simple systems.

2. Europe: the cemetery of complex systems

Scientists foresee the timing of an eclipse, but they cannot know whether they will see it. Metereology has been removed from history ... It is of interest only to those who are ignored by scholars: farmers and seamen ... Because it is the place of disorder and the unforeseeable ... Because it is the realm of clouds, where the head does not belong and which do not belong in the head. Michel Serres.

We may consider the so-called scientific revolution of the XVII century as a conflict between complex and simple systems, with the victory of the latter. Aristotle had also considered shape as one of the reasons for the buoyancy of bodies. As the boat moves, its shape becomes essential for its stability. Galileo, instead, reduced the problem to the force discovered by Archimedes. Aristotle had considered meteorites as bodies that dissolved as a result of friction in the atmosphere. Galileo denied this possibility. Thus the science of Aristotle was that of complex systems like the atmosphere, which presents the well-known butterfly effect. Also the movement of (or inside) a fluid like water is a typical complex case, because it usually takes place by means of vortices. [17, 18, 19]

On the side of complexity, we should place other characters who are seldom or never considered to be such. *De rerum natura* by Titus Lucretius Caro could be read as an explanation in verse of phenomena dominated by a *continuum* (in water?). The *primordia* would thus be vortices, and should not be translated, in the usual manner, as atoms. [11, 15, 17]

Leonardo da Vinci's studies on the movement of water also represent examples of complexity. In fact, the first mathematical law referring to vortices is his. But science historians should not ignore his numerous drawings of the shapes that water assumes as it flows. Furthermore, he conceived of the world as an organism in which phenomena mutually influenced each other. For him, explaining meant connecting, creating relationships, and pointing out analogies. [10, 17, 18, 19]

Descartes and Leibniz explained the movement of planets around the sun by means of the vortex in the ether. Newton eliminated the complexity by means of the attraction of the force of gravity. [23]

The history of sciences, at least the more orthodox kind, thus appears to be the cemetery of complex systems. The greatest successes of European scientists appear to be due to their peculiar ability to choose the phenomena that can be simplified. Explaining them thus means isolating them from others in the laboratory, ignoring certain aspects such as friction, and in the end, reconstructing them in a symbolic form in their own transcendent world. Those who tried to explain a phenomenon in other ways ended up by giving a complex vision of it. They believed that the various occurrences were all connected together, and that they had a mutual influence on each other. Thus it was necessary to add more and more new elements until the picture was complete. Otherwise the phenomenon would not be comprehended. 'To comprehend', from the Latin *comprehendere*, that is to say, to take together, and 'continuous', from *continere*, that is to say, to con-tain, or hold together, are the characteristic terms in this case. Whereas 'to grasp', from *capere*, that is to say, to seize, to master, to conquer, to attract (by means of deceit), gives the idea of the other kind of approach.

In the course of their evolution, European sciences have shifted the balance towards this second way of studying phenomena. In so doing, they have gained in the possibility of calculation, but they have lost in stability. [16] Yet in a different historical context, the balance between simple and complex had tipped over the other way. This is the case of China.

3. China: the apotheosis of complex systems

"...even the tiny noises of mosquitoes and flies would be able to arrive everywhere", *Huà Shu*, The Book of Transformations, A.D. 940. [14, p. 208]

In this text of the Tang period, attributed to the daoist, Tan Qiao, there are surprising references to the butterfly effect. The idea that something (a sound or another disturbance), however small it is, can arrive anywhere (and that we can perceive it) is so extraordinary that we cannot help wanting to find out where it comes from.

氣從聲 聲從氣 氣動聲發 聲發氣振

qì cóng sheng, sheng cóng qì, qì dòng sheng fa, sheng fa qì zhèn.[1]

"Qì comes from sound, sound from qì, moving the qì the sound is emitted, the qì shakes when it emits the sound."

[1] In the transliteration the pin yin is followed: the first tone - is not marked, the second is marked as ´ , the third as ˇ , the fourth as ` .

The generation of sound is described in the following way:

xíng qì xiang chéng. 形氣相乘
"bodies and qì move each other"

Qì forms terms such as:

qì fen 氣氛, atmosphere,
qì hòu 氣候, climate,
qì xìang 氣像, meteorology.

When preceded by the sign of water, qì becomes steam, vapour. What does qì mean? Qì means: gas, air, breath, smell, airs, spirit, make angry, get angry, insult, vital energy. For us Europeans, it may recall a kind of all-pervasive vapour, a classical ether where phenomena take place. However the European ether is abstract, heavenly, transcendent, a theoretical *medium*. The Chinese ether, qì, on the contrary, is wholly earthly and immanent. It surrounds us in the atmosphere, it pervades our breathing, giving us vital energy, it warms us up and (if necessary) it makes us angry.

Chinese science is deeply immersed in the culture of qì. This can be seen both in its characteristic way of considering certain particular phenomena, and in its general conception. [14, 3] Chinese scholars have concentrated on the movements of air and water. They have developed above all their own peculiar understanding of sounds, in particular considered together with music.

"The qì of heaven and earth combine and generate wind.
The windy qì of heaven and earth decide the right twelve lù,
zhèng shí èr lù dìng 正十二律定."

Qián Hàn Shu, History of the Former Han Dynasty, about a.D. 100 [14, p. 187]

The twelve lù are the Chinese musical notes, as shown in figure 1 and table 1, which are taken from Chéng Dàwèi. [4, 997–985]

	Lù Lǔ	length (cùn)	volume (fen)	Solar terms	Hòu	Earthly branches	Time
1)	Huáng Zhong	9	810	22	Winter solstice	1 zǐ	11pm–1am
2)	Dà Lǔ	$8\frac{104}{243}$	$758\frac{42}{81}$	24	Great cold	2 chǒu	1–3
3)	Dà Cù	8	720	2	Rain water	3 yín	3–5
4)	Jiá Zhong	$7\frac{1075}{2187}$	$674\frac{58}{243}$	4	Spring equinox	4 mǎo	5–7
5)	Gu Xǐ	$7\frac{1}{9}$	640	6	Grain rain	5 chén	7–9
6)	Zhòng Lǚ	$6\frac{12974}{19683}$	$599\frac{707}{2187}$	8	Grain full	6 sì	9–11
7)	Ruí Bin	$6\frac{26}{81}$	$568\frac{8}{9}$	10	Summer solstice	7 wǔ	11am–1pm
8)	Lín Zhong	6	540	12	Great heat	8 wèi	1–3
9)	Yí Zé	$5\frac{451}{729}$	$505\frac{55}{81}$	14	Limit of heat	9 shen	3–5
10)	Nán Lǚ	$5\frac{1}{3}$	480	16	Autumnal equinox	10 yǒu	5–7
11)	Wú Shè	$4\frac{6524}{6561}$	$449\frac{359}{729}$	18	Descent of frost	11 xu	7–9
12)	Yìng Zhong	$4\frac{20}{27}$	$426\frac{2}{3}$	20	Light snow	12 hài	9–11

Table 1.

The understanding of sounds and music was developed around the twelve lù lǚ, pitch pipes, by which the rules were fixed and instruments were tuned. Strings were used in Europe for the same purpose. [14, 3, 25] Right from the beginning, music has always been considered so important in Chinese culture that together with the famous Five Classics, there existed a sixth one dedicated to music, which is unfortunately no longer extant.

The rules for music spread everywhere, and were connected with a striking variety of the most varied phenomena. Music even succeeded in fixing the measurements of all things.

liù lù wéi wàn shì gen běn yan. 六律爲萬事根本焉
"Here the six lù act as the essential roots of the 10,000 things"
Simǎ Qian, *Shǐ ji*, Historical Records, about 90 B.C.
lù dù liàng héng yú shì hu sheng. 律度量衡於是乎生
"Hence originate the rules of pitch, length, capacity, weight".
Guó Yǔ, Discourses of the States, Zhou, Qín, Hàn periods.

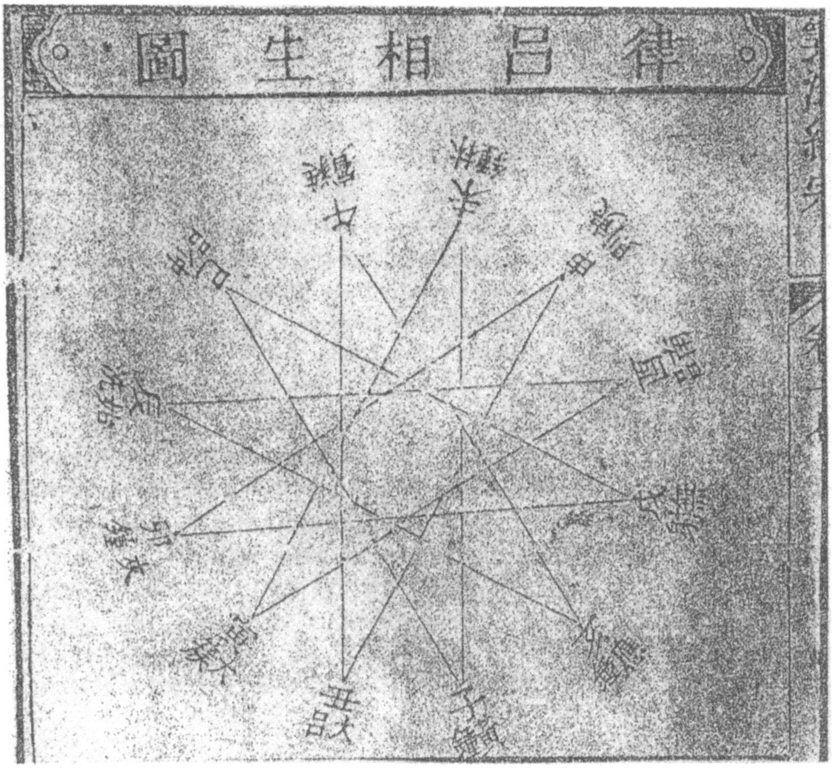

Figure 1. The twelve lùlǔ in a page of Chéng Dàwèi.

Where do they originate from? From a Zhong, a bell! The units of measurement of pitch, length, capacity and weight had been fixed in it. If the height of the note emitted was right, then the other units of measurement that depended directly on it were also right. Or else a pipe was taken, that emitted the first note of the lù lǔ, and the measurements of capacity (1,200 grains of meal) and weight (12 žhu) were fixed. This system of measurement based on notes has left traces that can still be seen today in the Chinese characters.

I do not know any other cultures in which sounds and music are so important as in China. The twelve notes indicated climatic events, solar terms, the hours of the day, and the earthly branches. If the music considered in Europe was that of the spheres, in the case of China we should speak of the music of the atmosphere. [25]

Those who have studied Chinese culture and sciences have offered widely varying interpretations, assisted in this by the numerous ambiguities of the language. And yet there is almost agreement over a couple of points. Chinese scholars appreciated correlations because they tended to explain every phenomenon by taking it together with others. Confucian scholars dominated the administrative and political structure of the empire. But it was rather the naturalistic events that had aroused interest among daoists. [9, 13] "As unfathomable as the sea, as majestic as the mountain, it generates the infinite cycle of the ends and the beginnings. It is the motor and the measurement of all beings. ... All beings draw their resources from this. It is inexhaustible. Is this not Dào?" And yet: "He who answers the one who enquires about Dào does not know Dào. ... Dào does not admit any questions, or answers to questions ...", *Zhuang Zǐ*, The Book of Master Zhuang, ch. XXII.

The principles of Yáng and Yin often pursue each other in a variety of situations, generating cyclic phenomena. However, let us also remember that Yáng and Yin do not create dualistic oppositions, but penetrate each other, and are transformed into each other, thus giving rise to the whole universe. Consequently, no distinction can exist here between the animate and inanimate world, between spirit and body, between man and nature, ..., between truth and error, ... For Chinese daoist scholars, everything is in movement and everything is transformed: their model appears to be water.

"Dì 地, the earth, is the origin of all things, the root and the garden of all life; it is the place where everything, the beautiful, the ugly, the good, the bad, the stupid and the intelligent are born. Now water is xuè qì 血氣, blood and qì of earth, which flows and communicates, as in nerves and veins, water is jù cái 具材, the matter that provides everything.", *Guǎn Zǐ*, The Book of Guan, Hàn period.

上善若水 水善利萬物而不爭

Shàng shàn rùo shuǐ. Shuǐ shàn lì wàn wù, ér bù zheng.

"The higher good is like water. Water-good benefits 10,000 things, but it does not contend"

chǔ zhòng rén zhi suǒ wū. Gù ji yú Daò. 處衆人之所惡故幾於道

"It is situated in the place that many men dislike. Hence pratically, in the Daò." *Daò Dé Jing*, The Classic of Dào and virtue, ch. VIII.

"The qì of Earth ascends above, the qì of Heaven descends below,

Yáng Yin xiang mó tian dì xiang dàng 陽陰相摩天地相蕩

Yáng and Yin rub each other, Heaven and Earth swing each other. Their drumming ... flourish." *Yuè Jì*, Records of Music, Zhou period [Needham, 1962, p. 205]

Chinese sciences are coherent with a "very organic world-view". [14, p. 7] "...the Chinese physical universe in ancient and medieval times was a perfectly continuous whole". [14, p. 8] The qì, the breath of the earth, the atmosphere embodied this *continuum*. Even in mathematics, one can find a geometrical *continuum* which forms the background of every construction and of every procedure, even of calculation. For this reason, Chinese mathematicians considered all numbers valid, without distinction, and have never condemned or forbidden the use of irrational numbers, as happened, on the contrary, in Europe. [13, 12]

"The jun 均 , the uniform is the greatest lǐ 理 , reason in the world. The lián 連 , the connection of all shapes and things is due to continuity", *Liè zi*, The Book of Liè, Han period.

Daoist apologues tell stories of mián, silk flosses which do not break. Mián means continuous too. Lián mián means continuous, unbroken. Other stories tell the effect of music on the climate. "Music assures continuity between human beings, and between man and the rest of nature". [14, p. 29]

In China, the balance between simple and complex had shifted towards complex. The history of sciences in China appears to be the apotheosis of complex phenomena. What the Europeans were gradually excluding, the Chinese were placing at the centre of their interests. However, Chinese science should not be taken for a holistic science. It might present some similarities to that, only if we exchanged it for a static science. On the contrary, it proves to be a process of transformation, precisely because its purpose was that of studying changes. It attempted to understand change by means of that which is unstable. European sciences, on the other hand, generally attempted to introduce order into that which changes, by fixing its static, eternal foundations.

The difference can best be seen in the different systems of musical notes. In the Chinese one, the notes are arranged in a circle which is not closed, and which never stops. In the European one, the notes are fixed in a motionless scale. [25]

4. Between sea and sand

Was that conversation held between the sea and the sand, that Juan Louis Borges, in his story "The Beginning", defined as the central fact in our history? [2] Were they two Greeks, or two South Chinese? In any case, they were meant to represent the conflict between the liquid, continuous, capricious, salty, corrosive element and the solid, granular, static, tasteless, inert element. The conflict between the geometry of the *continuum* and the numerical calculations of the computer, between complexity and reduction to a simple system. [18, 19]

Did the discussion between Albert Einstein ed Hermann Weyl continue on the beaches of Long Island Sound, during their exile in the United States? It started in an epistolary form in Germany after the Great War, and dealt with the possibility of describing the universe as a *continuum* represented by the gauge equations of a unified field. Aesthetically it would have been very nice, but wouldn't inflexible nature have laughed at such an attempt? [6, 18, 19, 20, 21]

And wouldn't our great mother have continued to laugh even more, today, at the General Omnipotent Device (a G.O.D. that is certainly incapable of laughing), who cannot succeed in simplifying the atmosphere so as to make reliable weather forecasts? "Quando Superga ha il cappello o piove, o fa bello [When Superga is wearing a hat, either it starts raining, or the weather is fine]", they say at Turin, in Italy. No approximation would then be sufficient.

The dynamics of the pendulum admits approximations relevant to the aims of description, but the dynamics of the atmosphere does not. It typically evolves through the formation of vortices, with a butterfly effect and therefore it represents the typical complex system. [7, 18, 19, 22]

René Thom composed the prelude to complexity when, in the glorious period of his Catastrophes, he pointed out that the network of the many causes could often not be reduced to a single force, as in Newton's parallelogram. At this point, ranges open up where the variations in parameters, however small they may be, produce observable effects in the qualitative evolution. [16, 26] It is likely, we hope, that the sun will rise again tomorrow. But why is it that this year (2001), the temperature in Italy in October was so close to the tropical level of Hong Kong? By chance, did a naughty butterfly move its wings? Or perhaps it was a mosquito.

Complex systems can even be created with the trusty, dear old pendulum. It is sufficient to connect two together by means of a magnet, and we obtain a quirky toy that will catch the eye in shop windows. [1, 18, 19]

Thus we, too, learn to laugh in the end, and we become more aware of our limitations, freeing ourselves from as much *hubris* as we can. As in Greek tragedies, *hubris* only leads to trouble. [24]

Jade Eyes walked along with her feet in the water, playing in the patterns of light created by the sun on the surface of the sea. She watched the waves breaking, the seagulls snatching up the fish, and she reflected on the shape of the rocks and the clouds. - Do you see how many different shapes there are in front of our eyes? Isn't your curiosity aroused by the way they keep changing? How beautiful the world is, with its unstable, capricious variety.

Nán Fǎ, on the contrary, was thinking about the length of the coastline. With every step he took on the sand, he could imagine the surveyor moving his instrument from one point to the next. As he followed the water's edge, he stumbled over a small rock which broke the regularity of the line created by the water with its rhythmic ebb and flow. Further on, the coastline became irregular, forcing them to proceed by jumping or swimming from one rock to the next.

- If the Emperor wants to know the length of his coastline, he will have to pay more surveyors than he thinks, because every rock and every irregularity will force them to shorten the standard of measurement. Who knows how many it will take, and how much it will cost.

- It will take an infinite number of surveyors, or else an infinite period of time, because they will end up by using infinitely short standards in order to measure the grains of sand, as well. Then they will need an enormous number of clerks to record the measurements, and an equal number of mathematicians to add them up on the abacus, Jade Eyes replied. In the end, the Emperor will be happy to learn that the shores of his empire are without limits. But what is the purpose of this senseless attempt to translate into numbers that which is continually changing?

- The Emperor's desire is sufficient for me. But for you ladies of the Court, who love intrigues and gossip, I may add that its use is of a military nature. Building a castle here, a tower there, with soldiers to guard them.

- A Great Wall along the seashore? An enormous expense, enormously useless, which will not save us from death, but will undoubtedly demand its tribute of blood for its construction. Happy are those populations who do not feel the need for greatness, and prefer beauty!

Acknowledgments

I would like to thank Dr. Ronald Packham for his help in the English styling of the manuscript.

References

[1] BERRY M., *Deterministic and chaotic motion as illustrated by billiards*, Colloqium Peyresq 1984.

[2] BORGES J.L., *Il principio*, in: "Atlante", Kodama M. ed., Editorial Sudamericana S.A., Buenos Aires 1984.

[3] CHEN C.-Y.(J.), *Early Chinese Work in Natural Science*, Hong Kong University Press, Hong Kong 1996.

[4] CHÉNG DÀWÈI, *Suàn Fă Tŏng Zong Jiào Shì(1592)*, Yán Dun Jié, Méi Róng Zhào, Lĭn Zhào Húa eds., An Hui Jiao Yù Chu Băn Shè 1990.

[5] DAÒ DÉ JING, J.J.L. Duyvendak ed., Librairie D'Amerique et D'Orient, Paris 1953.

[6] EINSTEIN A., *1923 May 26th, letter to Hermann Weyl, A. Einstein Duplicate Archive*, Princeton University Library, Princeton.

[7] EKELAND I., *Le Calcul, l'Imprévu*, Seuil, Paris 1984.

[8] GALILEO GALILEI, *Opere*, UTET, Torino 1964.

[9] GRANET M., *La pensée chinoise*, Éditions Albin Michel, Paris 1968.

[10] LEONARDO DA VINCI, *I codici di Madrid*, Giunti Barbera & McGraw-Hill, Firenze 1974.

[11] LUCREZIO TITO CARO, *De rerum natura*, Sansoni, Firenze 1969.

[12] MARTZLOFF J.-C., *A History of Chinese Mathematics*, Springer Verlag, Berlin 1997.

[13] NEEDHAM J., WANG LING, *Science and Civilisation in China*, Cambridge University Press, Cambridge 1954-....

[14] NEEDHAM J., WANG LING, ROBINSON K.G., *Sound (Acoustics)*, in: "Science and Civilisation in China", Volume 4 Part I, Physics, Cambridge University Press, Cambridge 1962, 126–228.

[15] SERRES M., *La naissance de la physique dans le texte de Lucrèce*, Les Éditions de Minuit, Paris 1977.

[16] THOM R., *Stabilité structurelle et morphogénèse*, Benjamin, Reading Mass 1972.

[17] TONIETTI T.M., *Elementi antigalileiani nelle scienze contemporanee: Eraclito, Aristotele, Lucrezio, Leonardo*, in "Atti del III Congresso nazionale di storia della fisica", Bevilacqua F. e Russo A. eds., Centro Stampa Facoltà di Ingegneria, Palermo 1983, 275–284.

[18] TONIETTI T.M., *Capricci di mare e calcoli di sabbia*, AMELA project, Paris 1986-87.

[19] TONIETTI T.M., *Disegnare la natura*, AMELA project, Paris 1986-87.

[20] TONIETTI T.M., *Four letters of E. Husserl to Hermann Weyl and their context*, in "Exact Sciences and Their Philosophical Foundations", Deppert W. et al. eds., Peter Lang, Frankfurt am Main 1988, 343–384.

[21] TONIETTI T.M., *Continuum's Continuous Conflicts*, in "Proceedings Conference on Foundations of Mathematics and Physics - Perugia", Bartocci U. et al. eds., Benjamin Wesley Publ., Blumberg 1990, 271–288.

[22] TONIETTI T.M., *Pisa-Como: andata e ritorno (Volterra, Waddington, Thom)*, in "Da Pisa a Como e ritorno", Cerrai P., Galleni L. eds., SEU, Pisa 1996, 57–74.

[23] TONIETTI T.M., *Verso la matematica nelle scienze: armonia e matematica nei modelli del cosmo tra Seicento e Settecento*, in "La costruzione dell'immagine scientifica del mondo", Mamone Capria M. ed., La città del sole, Napoli 1999, 155–219.

[24] TONIETTI T.M., *Tra armonia e conflitto: da Kepler a Kauffman*, in "La matematizzazione della biologia", Quattro venti, Urbino 1999, 213–228.

[25] TONIETTI T.M., *The Mathematics of Music During the XVI Century: Francesco Maurolico, Simon Stevin, Chéng Dàwèi, Zhu Zàiyù*, paper presented at the 9th International Conference of History of Science in China, Hong Kong 2001.

[26] TONIETTI T.M., *Catastrofi - Il preludio alla complessità*, Dedalo, Bari 2002.

[27] ZHUANG-ZǏ, Lion K.-h. ed., Gallimard, Paris 1980.

Index

A
AIC (Algorithmic Information Content), 44–46, 49–50, 221–222
Alexander S., 381
Andrade J., 201
Anti-reductionism, 339–340
Appert N., 369
Aristarchus, 126–127
ARMA (AutoRegressive Moving Average), 246, 248–250, 252
Arquès-Michel maximal circular code, 242
Arrow of time, 41–42, 46, 49–50, 105–107, 109, 113, 123
ART (Adaptive Resonance Theory), 5, 8, 23
Assimilation, 259–260, 263, 266–268, 293–294, 299–301
Atlas Project, 349
Auto-organization, 88, 202
Autopoiesis, 3, 339–340, 346
Average individual, 205

B
Babbage Ch., 352
Bacon F., 368
Bain A., 314
Bekenstein formula, 122
Bell inequalities, 15–16, 58
Bell J., 54
Bernard C., 340
Bernoulli Joh., 361
Betatronic
  motion, 27–28, 30, 37
  tune, 30, 36
Big computer, 117
Biogeography, 283, 286–287, 290
Biological
  indices, 171
  individual, 203–205
Birkhoff G.D., 327–334
Birkhoff theorem, 35, 331
Blanchard Ph., 57
Bodganov A., 355
Bohm D., 119, 130
Bohr N., 53–56, 127, 129, 375
Boltzmann L., 42, 47–48, 79–84, 145, 196, 329–330
Boolean
  logic, 15
  model, 153, 159–162, 165
Borelli A., 361
Bradley F.H., 319
Brahe T., 362
Brentano F., 315
Broad C., 381
Brownian particle, 89
Bunge M., 54–55, 258

C
Campanella T., 376
Carathéodory C., 330
Cardano G., 374
Cassini G.D., 125–127
CASToRe algorithm, 223–225
Causality, 105, 107, 113, 316–318
Cell system, 293–295
Cellular automata, 43, 49–50, 117, 120
Chaitin G., 121, 209, 221
  theorem, 46
Chaos, 3, 5, 19, 50, 99–101, 109, 121, 142, 191, 197, 346, 370
  See also Deterministic chaos
Chromosome, 208–209, 214, 216, 218, 224, 276–277, 279, 289
Church-Turing thesis, 16
CIC (Computable Information Content), 46–51, 222–224
Complexity, x, 3, ix, 45, 50–51, 87–88, 91, 93, 95, 101–102, 117, 119–121, 142, 146, 149, 199–200, 209, 211–212, 221–227, 273–274, 280, 285, 288–289, 293–294, 296, 298, 301, 307, 309, 324, 340, 345–346, 349–350, 352, 355–356, 377–378, 380, 382–383, 387–389, 395–396
Compression algorithm, 46–48, 121, 221–223
Computability, 117, 122, 222
Computational paradigm, 118, 120, 122–123
Coulomb Ch.A., 352

D
Dantzig G., 353, 356
Darwin Ch., 207, 232, 269, 284, 287–289, 314–315, 321
De Broglie L., 66
Decoherence, 16, 63–64, 71, 75–76, 186
De Maupertuis P.L.M., 193, 343, 361, 376
DEM (Diffusion Entropy Method), 87–89, 91, 93
Descartes R., 361, 365, 367, 374, 389
Determinism, ix, 105, 107–109, 133–134, 313, 316–317, 319, 324, 377
  dilemma of –, 109, 313, 315, 317–318, 323–324
Deterministic
  chaos, 99, 108–109, 111, 121, 234, 327–328, 333–334
  component, 245–248, 250
  order, 191, 197
  system, 105, 115, 135, 141–142
Digital system, 112
Dirac P.A.M., 10, 60, 65, 160, 192
Dupuit J., 352
Dynamical system, 4–7, 48, 99, 105, 141–142, 145, 149, 309, 327–334

E
EBI (Extended Biotic Index), 171–174
Ecosystem, 134, 172–174, 211–212, 219, 273
Eddington A.S., 192
EEG (ElectroEncephaloGram), 7, 182
Ehrenfest P. and T., 329–330
Einstein A., 42, 53–54, 58, 117, 129–130, 237, 395
Emergence, 3–4, 63, 141, 149, 183, 193, 273, 279, 289, 300, 305, 340–341, 345, 350, 376, 381, 385
Emergent property, 199–200, 207–209, 234, 285, 340, 346
Engel K.J., 133
Engineering systems, 349–350, 356
Entropy, 41–42, 44, 47–51, 87–91, 105, 114, 144–146, 196, 223, 258
EPR paradox, 58
Ergodic theory, 327–330, 334
  See also MET (Mean Ergodic Theorem)
Euclid, 127, 201
Euler
  equations, 82–84
  fluid, 81–82
Evolution, 4, 41–42, 48–49, 51, 61, 68, 70–72, 74, 80, 99–100, 108, 110–115, 130, 133–134, 141–143, 146–147, 149–150, 160, 197, 200, 202, 206–207, 214, 235, 238, 245–248, 254, 258, 264, 267–269, 274, 276, 280, 283–285, 287–290, 293, 295, 297–301, 309, 314–315, 319, 321, 349, 360, 364–365, 368–369, 376–377, 380–381, 383–384, 387, 390, 396

F
Fantappié L., 115, 193
Fayol H., 352
Feature binding, 7, 13, 22, 182
Fechner Th., 314
Fermat principle, 28
Feynman R.P., 54, 59–60, 122
Fliess W., 320, 322
FODO, 28, 30, 33, 35–38
Fokker-Planck equation, 145
Fouillée A.J.E., 319
Fourier
  series expansion, 158
  transform, 66–67, 250
Fractal borderland, 191
Freud S., 313–317, 319–324

G
Galileo Galilei, ix, 126, 128–129, 360, 362, 388–389
Galvani L., 361
Generation, 205–208, 211, 216–219, 293–294, 296–298, 300, 391
  distance, 206–207
Gesner K., 367, 374
GMO (Genetically Modified Organism), 211–212
Gödel K., 121, 209
GOD (General Omnipotent Device), 396
Gottfried K., 56
Green-Kubo formulas, 84
Green Th.H., 319
GR (General Relativity), 117, 119, 123
Grossberg S., 8

H
Hamilton function, 29
Hamiltonian, 29–32, 68, 72, 74
  dynamic, 28
Hamilton-Jacobi equation, 31
Hamilton W.R., 28, 31, 34, 42, 142
Heisenberg
  inequality, 61
Heisenberg M., 65, 117, 130
Heisenberg
  principle, 57, 197
Helfrich model, 88
Helmholtz (von) H., 314, 361
Hilbert D., 67–68, 70, 79, 83–84
Hilbert sixth question, 79–80
Holism, ix, 339–342, 344–346, 379–384
Hox genes, 289
H-theorem, 113
Hughes Th.P. and A.C., 350
Hume D., 314
Husimi measures, 67
Hutton J., 237

Huxlyey A., 359, 362–363
Hysterical behaviour, 99

I

IFF (Index of Fluvial Functionality), 171–175
Industrial revolution, 351
Information, 3–5, 9, 14, 35, 38, 41–44, 46–51, 70–71, 99, 108, 121–123, 136, 142, 144–146, 154, 173, 182, 186, 197, 221–223, 225, 255, 267, 273, 283, 290, 293–295, 298–299, 301, 307, 350, 375
    See also Sinai-Kolmogorov information
Instability, 5, 88, 96, 141–142, 236
  Faraday's –, 94
Irreversibility, 41–42, 50, 80–81, 215

J

James W., 313–315, 317–319, 324
Jammer M., 55

K

Kantorovic L., 353
Kauffman S.A., 121, 305, 308, 346, 361
Kepler J., 126, 130, 362
Kolmogorov A.N., 221
  entropy, 144
  probability, 15
  theorem, 45
    See also Sinai-Kolmogorov information

L

Lagrangian representation, 35
Lanford O.E., 83
Langevin equation, 145
Laplace P.S., ix, 107–109, 377
Launhart W., 352
Lebesgue H.L., 65, 161, 163, 165–166, 168
Leonardo da Vinci, 389
LGN (Lateral Geniculate Nucleus), 5, 7
Linneo C., 368
Lister J., 369
Lobatchevsky N.I., 201
Lorentz equation, 29
Loschmidt J., 113–114
Lucrezio, 141, 389
Lyell Ch., 237

M

Mach E., 84, 118, 380
Macroscopic description, 79, 81, 85, 142
Management, 246, 349–354, 356
Markov processes, 136
Mathématique sociale, 351, 354
Maxwell equations, 29, 42
Maxwell J.C., 130
Merli law, 172
Mesoscopic description, 142
MET (Mean Ergodic Theorem), 330
Microscopic description, 79–80, 82, 85, 142
Mill J.S., 314–315
Minimal life, 257, 263, 265–267
Modern Synthesis, 284–285
Morgan C.L., 381

N

Nagel E., 233, 323
Nagel R., 133
Nán Fă, 396
Natural science, 134, 313, 320–321, 359, 361, 364, 366–367, 377
Navier-Stokes
  equations, 83–84
  fluid, 81–82
Necker cube, 15
Neocortical interneurons, 181–182
Neurodynamics, 5–6, 10
Neurophysics, 3–4, 6, 16
Newton
  equations, 80
Newton I., 127, 129–130, 351, 389, 396
Newton
  law, 80–81, 83–84
Nietzsche F., 315
NLC (Nematic Liquid Crystal), 87–89, 91, 93–94
Non-target organism, 211–213, 216, 219

O

Occam G., 61
Optical FODO cell
    See also FODO
Optics, 31–32, 34
  geometric –, 27–28, 30–31, 34
Organicism, 339–342, 344, 346–347, 380–381
Organization, 14, 141, 202, 234, 261–263, 273–275, 293, 296, 346, 350–353, 355–356
    See also Auto-organization
Overproduction, 257–259, 262–263, 265

P

Particle accelerator, 27–28, 30, 37–38
Pasteur L., 369
PCF (Pre Frontal Cortex), 13–14
Peirce Ch.S., 319
Penrose R., 16, 118
Perturbation, 8, 12, 28, 99, 144, 197, 246
Phylogeny, 280, 283, 285, 287
Planck M.K.E.L., 118, 123, 130
Planck's constant, 14, 25, 57, 66, 192, 196
Poincaré H., 81, 143, 327, 329–331, 333–334
Poisson S.D., 159, 329
Poisson's functional equation, 202
Pollen dispersal, 213–214
Popper K.R., 323, 359, 363–366
Pragmatism, 314, 319
Predictability, 105, 107–108, 113, 142, 370
Probability, 16, 42–43, 50, 53, 55, 57, 59–61, 65–66, 69–70, 73, 76, 81–84, 90, 92, 114, 142, 147–149, 156, 158, 160–161,

168, 213, 215, 246, 248, 327-332, 334, 354
See also Kolmogorov probability
Properties of synthesis, 200
Psycho-analysis, 109, 313, 315, 320, 322–324
Psychology, 313–322, 324, 377
Psycho-pathology, 315–316
Purpura K.P., 10, 13–14

Q

QFT (Quantum Field Theory), 117–118, 123, 192
QM (Quantum Mechanics), 53–56, 58–60, 63–65, 67, 69–71, 80, 118–119, 127, 129–130, 236, 332–333, 379–380
Quantum, 3–5, 10, 13–14, 16, 53–61, 63–64, 68, 70–71, 76, 109, 115, 117, 120, 123, 125, 127, 130, 191–192, 194, 196–197, 333
Quantum vacuum, 122
Quaternion, 27–28, 34–35, 38

R

Redi F., 368
Reductionism, x, 231–235, 238, 339–341, 343–347, 377, 379
Remak R., 368
Renouvier Ch., 317, 319
Reproduction, 257–258, 262–263, 278–280, 294, 373, 376, 381
Reticular dynamics, 118, 120, 122
Reversibility, 42, 80–81, 144
Riemann B., 202
Ritter W., 381
Rossum metric, 14
Russel B., 359, 362–367, 375, 378

S

SAGE (Semiautomatic Ground Environment), 349, 355
Scaling laws, 79
Schiaparelli G.V., 207
Schopenhauer A., 315
Schrödinger E., 15–16, 56, 65, 67, 115, 129–130, 364, 373
Schrödinger equation, 42, 69, 72
Self-organization, 285, 305, 307–309, 346, 379
Self-organizing process, 9, 309
Semigroup, 133, 135–138
Shannon C.E., 42, 90, 223
Sherrington Ch.S., 320
Sinai-Kolmogorov information, 122
Singer W.E., 8
Smale-Birkhoff theorem, 333
Smith A., 351
Smuts J.C., 379–385
Snell's law, 32
Soft science, 324, 359, 361
Sokal affair, 125
Solomonoff R., 121
Speciation, 206, 286–287, 298
Species, 150, 204–207, 213–215, 218, 245, 258–260, 263–265, 279, 284–287, 289, 299–300, 341, 360, 366, 370–371, 373
Stability, 28, 30, 82, 96, 142, 287, 328, 332, 388–390
Stanley Hall G., 313
Stream of consciousness, 317–318
Synapse, 5, 181–186, 189–190, 278, 320
Synchronization, 3–4, 7–8, 21, 181–182, 184–186

T

Taylor F.W., 352
Terminator technology, 211, 216, 218
Test-tube, 257, 265
Thermodynamics, 41, 51, 113, 142, 236, 377
Thom R., 396
Time arrow
See also Arrow of time
Trinucleotide, 241–242
Tubulin, 273–275, 277, 279–281
Turing
halting problem, 46
machine, 50, 117
question, 16

U

Unpredictability, 50, 108, 211, 216, 346

V

Victor J.D., 10, 13–14
Virchow R., 368
Vitalism, 339–340, 342–344, 381
Volterra V., 192, 200, 353–354
Von Bertalanffy L., 202, 340, 342, 346, 381
Von der Malsburg C., 8
Von Neumann J., 56, 65, 330–331, 353–354

W

Walras L., 353–354
Weinberg S., 125, 128–130
Weyl H., 192, 194, 395
Wigner-Feynman
equations, 60
functions, 119
Wigner function, 15, 25, 59–60, 67
Wise N., 129–130
Wolfram S., 120–121
Wundt W., 314

Z

Zermelo E., 81
Zygote, 204, 208–209

GPSR Compliance

The European Union's (EU) General Product Safety Regulation (GPSR) is a set of rules that requires consumer products to be safe and our obligations to ensure this.

If you have any concerns about our products, you can contact us on ProductSafety@springernature.com

In case Publisher is established outside the EU, the EU authorized representative is:

Springer Nature Customer Service Center GmbH
Europaplatz 3
69115 Heidelberg, Germany

Batch number: 10800094

Printed by Printforce, the Netherlands